Drone Technology

Scrivener Publishing
100 Cummings Center, Suite 541J
Beverly, MA 01915-6106

Publishers at Scrivener
Martin Scrivener (martin@scrivenerpublishing.com)
Phillip Carmical (pcarmical@scrivenerpublishing.com)

Drone Technology

Future Trends and Practical Applications

Edited by
Sachi Nandan Mohanty
J.V.R. Ravindra
G. Surya Narayana
Chinmaya Ranjan Pattnaik
and
Y. Mohamed Sirajudeen

Scrivener
Publishing

WILEY

This edition first published 2023 by John Wiley & Sons, Inc., 111 River Street, Hoboken, NJ 07030, USA and Scrivener Publishing LLC, 100 Cummings Center, Suite 541J, Beverly, MA 01915, USA
© 2023 Scrivener Publishing LLC
For more information about Scrivener publications please visit www.scrivenerpublishing.com.

Wiley Global Headquarters
111 River Street, Hoboken, NJ 07030, USA

For details of our global editorial offices, customer services, and more information about Wiley products visit us at www.wiley.com.

Limit of Liability/Disclaimer of Warranty
While the publisher and authors have used their best efforts in preparing this work, they make no representations or warranties with respect to the accuracy or completeness of the contents of this work and specifically disclaim all warranties, including without limitation any implied warranties of merchantability or fitness for a particular purpose. No warranty may be created or extended by sales representatives, written sales materials, or promotional statements for this work. The fact that an organization, website, or product is referred to in this work as a citation and/or potential source of further information does not mean that the publisher and authors endorse the information or services the organization, website, or product may provide or recommendations it may make. This work is sold with the understanding that the publisher is not engaged in rendering professional services. The advice and strategies contained herein may not be suitable for your situation. You should consult with a specialist where appropriate. Neither the publisher nor authors shall be liable for any loss of profit or any other commercial damages, including but not limited to special, incidental, consequential, or other damages. Further, readers should be aware that websites listed in this work may have changed or disappeared between when this work was written and when it is read.

Library of Congress Cataloging-in-Publication Data

ISBN 978-1-394-16653-4

Cover image: Pixabay.Com
Cover design by Russell Richardson

Set in size of 11pt and Minion Pro by Manila Typesetting Company, Makati, Philippines

Printed in the USA

10 9 8 7 6 5 4 3 2 1

Contents

Preface

This book provides a holistic and valuable insight into the revolutionary world of unmanned aerial vehicles (UAV). The book reflects on the dependence on smart surveillance for drowning, theft detection, an emerging trend in precision farming, land mine detection, and illegal migration surveillance support in military applications. With the amalgamation of multiple chapters, this book elucidates the revolutionary and riveting research in the ultramodern domain of drone technologies, drone-enabled IoT applications, and artificial intelligence-based smart surveillance. Each chapter gives a concise introduction to the topic. The book explains the most recent developments in the field, challenges, and future scope of drone technologies. Beyond that, it discusses the importance of a wide range of design applications, drone/UAV development, and drone-enabled smart healthcare systems for smart cities. The book describes pioneering work on mitigating cyber security threats by employing intelligent machine learning models in the designing of IoT-aided drones. The book also has a fascinating chapter on application intrusion detection by drones using recurrent neural networks. Other chapters address interdisciplinary fields like artificial intelligence, deep learning, the role of drones in healthcare in smart cities, and the importance of drone technology in agriculture.

This book covers almost all applications of unmanned aerial vehicles and will provide you with an encompassing knowledge about this wide field and its potential offshoots in the digitalized era. It is an ideal book for newcomers in the field of drone technologies with simple and lucid language for better understanding. This book provides updated knowledge on different types of drone technologies that will leave you flabbergasted.

The Editors
February 2023

Drone Technologies: State-of-the-Art, Challenges, and Future Scope

Arun Agarwal[1]*, Chandan Mohanta[1] and Saurabh Narendra Mehta[2]

[1]Department of Electronics and Communication Engineering, ITER, Siksha 'O' Anusandhan Deemed to be University, Bhubaneswar, Odisha, India
[2]Department of ECE, Vidyalankar Institute of Technology, Mumbai, India

Abstract

Drones, also known as unmanned aerial vehicles (UAVs), have become increasingly vital in the recent times. They have managed to find their sweet spot in rigorous and sensitive arena of military, agro-based applications, logistics, and supply chain and in observation and security. This sophisticated equipment has found its fancy uses too. In areas like DIY hobby crafts, amateur and professional photography, race tracks, drones sports etc. In recent times one of the best uses that the users could make out of drones was swarm drones show. Be it local festivals or national events swarm drones show are a delight to witness. Show like these are now on their nascent phase and have huge potential. Moreover, they are eco-friendly when compared to the fire crackers. Drones, due to their small size and high maneuverability provides a great wide coverage view of any large area. This enables the user to get a detailed holographic view of the entire space. The possibilities of integration of drones with the meta-verse are literally infinite and would definitely add so much of dimensions of usage. In order to harness these capabilities, the hobbyists and industry personnel alike are finding their own ways to accommodate drones in their own vicinity; altering dramatically many businesses and creating new avenues. By properly implementing the technology in UAVs, many emergency situations can be mitigated in both the civilian and military applications. The exploits of implementation of drones can find its capabilities in emergency medical supplies that could be made possible irrespective of transport feasibilities. Also military and enforcement agencies would have an upper hand in monitoring movements efficiently. In desperate times such as those following a

---

Corresponding author: arunagrawal@soa.ac.in

Sachi Nandan Mohanty, J.V.R. Ravindra, G. Surya Narayana, Chinmaya Ranjan Pattnaik and Y. Mohamed Sirajudeen (eds.) Drone Technology: Future Trends and Practical Applications, (1–20) © 2023 Scrivener Publishing LLC

natural disaster or a terrorist attack, the concept of surveillance drones allows for accurate tracking of people without putting more lives at risk. In order to briefly understand the military usage let us consider the following example. If we go back in time and look at the history of Japanese Kamikaze pilots; the brave pilots literally amalgamated themselves with the aircraft in order to become a guided missile. At present, the modern day fighter pilots have guided weapons systems. However, with time the air defense systems have also substantially evolved. The air defense systems have longer range, highly maneuverable nearly hypersonic missile systems which create an area-denial zone for any adverse aircraft/hostile missile system. In situation like these drones can be really useful. The operating bandwidth of ground based systems could be overwhelmed by use of cheap and disposable swarm drones. Meanwhile, another variety of drones known as loitering munitions could be made to sneak into the zone to clear lesser hostile targets. After this entire chaos has been created; the final attack from main assault party could be launched by both surface and air assault systems. This process puts no human life in danger while enabling the user complete control of the situation. Despite having immense possible benefits, there is hysteria of mass adoption of these systems due to the frequent security related incidents. UAVs could possibly be targeted by nefarious groups and if exploited, life-threatening compromising situation may arise. This has given rise in increasing number of regulations for drone applications.

Keywords: Drones, UAV, surveillance, applications, safety, protocols, guidance system

1.1   Introduction

Unmanned Aerial Vehicle (UAV) commonly referred as Drone is a system of flight equipment that can fly and maneuver without a human pilot on board, and sometimes fully autonomous. The development of UAVs for many uses has advanced greatly. Self-controlled drones are enjoyable to fly, which has made them popular enough that they are available all over the world and are used by kids and teenagers. However, there are a lot of practical applications of it. By the known and accepted history, the military has first deployed UAV in 1849, when Austria attacked Venice with an unmanned air balloon. There has been continuing research into every facet of mobile security and the development of autonomous UAVs. Current research efforts target different UAV control issues as well as applications of AI research. They have recently attracted a lot of attention since they can be employed for both military and civilian purposes, such as remote sensing, agriculture, border security, state intelligence gathering, and military attacks. In addition, industries and amateurs are constantly coming up with innovative methods to employ UAVs, which is transforming numerous industries and opening up new opportunities.

1.2   Forces Acting on a Drone

This is presented in Figure 1.1.

Lift: Lift is the force on the aerial vehicle that gives it altitude. It can be generated through propulsion, body shape, inner materials etc. [16].

Gravity: Gravity is the pulling force of the earth exerted on the aerial vehicle [16].

Drag: Drag is the force that acts opposite to the indented motion of the aerial vehicle [16].

Thrust: Thrust is the net force provided by the propulsion unit of the aerial vehicle [16].

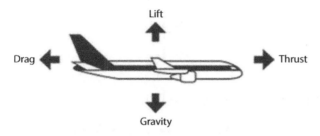

Figure 1.1 Representation of forces on an aerial object [16].

1.3   Principal Axes

This is presented in Figure 1.2.

1. Roll: The wings of the drone will turn to the right or left around the fuselage during this movement [17].

2. Pitch: This enables the nose of the drone move upwards or downwards [17].

3. Yaw: This turns the nose of the drone left or right without tilting the entire body [17].

1.4   Broad Classification of Drones

1.4.1   Fixed-Wing Drones

These are the most basic type of drones. A fixed-winged drone has a fixed rigid wing which generates lift. Here, main generator of lift is the wings instead of propellers [1]. This is presented in Figure 1.3.

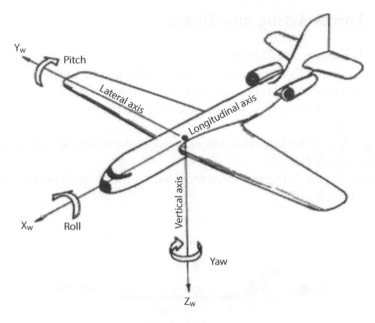

Figure 1.2 Representation of principal axes [17].

Figure 1.3 Representation of a fixed-wing drone [16].

1.4.1.1   Advantages [16]

i) They have a simpler structure and do not require much airframes and slots for avionics and other mechanical equipment.
ii) The high-winged and mid-winged configuration is aerodynamically stable and has great gliding capabilities.

iii) Since the parts involved are lesser and the body is less complicated they are easy to repair and maintain.

iv) Better aerodynamics and stable body enables a better battery/fuel life.

v) By considering above points, we can safely conclude that they can remain air worthy for a longer duration of time.

1.4.1.2   Disadvantages

i) Majority of them are either propeller or jet driven and hence require a dedicated runway or launcher and recovery system to operate and recover.

ii) Since wings are the primary driver of lift, they need to be in continuous motion to maintain altitude and hence cannot hover.

1.4.2   Lighter-Than-Air Systems

These refer to the aerial platform which could fly due to its buoyancy in air. This is achieved by encasing lifting gases (e.g., Helium) within a membrane and on that the entire structures and components are mounted [15]. This is presented in Figure 1.4.

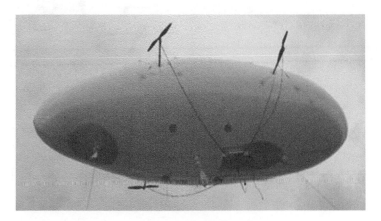

Figure 1.4 Representation of a lighter-than-air system [19].

1.4.2.1   Advantages

i) Irrespective of their size and individual weight of the sub-systems, the overall weight is too less. This enables them to maneuver with minimal effort.

ii) The net light weight of the entire system makes them capable of quickly changing direction both during motion and stationary.
iii) The energy consumed to operate them is low.
iii) This makes them ideal for stationary hovering (e.g.: sports, monitoring, agriculture, mining areas, etc.) [19].

1.4.2.2   Disadvantages

i) The net light weight makes them difficult to operate in bad weather or vulnerable to any external forces.
ii) They are bigger in size. This makes them non ideal for military or civilian use if user wants to maintain a low profile [19].

1.4.3   Multi-Rotor Configuration

Multi-rotor drones are cheapest means to get an aerial view. They are easiest to operate and transport and hence are ideal for personal and industrial usage. The most common platforms are tricopters (three rotors), quad copters (four rotors), hex copters (six rotors) and octacopters (eight rotors) [20]. This is presented in Figure 1.5.

Figure 1.5 Representation of multirotor platform [20].

1.4.3.1   Advantages

i) The rotor configuration provides enough lift and control to accommodate the ability of vertical takeoff and landing.

ii) They can be built over a very compact airframe and can be accurately maneuvered. This enables the user to operate within smaller vicinity.
iii) Since they have enough thrust to sustain by themselves they do not need a dedicated runway/helipad for takeoff and landing.
iv) Although they have an unstable design, they are highly and precisely maneuverable in presence of a good on board flight controller [20].

1.4.3.2   Disadvantages

i) The build and structure is sophisticated to make when compared to other models.
ii) They are aerodynamically unstable and on-board flight controller is required to perform basic maneuvers.
iii) They do have more number of propulsion units and other supporting peripherals and hence require a bigger energy pack [19].

1.5   Military Necessity of Drones

1.5.1   Features of Sixth-Generation Fighter Planes

This is presented in Figure 1.6.

1.5.1.1   Introduction

All other fighters will be obsolete as a result of the sixth generation fighters. There will be a future for stealth and beyond visual range missiles. Several

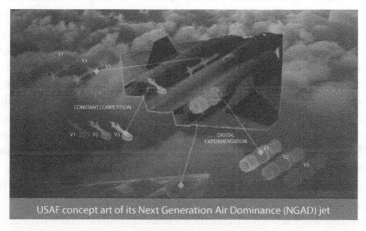

Figure 1.6 Representation of a sixth-generation fighter plane [19].

of the six-generation concepts share many of the same features. Stealthy airframes and long-range missiles will remain major characteristics of fifth-generation fighters [9].

In order for stealth aircraft to penetrate deeper into an anti-access/area-denial bubbles and neutralize the potential air defense batteries or regiments from a safer distance in light of economically low cost ground based air superiority systems such as the S-400 and Patriot Missile Systems, they must be capable of neutralizing anti-access and area-denial bubbles. Moreover, stealth jets perform better in aerial war games than non-stealth aircraft. The development of sixth-generation fighters will therefore require low radar cross-sections and radar-absorbing materials. As advanced sensor technology advances, stealthy airframes may become obsolete, and that stealthy airframes aren't as easily upgradeable as avionics and weapons packages. The use of jamming, electronic warfare, and infrared obscuring defenses will also become increasingly important in this regard. There will remain a strong presence of beyond visual range (BVR) missiles like the AIM 120, Astra, Python, etc. in the future which can already hit targets over 100 km of distance, but realistically that needs to be fired from a much closer range to have a good probability of acquiring target against a fast moving and agile fighter-sized target [9, 19].

However, the future air warriors may mostly fight in great distances from their opponents with new ramjet-powered, high-speed air-to-air missiles like the British Meteor and Chinese PL-15. Using helmet-mounted displays, the F-35 provides superior situational awareness by being able to see through the aircraft. Key instrument data and target missiles can be viewed through the cockpit and mounted on a helmet mounted display. Despite their teething issues, these helmets are expected to be adapted as standard part of future combat jets, perhaps superseding the cockpit or the entire instrument panel in some cases. As a result, fighter pilots might be able to operate more efficiently through voice-activated command interfaces. The larger airframes with better engines would allow combat jets to fly long endurance missions and lift heavier weapons as air bases and carriers become more vulnerable to any enemy attacks. When stealth jets are dependent solely on internal fuel tanks and weapons, it is difficult to do so as a stealth jet only has internal fuel-tanks and weapons bay. The obvious answer is a plane bigger in volume because the air forces anticipate that aerial dogfights within visual range will be uncommon and possibly fatal to both sides. As a result, there is a greater willingness to compromise on maneuverability in favor of higher sustained speeds and a heavier payload [9].

G-variable engines have become increasingly advanced with the development of advanced adaptive cycle engines, which could adapt to a required configuration mid-flight to operate better at higher speeds like a turbojet or with more fuel efficiency at lower speeds like a high-bypass turbo fan, these design requirements may work well together. Air power experts predict the shift to unmanned combat jets, which won't have to shoulder the additional responsibility and risk to life imposed due to a human pilot, after being optionally manned for several decades. The idea of an optionally manned aircraft that may fly with or without a pilot on board is consequently being advanced by sixth generation designs. This has the drawback that it will take more design work to create an airplane that will still have the drawbacks and high training costs of a manned airplane [4, 9].

However, the optionally manned pilot might make the transition to an unmanned fighter plane more doable and, in the middle term, allow the military officials to send planes on dangerous missions without endangering the lives of the pilots. Sensors fusion with allies on land, in the sea, and in space. The F-35's capacity to take in sensor-data and transmit it via data links to friendly and allied forces, producing a comprehensive image, is one of its significant achievements. As friendly forces move into vantage positions and launch strikes from further behind without even enabling their radars, a stealth aircraft may ride point and drive away enemies.

It is certain that sixth-generation jets will incorporate fusion sensors and cooperative engagement because this tactic promises to be such a force multiplier. Integration of satellites and drones alongside jet fighters will likely deepen the fusion [7].

1.5.1.2   Cyber Warfare and Cyber Security

Although the F-35's design promised a major increase in efficiency, sensors fusion and optional manned piloting suggest that sixth generation jets will rely significantly on data linkages and networks that could be jammed or even attacked through hacking. However, it also makes even landed airplanes vulnerable to future cyber-attacks. The sixth generation of avionics equipment may be able to launch such assaults on adversaries in addition to being intended for resilience against electronic and cyber warfare [3, 6].

1.5.1.3   Artificial Intelligence

One issue is that the complexity of all these comprehensively packed sensors communication and weapon systems has grown to the point where they may be too much for the human brain to absorb, while some fourth-generation aircrafts had a weapons system officer in the backseat for assistance. At present, all the available fifth-generation stealth fighters have single-seater configuration, the capable air forces are turning towards artificial intelligence to control more routine and mundane fighter operations and organize which information should be fed to the pilot. Drone coordination may also make use of artificial intelligence and machine learning [4, 6].

1.5.1.4   Drones and Drone Swarms

In a test over China Lake in October 2016, two FA-18 Super Hornets flew 103 Perdix Drones. Drones that have been activated by an artificial intelligence hive mind descend over a chosen target spot like a cloud of locusts. Although kamikaze drones have already been used in combat, it's simple to understand how inexpensive, little drones may develop into a particularly deadly weapon. Future warfare experts say that a few pricey and well-defended weapons systems and missiles may prove to be much harder to defend against than a cheap and disposable network of drones. But it's also conceivable that sixth generation fighters will collaborate with bigger, faster drones to act as decoys, weapons platforms, and scouts with sensors [6, 9].

1.5.1.5   Directed Energy Weapons

Threatening to overwhelm and enhance the offensive and defensive capabilities of stealth jets are swarms drones, precision missiles, and even decommissioned decoy fighter jets. Directed energy weapons like lasers or microwaves, which could be launched immediately, precisely, and with a nighttime magazine capacity delivering enough electricity, are one frequently mentioned countermeasure. In order to disrupt or harm enemy sensors and seekers, the US air force envisions three different types of airborne directed energy weapons: low-powered lasers; mid-level category that could neutralize an approaching air-to-air threat out of the sky; and high-powered lasers that can destroy aircraft and ground targets. Programs for sixth-generation fighters are still purely theoretical. Especially in light of the significant costs and time spent ironing out the issues in the fifth generation [9].

A few of the component technologies that are already well under development include lasers, cooperative interaction, and unmanned piloting. However, fitting them into the same airframe would be a critical job to address. Sixth generation fighters may appear at their earliest in the 2030s or 2040s, by that time the strategies and tactics of the air warfare would have become more and more sophisticated [6, 7].

1.5.2 Pseudo Satellite of HAL

The first flight of the Combat Air Teaming System (CATS) warrior drone, a low-observable, semi-stealthy unmanned wingman controlled from a CATS-max aircraft like LCA Tejas, is anticipated to take place in 2022, and it will be ready by 2024 or 25. CATS warrior has begun its made-in-wind-tunnel testing and is progressing well on schedule. There has been an update regarding a further CATS program component known as the High Altitude Pseudo Satellite (HAPS). As per the latest update the work on HAPS has started. Creating is not anticipated is working with private companies to improve the nation's military strike capabilities through the development of a futuristic high-altitude pseudo satellite. This innovation is regarded as a significant technological advance [22]. This is presented in Figure 1.7.

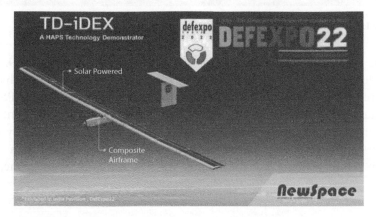

Figure 1.7 Graphic representation of HAL pseudo satellite [21].

HAPS will be a false satellite fuelled by solar energy at extremely high altitudes. It will be close to 500 kg in weight, capable of sustained flight for more than two to three months, and able to cruise in the stratosphere at an altitude of more than 70,000 feet. It can power itself thanks to its array of solar cells. To power nocturnal flights, the secondary batteries will be charged throughout the day. Due to its extremely high altitude operation and lack of propulsion it relies on solar cells to move—this allows soldiers to go over hostile area without worrying about being discovered. A manned aircraft will operate within the perimeter while an unmanned aircraft enters the hostile zone and can launch an attack deep into the enemy territory. This is a first-of-its-kind initiative worldwide. The intelligent surveillance and reconnaissance (ISR) capability of the Indian armed forces will be significantly improved by this technology [10]. This is presented in Figure 1.8 below.

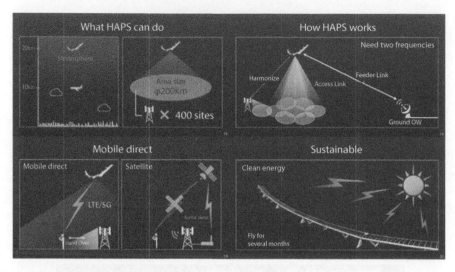

Figure 1.8 Capabilities of HAPS [21].

The concept of air teaming system is presented in Figure 1.9. It is intended to belong to the same class of unmanned aerial systems that are powered by fire, sunlight, and electricity. During operating, HAPS can also transmit photos and a live video feed to the ground station. Synthetic aperture radar would be added to the new drone to detect activity deep within enemy territory. The drone will be able to coordinate with other Indian prone systems like the Hunter missile, Alpha S Swan drone, and

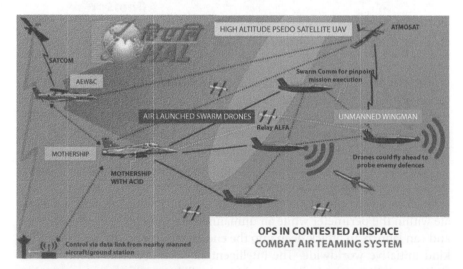

Figure 1.9 Air teaming system [21].

warrior-loyal Wingman drone using its superior sensor. The HAPS can be utilized for humanitarian aid and disaster relief efforts in addition to military purposes. Additionally, it can help 4G and 5G connection in isolated locations, particularly in difficult terrain and high altitudes [22].

1.5.3   Surface to Air Missile vs. Modern Fighter Aircraft

The Dassault Rafale's Spectra self-protection technology, which defends against threats to the fire aircraft, is one of the characteristics of modern fighter aircraft that make them such effective fighters. For the Rafale combat aircraft, Thales Group and MBDA jointly developed this technology. This is an extremely complicated system that combines a number of different sensors and radars, including a phased array radar jammer, a missile approach warning system, an infrared missile launch detector, a radar warning receiver, a laser warning, and a decoy dispenser. Based on the information obtained from its sensor and its sizable database, Spectra, which acts like a central processing unit, receives data from these sensors and feeds it. A spectrum provides exceptional situational awareness and even makes recommendations for the best course of action. In fact, spectra are as good as advanced sensor fusion using F-35 jet fighter [9].

The element that sets powerful air defense systems like the S-400 or Patriot air defense apart from other defense systems is their strong radar. Within a 600 km range, the S-400's radar is capable of detecting and tracking large aircraft, rotorcraft, cruise missiles, guided missiles, drones, and ballistic rockets. The declared anti-stealth targeting range of the 91N6E panoramic radar is 150 kilometers. The detection range of a ballistic target travelling at Mach 14 with a radio cross section (RCS) of 0.4 m^2 is 230 km. And the distance is 390 kilometers for a target with an axis of 4 m^2. Rafale has a RCS of 0.5 m^2. Consequently, the S-400 system could readily track the aircraft from several hundred kilometers away [11].

How could Rafale carry out SEAD (suppression of enemy air defense) missions is then the question. First, we need to use Signal Intelligence platforms like Phalcon AWACS, reconnaissance and patrolling UAVs, and satellite networks to locate and identify Surface to Air (SAM) threats inside the enemy's territory and proper planning of the flight path for the fighter aircraft should be calculated and measured to dodge active SAM site's radar coverage. Second, the fighter aircraft has to fly low hugging the terrain functionality integrated with flight control system actually controls and maneuvers the aircraft closer to the ground or any other surface that would be acceptable for the crew members flying in a manual mode. Rafale would be tough for the S-400 radar to detect from such a distance away and at such a low altitude. Third, the fighter aircraft's threat library system needs to be updated with S-400 radar systems wavelength frequencies. S-400 system consists of three different types of radars; therefore in a SEAD mission there must be three aircrafts, each carrying jamming equipment tuned to jam the frequency of three different radars of the S-400. Once the S-400 attacks are jammed, they can be destroyed because they are rendered absolutely blind. With the help of spectra's active cancellation technology, the attacking aircraft could deceive the radars by counter echo returns. In active cancellation technology the mirror image of incoming wave received from adversary radar is produced and then both cancel each other out. A fighter jet can hover directly in front of the radar without being detected if the aircraft has a fast enough computer and a powerful enough emitter. This strategy has been demonstrated in a Slovakian drill. Fourth, self-protection system should be improved on Rafale. The next generation jammers like Bright Cloud should be installed. Platforms can now be protected from these modern tracking technologies by using Bright Cloud. Fired from a standard 55 mm flare cartridge, bright clouds has been programmed to draw threads away from the host platform creating room for large evasive distances. Fifth, aircrafts can also use Storm Shadow or Scalp missiles for attacking S-400 system from long standoff range. It won't be simple to use the file with the set of S-400. It requires precise planning and execution. The real world war scenario could be different and there are chances that profiles might get detected even after applying all the counter measures [11].

1.5.4   Drones as Weapons of Mass Destruction

In the context of weapons control, a "weapon of mass destruction" is a device that can cause widespread destruction and is governed by international agreements. Analyst Zach Kallenborn argues in a recent study for the US Air Force Center for Strategic Deterrent Studies that some drone swarm configurations could be treated as WMD. Modern drones, such as the MQ-9 Reaper, are operated remotely by a pilot who flies the aircraft while a payload operator fires warhead carrying missiles. Over their shoulders, a phalanx of additional specialists, such as military attorneys and image analysts, debate what constitutes a legitimate target. Drones in the future may operate autonomously and fight without much human supervision, especially when they are swarming. Kallenborn, an expert in unmanned systems and WMD describes one configuration of swarm that he refers as an Armed Fully Autonomous Drone Swarm (AFADS) [13].

Once released, aphads will locate, recognize, and attack all the targets without the need for human assistance. According to Kallenborn, an AFADS-type swarm does qualify as a weapon of mass destruction due to the amount of damage it is capable of causing and its inability to distinguish between military and civilian targets. The Cluster Swarm project is working on creating a missile carried warhead that would release a swarm of tiny drones that will fan out and seek out and destroy vehicles equipped with explosively produced penetrators (EFP). An EFP strikes an armor-piercing metal slug moving at a high speed. The CBU 105 bomb, 1000 pounds category ammunition that drops 40 skeet sub-munitions over the target area, is analogous to this in concept. Each one parachutes down and uses the seeker to scan the terrain until it spots a tank and shoots an EFP. The strength of the cluster swarm would be enormous [20].

The army operating GMLRS rockets could deliver a 180 pound payload and over a range of 70 km, or tax missiles, which comprises of 350 pound payload and a range of over 270 km, were used in the cluster swarm. The first plan was for the missile to carry a payload of quad-copter drones that would be dispersed throughout the target area by an aerodynamic shell. However, the challenges related to the unfolding of quad-copters mid-air may have been significant. Avid is best known for its work with Honeywell on the T-hawk drone systems, a tactical VTOL capable airframe deployed in Iraq to help detect IEDs in 2007. The T-hawk was propelled by duct fans located inside the fuselage and lacked external rotors. Later on, Avid created the EDF 8, a smaller electrically driven duct fan drone with a one pound payload [14, 20].

For two reasons, the cluster of swarm drones would be significantly more potent than the presently operating CBU 105 water canisters. Targets can only be hit by a CBU 105 in an area that is a few hundred meters broad. The cluster swarm may search for cars scattered across a large area. The other advantage is efficiency. A few warheads will not have locked onto a target, and where there could be an overlap, two or more may hit the same target while ignoring others. In the search zone for each warhead, CBU 105 offers just a small amount of overlap. An actual swarm that cooperates will engage in conflict, therefore 40 drones will always launch 40 separate attacks. Each MLRS missile unit would release roughly 10 cluster swarm drones if they were equipped with EFP warheads identical to current weapons. 12 missiles are fired by each M-270 MLRS vehicle in a salvo for 120 drones. Therefore, a battery of nine such launch vehicles might theoretically be able to engage an entire armored column in its tracks by dropping thousands of smaller sized killer drones over the target region [18].

Would a swarm like that qualify as a WMD?? It is conceivable that the weapon qualifies as a weapon of mass destruction. The quantity and payload of armed UAVs in the swarm would determine this, though. A swarm carrying the same amount of ammunition as a thousand M-67 hand grenades would probably fall under the WMD category. If the swarm reaches this mark, international arms control law may apply to it. Uncontrolled drones would undoubtedly have the ability to cause serious harm if they mistakenly believed that military vehicles were civilian ones. It is simple to understand how an attack on a column of armored vehicles could wind up misfiring on a neighboring refugee caravan with disastrous repercussions. This ought to be unacceptable and the DoDD 3000.09 aims to prevent unintended engagements caused by malfunctions in autonomous or semi-autonomous weapon systems. Errors do, however, occur in war. It might be challenging to demonstrate whether or not it is a WMD. How to identify if we are dealing with a single swarm of WMD-scale? If swarms can behave as WMD, it may be necessary to determine if a group of drones constitutes a single swarm or numerous swarms. It would be difficult to determine whether the swarm is entirely autonomous. Any swarm with this kind of potential raises question about how much autonomy is acceptable even if it is not classified as a WMD. The weapon serves as an example of the necessity to carefully evaluate the risks that the US is prepared to take. The phase 2 development was finished in March and comprised a different type of demonstration of the efficacy of the EFP warheads as well as deployment, powered flight, neutralizing the target, autonomous nav system and gliding on the target [15].

The armed forces may be able to deploy the drones quickly if Phase 3 is completed and the drones are integrated into missile warheads. Unlike other WMDs drone swarms can be acquired at low cost and require relatively far less technical skills if there is a military. If other parties use them and they begin resulting in a lot of casualties, the scenario might change [4].

1.6   Conclusion and Future Scope

This chapter discussed all aspects of Drones like components, architecture, classification, network layers etc. Future is UAV in different fields like civil, farming, surveillance etc. Coming to the usage of civilian drones, there exist a vital risk with respect to the rights and privacy of citizens. But battery issue remains a focus point for future research. Work on Autopilot system including GPS module for better position data remains a potential field of research along with data speed as well.

Drones were traditionally used to for surveillance and gain real-time imagery and sensor data from a designated area. During the subsequent years of development and further iterations they were instilled with payload carrying capacity and then got attack and strike capability. The current scenario of air conflict is very tricky and risky. Nations have allocated and spent huge budgets to protect themselves from any aerial threat. There is a constant tussle going on between air defense systems and electronic warfare suite installed on-board an aircraft. This tussle has put risk on life of personnel operating both the ground stations as well as the pilot flying the aircraft. Drones have traditionally played a supportive role and now they are being developed to play more and more offensive roles. Additionally, a smart fully autonomous drone capable of controlling multiple decoy drones would be game changing. Whether at the times of conflict or at peace, drones are useful in every scenario. Be it providing assistance in moving logistics over short ranges or providing aerial surveillance capability for more than 20 hours per day, drone technological development has come long way. Future of drones looks promising and they are capable of delivering capabilities that are not even though of at the moment.

References

1. Masum, M.A. *et al.*, Simulation of intelligent unmanned aerial vehicle (UAV) for military surveillance. *2013 International Conference on Advanced Computer Science and Information Systems (ICACSIS)*, pp. 161–166, 2013.
2. Sehrawat, A., Choudhury, T.A., Raj, G., Surveillance drone for disaster management and military security. *2017 International Conference on Computing, Communication and Automation (ICCCA)*, pp. 470–475, 2017.
3. Dufrene, W.R., Mobile military security with concentration on unmanned aerial vehicles. *24th Digital Avionics Systems Conference*, vol. 2, p. 8, 2005.
4. Tahir, M., Shah, S., II, Zaheer, Q., Aircraft system design for an anti terrorist unmanned aerial vehicle. *2019 International Conference on Engineering and Emerging Technologies (ICEET)*, pp. 1–8, 2019.
5. Hartmann, K. and Giles, K., UAV exploitation: A new domain for cyber power. *2016 8th International Conference on Cyber Conflict (CyCon)*, pp. 205–221, 2016.
6. Pauner, C., Kamara, I., Viguri, J., Drones. Current challenges and standardization solutions in the field of privacy and data protection. *2015 ITU Kaleidoscope: Trust in the Information Society (K-2015)*, pp. 1–7, 2015.

7. Muneem, I.A., Fahim, S.M., Khan, F.R., Emon, T.A., Islam, M.S., Khan, M.M., Research and development of multipurpose unmanned aerial vehicle (flying drone). *2021 IEEE 12th Annual Ubiquitous Computing, Electronics & Mobile Communication Conference (UEMCON)*, pp. 0402–0406, 2021.

8. Iqbal, S., A study on UAV operating system security and future research challenges. *2021 IEEE 11th Annual Computing and Communication Workshop and Conference (CCWC)*, pp. 0759–0765, 2021.

9. US Military News, 6th generation fighters are coming, Apr. 25, 2021. [Online]. Available: https://www.youtube.com/watch?v=8YkQs3lm85k&list=PLNGzO_h3wEbNzjJmmQ7q3m7aOQVJBTOJl&index=16. [Accessed: 29-Jun-2022].

10. Indian Defense Analysis, Development started on HAPS-infinity Pseudo satellite of HAL CATS (combined air teaming system), Feb. 11, 2022. [Online]. Available: https://www.youtube.com/watch?v=2d1fYEBIUkI&list=PLNGzO_h3wEbNzjJmmQ7q3m7aOQVJBTOJl&index=14. [Accessed: 29-Jun-2022].

11. Indian Defense Analysis, S-400 vs rafale|can rafale take down s-400 SAM?, Apr. 04, 2022. [Online]. Available: https://www.youtube.com/watch?v=W-br5aTDX6Xg&list=PLNGzO_h3wEbNzjJmmQ7q3m7aOQVJBTOJl&index=14. [Accessed: 29-Jun-2022].

12. Military Update, U.S. army announced new drone swarm would be a weapon of mass destruction, Mar. 07, 2021. [Online]. Available: https://www.youtube.com/watch?v=KLmmPnMvwNY&list=PLNGzO_h3wEbNzjJmmQ7q3m7aOQVJBTOJl&index=13. [Accessed: 29-Jun-2022].

13. User, 179. A new age of terror: New mass casualty terrorism threats. *Mad Scientist Lab.*, Sep. 26, 2019. [Online]. Available: https://madsciblog.tradoc.army.mil/179-a-new-age-of-terror-new-mass-casualty-terrorism-threats. [Accessed: 29-Jun-2022].

14. Defense.gov. [Online]. Available: https://media.defense.gov/2020/Jun/29/2002331131/-1/-1/0/60DRONESWARMS-MONOGRAPH.PDF. [Accessed: 29-Jun-2022].

15. Researchgate.net. [Online]. Available: https://www.researchgate.net/publication/331227634_LIGHTER-THAN-AIR_LTA_UNMANNED_AERIAL_SYSTEM_UAS_CARRIER_CONCEPT_FOR_SURVAILLANCE_AND_DISASTER_MANAGEMENT/fulltext/5c6d566092851c1c9defd80f/LIGHTER-THAN-AIR-LTA-UNMANNED-AERIAL-SYSTEM-UAS-CARRIER-CONCEPT-FOR-SURVAILLANCE-AND-DISASTER-MANAGEMENT.pdf. [Accessed: 29-Jun-2022].

16. 911 Security, Types of drones-fixed wing. 911Security.com. [Online]. Available: https://www.911security.com/learn/airspace-security/drone-fundamentals/types-of-drones-fixed-wing. [Accessed: 29-Jun-2022].

17. Axis-the basics of aviation. Google.com. [Online]. Available: https://sites.google.com/site/thebasicsofaviation/axis. [Accessed: 29-Jun-2022].

18. Plane flight controller-veronte autopilot-products. *Embention*, Sep. 05, 2019. [Online]. Available: https://www.embention.com/product/plane-flight-controller/. [Accessed: 29-Jun-2022].

19. Researchgate.net. [Online]. Available: https://www.researchgate.publication/net/331227634_Lighter-Than-Air_LTA_Unmanned_Aerial_System_UAS_Carrier_Concept_for_Survaillance_and_Disaster_Management/Fulltext/5c6d566092851c1c9defd80f/Lighter-Than-Air-LTA-Unmanned-Aerial-System-UAS-Carrier-Concept-for-Survaillance-and-Disaster-Management.pdf. [Accessed: 29-Jun-2022].

20. Tyrrell, M., Has lockheed Martin unveiled a sixth generation fighter factory? Aero-mag.com, Aug. 13, 2021. [Online]. Available: https://www.aero-mag.com/next-generation-air-dominance-ngad-13082021. [Accessed: 29-Jun-2022].

21. Raksha Anirveda, HAPS: India's stratospheric foray. *Self Reliance Defence*, Mar. 22, 2022.

22. HAL's HAPS program gets Government approval. Idrw.org. [Online]. Available: https://idrw.org/hals-haps-program-gets-government-approval/. [Accessed: 29-Jun-2022].

2

Introduction to Drone Flights— An Eye Witness for Flying Devices to the New Destinations

S. Venkata Achuta Rao[1], P. Srilatha[2], G.V.R.K. Acharyulu[3] and G. Suryanarayana[4*]

[1]*Department of Computer Science & Engineering, Sree Datta Institute of Engineering and Science, Hyderabad, India*
[2]*Department of Computer Science & Engineering, Sreyas Institute of Engineering and Technology, Hyderabad, India*
[3]*Master of Business Administration, SCM Studies, University of Hyderabad, Hyderabad, India*
[4]*Department of Computer Science & Engineering, Vardhaman College of Engineering, Hyderabad, India*

Abstract

An unmanned aircraft is called a drone. Significantly known as Unmanned Aerial Vehicles (UAVs) or Unmanned Aircraft Systems (UASs) without a pilot or also called a Flying Robot (FR), which can be remotely controlled or fly autonomously using embedded software-controlled in conjunction with onboard sensors, signals, a global positioning system (GPS), including IoT and a much larger number of electronic components based on their type and its functionality. The first pilotless Vehicles developed in Britain were tested in March 1917 and the American aerial torpedo first flew in October 1918. From then onwards to till date there is a rapid evolutionary development in this technology, especially the two decades of 1990-2010 a pivotal significant time period for military and civilian drone development, and also during 2010 till today is the Golden Age of drones. The description of the introductory chapter provides about drone technology evolves drastically day by day using different types of drones, the working principles of drones, how these are advantageous harmful to the society and their anatomy, different career options, safety, and security are discussed. Drones play a significant

Corresponding author: sreedatthaachyuth@gmail.com

Sachi Nandan Mohanty, J.V.R. Ravindra, G. Surya Narayana, Chinmaya Ranjan Pattnaik and Y. Mohamed Sirajudeen (eds.) Drone Technology: Future Trends and Practical Applications, (21–52) © 2023 Scrivener Publishing LLC

role in collecting waste under the sea and on the earth across the globe surprisingly, knew that every year the world generates around 2.01 billion tons of municipal solid waste; it may be projected to rise by 70% to reach 3.4 billion tons by 2050. In this chapter the important applications and significant uses and scope of drones and explained that like a cell phone, it turned it as the handheld device in every hand. The researchers and engineers are having a significant responsibility to see that drones and their disruptive technology should not harm the mankind and it should not be an aid for terrorists and vulnerabilities.

Keywords: Unmanned aircraft systems, global positioning system, autonomous, sustainability, drone identification system

2.1   Introduction

In a technical sense, the aircraft may be remotely operated or fly autonomously due to embedded technology that uses software-controlled flight plans in combination with onboard PCB with sensors and GPS to fly. UAVs were most often operated in the Department of Defense (DOD) in the recent past, when they were used as hard real-time embedded systems for anti-aircraft target shooting, intelligence massive gathering, and, more controversially, as weapons operating devise systems. Drones are currently operated for a wide variety of civil applications such as GIS. GPS includes search and rescue, operations in surveillance, sudden calamities like sand storms, high-speed catastrophic wave transformations to the people's lands, traffic monitoring, weather forecasting, monitoring, and firefighting, as well as personal drones for commercial purposes and commercial drone-based photography, videography, agricultural, and even deliveries. Special procedures can be used to trace the source of a drone and give relevant information to the department of defense. The basic and first airplane with a reusable type radio control mechanism was invented in the 1930s, and it served as a basis to cater to all the subsequent technological improvements [1, 2]. Military purpose drones were later designed and deployed with the required types of sensors and necessary pixelated camera units, and they are currently installed within missiles. With so much advancement in technology, there is now a diversity of drone models to choose from. Few are employed for military purposes, while others are trying to find a place in many large corporations.

The impact of disruptive technology could be devastating for the business, environment, society, and government (ESG) concern. As we see how terrorists are using drones as many consequences are happening and citizens appeal to all diversified countries presidents to make a law and enforcement body across jurisdiction continue to hamper to control that with license or testing then only drone shall allow flying otherwise, with the help of Drone Identification System for Sustainability (DISS) is to be planned. This DISS System must provide safety, security, privacy, monitory of misuse, control and regulate its intensity for the well-being of society.

2.1.1   Brief History

Abraham Karem is a pioneer in innovative fixed and rotary-wing unmanned vehicles and is regarded as the founding father of UAV (drone) technology. By the 1970s he had built for the Israeli Air force. The earliest attempt at a powered unmanned aerial vehicle was A.M. Lows's "Aerial Target" of 1916. After World War 1, the first scale remote piloted vehicle was developed by the film star and model airplane enthusiast Reginald Denny in 1935. The birth of U.S UAVs began in 1959 when U.S. Air Force officers were concerned about losing pilots over hostile territory. On February 4, 2002, the CIA first used an unmanned predator drone in a targeted killing. In 2013, John Horgan reported in National Geographic that at least 50 countries used UAVs, with China, Iran, Israel, and others deployed and building their own varieties. On September 7, 2015, the UCAV was used for the first time in a live military operation.

2.1.2   The Indomitable Significance of Drone Technology

For centuries, maps have played a pivotal role in human existence. From the map murals in the Vatican to manual surveying to satellites and eventually drones, the significance of maps, locations, and positioning services since the beginning of civilizations have been incomparable [3, 4]. Drones have proved to have unexpected growth in demand in India over the past five years. Various statistics published by the Ministry of Civil Aviation corroborate that India will witness an estimated investment of INR 5,000 crores for the manufacturing of drones, which in turn will see the drone industry clock an annual turnover of INR 900-crore by fiscal 2024. So what is the various device pilot or drivers behind the usage of drones in India? [14].

As the Drone Industry Insights Report 2020, the worldwide drone industry is predicted to increase at a 13.8% CAGR to $42.8 billion by 2025. With Ministry of Civil Aviation updating the Drone Rules 2021, efforts are to make India a global drone hub by 2030. According to the results, by 2025, India is anticipated to be the world's third-largest drone market. The UAV market in India is expected to grow at a CAGR of 20.9% between 2020 and 2026 which is future estimation on the investments from industrial conglomerates, chip companies, IT consulting firms. The global unmanned aerial vehicle (UAV) drones market size is predicted from 2022 to 2030 as a worth around US$ 102.38 billion by 2030 and expanding growth at a CAGR of 18.2%.

Drones were used in the COVID-19 pandemic in many ways. Some of the use cases are mentioned below. The reports are taken from the media and other sources and specifically three key use cases of drones are given as [7].

- Lab sample pick-up and delivery and transportation of medical supplies in order to reduce the transportation times and minimize the exposure to widespread infection
- Aerial spraying of public areas in order to disinfect potentially contaminated places
- Public space monitoring and guidance during lockdown and quarantine.

2.1.3   Trends

Majorly being used by the law enforcement agencies, drones have observed more usage by public sector utilities and government; non-government agencies for land surveying, mapping, and monitoring. To cite an example, the National Highways Authority of India (NHAI) has installed current operations the use of drones for 3D printing technology mapping for their various detailed project reports. The data collected is being utilized for the calculation of compensation for citizens with property rights along the highway. What's more? The Indian Railways is using drones to monitor the construction of its railway lines by 3D mapping a dedicated freight corridor network of a 3,360-km project. The entire project will be mapped using drones [9]. Meanwhile, the state-owned Power Grid Corporation of India has already started monitoring its 15,000-20,000 km long transmission network with the aid of drones, which are equipped with high-definition pixelated cameras. There is a real scenario is efficiently executing for automation of tasks in the mining department. The drones are installed and operations are going on to the central data monitoring center at Manesar. That said, Coal India, India's biggest government-owned miner has started with the purpose of aerial surveys of coal blocks for evaluation and assessment of greenery restoration post-excavation from mines [14].

Already, the SVAMITVA Initiative—a central sector scheme of the Ministry of Panchayati Raj that relies on the usage of drone technology for mapping of land parcels in rural inhabited areas, and creating accurate land records for rural planning—has given a boost to the entire ecosystem, making businesses and varied state governments more receptive to the idea of using drones. Through SVAMITVA, the Survey of India (SOI) is focusing on high-resolution, high-accuracy mapping, and using only drones to map about 6.6 lakh villages in India. Despite the pandemic, idea Forge dedicated a big team of research and development engineers to develop class-leading mapping products to help SOI meet this audacious role. Currently, idea Forge's RYNO is the micro category drone because of its high accuracy in mapping applications, which has been approved by SOI for the SVAMITVA Yojana [14].

2.2   How Drones Work and Their Anatomy

A typical unmanned aircraft is developed and deployed with low-weight composite materials to ease fly and increase maneuverability. UAV drones are integrated with state-of-the-art technology such as infrared cameras, GPS, and laser for consumer, personal, commercial, and departments of defense like military, navy, and air force-related UAVs. Drones are governed by remote ground control systems and are also referred to as a ground cockpit. A UAV has two parts, the drone itself and the control system. The nose of the UAV is where all the sensors and navigational systems are available. The remaining part is full of drone technology systems since there is no need to accommodate humans in the body. The design of the embedded circuit is fabricated with necessary core processors and required engineering materials are used to build highly complex composites which are used to absorb vibrations and decrease the sound produced [9].

2.2.1   Anatomy of a Drone

2.2.1.1   Propellers

The propellers of quadcopters are normally seen in the front. In terms of size and materials, propellers come in a wide range of sizes and materials. Many of them, especially the smaller ones, are constructed of plastic, while the more costly ones are made of carbon fiber. Propellers are continually being created, and technical research to generate more effective propellers for both tiny and large drones is currently underway [7]. The drone's orientation and movements are controlled by its propellers. As a result, before taking your drone out for a flight; make sure that each of the propellers is in good working order.

2.2.1.2   Brushless Motors

Brushless motors, as opposed to brushed motors, are used in all new drones. The motor's design is just as crucial as the drone's. This is due to a more efficient motor that will allow you to save money on both the buy and operational maintenance of the machine. Additionally, you will preserve battery life, allowing you to fly your drone for extended periods of time. The drone motor design manufacturing sector is now rather dynamic, as businesses compete to build the most efficient and well-designed motors.

2.2.1.3   Landing Gear

Drones are equipped with landing gear that resembles that of a helicopter. Drones that require a lot of ground clearance during landing will need to have their landing gear modified to land safely. Moreover, delivery drones carrying goods or things may require a large landing gear due to the area needed to hold the objects when they strike the ground. However, the landing gear is not required for all drones. Some smaller drones can fly without landing gear and land safely on their stomachs once they contact the ground. Most drones with longer flight times and larger ranges feature fixed landing gear.

2.2.1.4   Electronic Speed Controllers [ESC]

It is an electronic circuit whose primary function is to monitor and alter the drone's speed while in flight. It is also in charge of the drone's flight direction and variations in braking. The ESC is also in charge of converting DC battery power to AC power, which is used to power brushless motors. For all of their flight demands and performance, modern drones rely totally on the ESC. The DJI Inspire 1 ESC is the most recent example of a higher-performing ESC that reduces power consumption while increasing performance. The ESC is mostly found inside the drone's mainframe.

2.2.1.5   Flight Controller

The flight controller serves as the drone's motherboard. It is in response to all the commands that the pilot delivers to the drone. It decrypts data from the receiver, GPS module, battery monitor, and onboard sensors. The flight controller is also in charge of regulating motor speeds via the ESC and controlling the drone's direction. The flight controller is in charge of all commands, including camera triggering, autopilot mode control, and other autonomous operations.

2.2.1.6   Receiver

The receiver is the component in charge of collecting radio signals delivered to the drone via the controller. The minimum number of channels required to operate a drone is generally four. It is recommended, however, that a maximum of five channels be made available. There are numerous types of receivers available on the market, and any of them may be utilized to build a drone.

2.2.1.7   Transmitter

The transmitter is in charge of transmitting radio signals from the controller to the drone in order to give flight and navigational directives. The transmitter, like the receiver, needs to have four channels for a drone, but five is commonly recommended. Drone manufacturers can pick from a variety of receivers available on the market. To communicate with the drone while being in flight, the receiver and transmitter must use a single radio signal.

2.2.1.8   GPS Module

The GPS system is in charge of determining the drone's longitude, latitude, and elevation. It is a very huge component of the drone. Drones would not be as essential as they are now if not for the GPS device. The components assist the drone in navigating greater distances and capturing data of land areas.

2.2.1.9   Battery

The drone's battery is the component that allows all motions and reactions. The drone would not be able to fly without the battery because it would be powerless. The battery needs for different drones vary. Due to their lower power requirements, smaller drones may require smaller batteries. Larger drones, on the other hand, may need a larger battery with a greater capability in order to power all of the drone operations.

2.2.1.10   Camera

Many drones feature an installed camera, while others have a camera that can be removed. The camera aids in the shooting of photographs and images from the air, which is a common function of drones. There are a number of camera kinds and qualities to pick from on the market.

2.2.2   Types of Drones

Drones generally are of two types (Figure 2.1) [12]:

    a.   MQ-9 REAPER, a hunter-killer surveillance UAV which is also called a killing machine. They are specifically designed and deployed for military and special operation applications.
    b.   DJI PHANTOM are designed and deployed for commercial and recreational aerial photography which are popularly named educational and research purpose drones.

Some other types:

- Micro UAVs are called as small-scale flying vehicles, miniaturized scale air vehicles (MAV).
- Biomimetic UAVs are specified for the Art of planning and building biomimetic usage with the help of mechanical assembly.
- Biomimetic precisely to human-made procedures, substances, gadgets, a framework that impersonate nature.

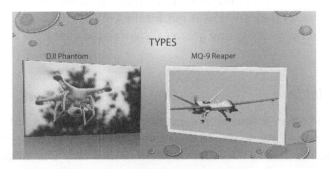

Figure 2.1 Two general types of drones [11].

- Blimps and Balloons UAVs are zeppelin or non-unbending aircraft is a carry airship without an inward backup system or bottom depends on the weight of the lifting gas.

2.2.2.1   Sub-System of UAVs

2.2.2.1.1   Communications

a. Ultra-High Frequency based operability
b. With Ku-Band Communication System.
c. Uplink Frequencies from 15.15 GHz to 15.35 GHz and Downlink Frequencies from 14.40 GHz To 14.83 GHz with all operations.

2.2.2.1.2   Navigation

a. GPS and WAA are on Avionics Use Satellite-Based System.
b. Automatic calculate the position.

2.2.2.1.3   Monitoring

a. GPS like a System
b. Video Camera with High-Resolution.
c. Still Camera with Super High Resolution

2.2.2.1.4   Collision Avoidance System

a. Traffic & Collision System Purpose
b. Simple Traffic Alerting System by Smaller Aircraft Use
c. Proximity Warning System for Aircraft Use Ground

2.2.2.1.5   Weather System

a. Light Detector System by weather radar usage.

2.2.2.2   Other Specific Types of Drones

2.2.2.2.1   Multi-Rotor Drones

Above their tiny form, these designs have several fans. The many propellers provide powerful lift and precise control to the pilot. Multi-rotor drones

Figure 2.2 Multi-rotor drone [11].

are a terrific alternative for aerial photography because of their compact size and superb handling. They can also hover and take off vertically, giving them even more mobility. Quadcopters, hex copters, and octocopters, on the other hand, have drawbacks. Increasing the number of rotors on a drone makes it more difficult to understand and operate. All those moving bits use more energy, causing the batteries to deplete quicker (Figure 2.2) [12].

2.2.2.2.2   Fixed-Wing Drones

Fixed-wing drones, as their name implies, resemble traditional planes and must be launched from a runway or a launcher. They don't have the same vertical take-off capability as quadcopters and single-rotor drones, and they can't hover [12]. A wing-based design offers several significant advantages. Static wings keep the drone in the air and improve the drone's aerodynamics, making it more efficient. As a result, you may run for longer periods of time without having to charge or switch the batteries (Figure 2.3).

Figure 2.3 Fixed wing drones [11].

They're also faster than rotor-based drones, given to the plane's speed advantage over a helicopter.

These qualities make them perfect for long-distance travel. They don't even have to use batteries; many fixed-wing drones use gas engines instead.

2.2.2.2.3   Single-Rotor Helicopter Drones

These drones combine the greatest features of both small multi-rotor and big single-wing aircraft.

These gadgets can hover and launch vertically thanks to their reliance on rotors. They are often bigger than their miniature counterparts, allowing them to transport heavier cargoes. They are more efficient than multi-rotor variants since they do not have several motors. Because of the increased efficiency, single-rotor machines frequently employ gas engines rather than batteries [12]. Their range is significantly increased when they choose the gas option. These drones, especially multi-rotor variants, are typically bigger and more complicated than other types of drones. This raises their prices and makes them more difficult to learn to operate. Because the bigger blades might make them more hazardous, it's a good idea to acquire some training before setting them down (Figure 2.4).

2.2.2.2.4   Fixed-Wing Hybrid VTOL Drones

The most recent drone technology to hit the market is fixed-wing hybrid aircraft. They combine the extended range and flying endurance of fixed-wing drones with the ability to take off vertically from a rotor-based device. This hybrid hardware's adaptability makes it an excellent choice for commercial drone operation. Some businesses are currently utilizing these approaches for delivery (Figure 2.5).

Figure 2.4 Single-rotor helicopter drone [11].

Figure 2.5 Fixed-wing hybrid VTOL drones [11].

Fixed-wing drones with vertical lift rotors make up several hybrid units. Others shift themselves after take-off to fly straight up, and then move to a horizontal position once in the air. There are also versions with wings and rotors that can swivel one direction for vertical take-off and another during flight. Hybrid VTOL drones are extremely adaptable, making them an excellent choice for long-distance flights [12].

2.2.3   Components of Drones

The central hub (control unit) consists of

- Actuator
- A Receiver
- A wide array of electronic sensors
- Battery packing with an electronic kit
- GPS
- Gyroscope
- Motor or Motors
- Processing units

2.2.3.1   Hardware

- Components
- Counter Drones

- Platform Manufacturers and
- System Manufacturers

2.2.3.2 Software

- Data Analytics Software
- Workflow
- Computer Vision and AI
- Flight Planning
- Fleet and Operation Management
- Navigation
- Open Source Infrastructure
- SDK
- Unmanned Traffic Management (UTM)

2.2.3.3 Other Specific Components

Many specific intensive purpose UAVs are designed with highly complex technological gadgets, drones with following other several main component.

2.2.3.3.1 Body
The body of a drone is integrated with the fuselage, plane wings, tail rotor and canopy, multi-rotor frame, and other arms.

2.2.3.3.2 Power Supply with the Corresponding Platform
Smaller drones are mostly designed with fly-on batteries while the larger ones use fuel or even solar power.

2.2.3.3.3 Computer Operations
Drones' hardware systems are firmly being specialized, and operation numbers accumulated and accelerated in order to support operating systems with no failures.

2.2.3.3.4 Sensors
There are three types of sensors that are widely used in UAVs: Proprioceptive, Exteroceptive, and Exproprioceptive.

2.2.3.3.5   Actuators
The Actuators which are installed are determined by the type of drones equipped with several electronic controllers, engines, propellers, and other components.

2.2.3.3.6   Software
Installed Software safely enables uninterrupted flight to lead the drone on its way and gives the information where to go and when to react.

2.2.3.3.7   Loop Principles
More sophisticated systems are designed with open loops for elementary types and closed loops for larger ones.

2.2.3.3.8   Flight Controls
However, automatic flight control is much more demanding. They are similar to a regular, aerial vehicle flown by a pilot.

2.2.3.3.9   Telecommunications
Telecommunications is one of the main components which gives the connection is quite regular and the conversion antenna is analog-digital and enables the transmission of data needed for the flight to proceed. Radio signals are used to transmit from ground control, a remote system as well as another manned aerial vehicle.

2.2.3.3.10   UAV Software
As drone software is the brain of the drone. It tells the drones, where to go and what to do while flying from source A to Destination B. software is designed to tell the drone where to go and what to do while flying from A to B. All the necessary information is mapped and thus the part of the drone becomes a very complex system. The necessary mapping structure which installed and kept as embedded software, then it operates as a system [18]. Further, all the layers are interconnected in the form of a framework and combined properly into tiers that perform various time slots after combined with meticulous design to work and act to control flight patterns, altitude, and other important information for the drone to work accurately with more efficiently.

We can say the framework of layers is called the flight stack or autopilot. There are many surveys and inquiries that have been successfully conducted and got a firm indication as that it doesn't matter if the drones have different efficiency or mission complexities, they all necessary the effective operating components. The received information has analyzed during the flight. This is an operative procedure to achieve unified components communication a generic architecture must be designed and promoted. We can conclude that the onboard system is not only sufficient but also required middleware and operating systems and emerging drivers also required [19].

2.2.3.3.11   UAV Software Layers
The following Software Layers are used in drone technology:

- Operating system
- Firmware
- Middleware

The architecture of software and its requirements in firmware and middleware are time-critical and come under hard real-time systems. These are intensive and if fail in accuracy and precision, there will be a massive catastrophic impact would be possible, which may not recoverable. Firmware operations are with machine code to the processor and then forward to memory access. Middleware acts as a common interface that organizes flight control, navigation, and telecommunication. The operating system acts as an administrator or resource provider for monitoring optic flow, avoids interference while SLAM searches for the solution and decides what are the actions according to the received information would be. Open source stacks would consider and assist in finding solutions for the drone industry, globally speaking. There are a number of innovative new enhanced applications that will have to be reliable, rapid and flexible ordered to meet all the consumers, personal and commercial requirements of the UAV Industry.

Listed below are some of the civil-use open-source stacks:

- ArduCopter
- Base Flight
- Beta Flight
- Clean Flight
- Clean Flight
- CrazyFile
- Drone Code
- Race Flight
- Paparazzi

- KKMultiCopter
- MultiWii
- OpenPilot Copter Control
- Taulubs, etc.

The following are popular UAV software serious pilots and UAV service providers should on sider using:

- Pix4D
- 3DR
- Microdrones
- mdCockpit
- mdFlightSim

2.3   Salient Features and Important Codes with Public Awareness with Respect to Safety and Necessary Precautionary Points

2.3.1   Safety and Legal Note

FAQ webpage: https://www.faa.gov/uas/faq/#qn24
Details of that can be found at this website: https://www.faa.gov/uas/civil_operations more details on issues of authorization can be found here [5, 6]: https://www.faa.gov/uas/public_operations/

2.3.2   Public Perception

It is common for some people to become very uncomfortable with the presence of drones in the sky.

2.3.3   Crew

A drone crew should consist of a minimum of two people:

    2.3.3.1   The Pilot in Command (PIC)
    2.3.3.2   The Spotter

2.3.4   Know Before You Fly

The FAA has partnered with several industry associations to promote Know before You Fly, a campaign to educate the public about using unmanned aircraft safely and responsibly.

2.3.5   Simulation Training

Although the 3DR Iris+ is a very capable semi-autonomous platform, I always recommend that prospective pilots spend some of their spare time using a flight simulator.

2.3.6   Mapping Configuration

For mapping, mapping camera is the only required equipment, the mapping camera mount and the remote camera trigger as shown in the Figures 2.6 and 2.7 [5]. The mapping camera is mounted on the front via the mapping camera mount; it faces downward so that its photographic plane is more or less parallel to the ground plane. The mapping camera is triggered remotely by the remote camera trigger, which is connected to the power output from the drone and to receiver output channel 7.

2.3.7   Mapping BFS Camera and Mapping Camera Mount

- Optional: First-person view equipment (see Figure 2.8). Required for real-time video downlink, but dramatically increases flight time.
- GoPro Hero4 Silver

Figure 2.6 Mapping camera and mount [11].

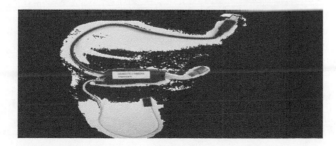

Figure 2.7 Remote camera and trigger [11].

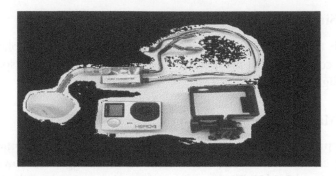

Figure 2.8 First person view equipment [11].

- GoPro Simple Mount
- Video Transmitter

2.3.8   Equipment to Remove

- Mapping Mode (assumes no FPV setup)
- Video Mode

2.3.9   Flight Planning

It is mandatory to obtain as much information as possible about a survey site before conducting any kind of drone-related mission (Figure 2.9) [6].

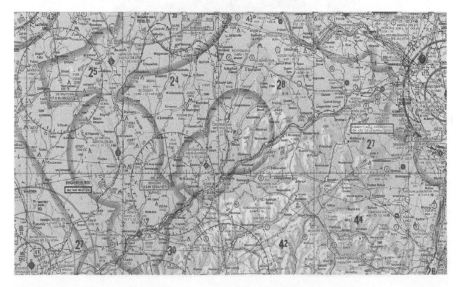

Figure 2.9 Sectional chart showing Cooperstown Westfield airport [10].

Initially, the drone operator should consult the 'Know before you fly" site is available at http://knowbeforeyoufly.org/ for helpful tips regarding how to maintain safety, privacy and ethical operations. There is an interactive map available at the following website: https://skyvector.com/

2.3.10   Post Processing Data

After a mission, the operator is left with a number of pictures of their survey area, none of them containing the whole picture. It is necessary to use post-processing to turn these data sets into something useful. To know all the necessary post-processing ideas by go through the following YouTube channel. http://youtube.com/watch?v=Vth6LXCifOs

2.4   Top 10 Stunning Applications of Drone Technology

Drone technology, now inexpensive and accessible, is continuously evolving and being put to several novel uses around the world. Here there are top 10 stunning applications of drone technology (Figure 2.10) [18, 19].

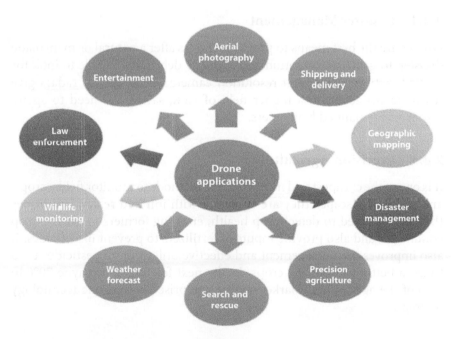

Figure 2.10 Applications of drone technology [14].

2.4.1   Aerial Photography

Drones are now being used to capture footage, Fast paced action and scientific fiction scenes with cinematography films, real estate and sports photography and further more journalists are used in an innovative to identify footage and information in live broadcasting.

2.4.2   Shipping and Delivery

Major service-based and oriented companies like Amazon, DHL, and UPS companies are in favor of drone delivery, saving a lot of manpower and avoiding road traffic, and many emergency deliveries to the hospitals, food, beverages, and many shipping's.

2.4.3   Geographic Mapping

Many available to amateurs, professionals are using drones for acquiring high-resolution data and also can reach where a man cannot reach locations such as coastal lines, mountains, and islands; to create 3D Maps and contribute to crowd-sourced mapping applications

2.4.4   Disaster Management

Drones are the best means to provide services after a natural or man-made disaster, to gather information and navigate debris and rubble to look for injured victims. Their high-resolution cameras, sensors, and radars give rescue teams access to a higher field of view, saving the need to spend resources on manned helicopters.

2.4.5   Precision Agriculture

It is one of the cheapest and most effective methods to monitor fields, crops, and paddy landscape. They are equipped with infrared sensors in drones that can be tuned to detect crop health, enabling formers to get the crop conditions, and also provide inputs of fertilizers to prevent insecticides. It also improves the management and effective utilization of pesticides used to get a better yield of the crops. In the next few years, it maybe 70% to 80% of the agricultural market will be comprised with drone technology in usage.

2.4.6   Search and Rescue

Especially, with the presence of thermal sensors provide drones' night vision and makes them a powerful aid for surveillance information. For example, a drone can be utilized to lower a walkie-talkie, GPS locator, medicines, food supplies, clothes, and water to stranded victims before rescue crews can move them to someplace else. Drones are able to discover the location of lost persons and unfortunate victims, especially in critical conditions or challenging terrains. Besides locating victims, a drone can drop supplies to unreachable locations in war-torn or disaster-stricken countries.

2.4.7   Weather Forecast

Drones are being best developed to monitor dangerous and unpredictable weather. Due to their being cheap and unmanned, drones are used to send into hurricanes and tornadoes, hence many researchers, engineers, scientists, and weather forecasters acquire new insights into their behavior and trajectory. Their specialized sensors would be used to detail weather parameters, collect data and mappings and also forecast weather and prevent mishaps.

2.4.8   Wildlife Monitoring

Drones are serving as a deterrent to poachers. They provide utmost protection to animals, like elephants, rhinos, and big cats, a favorite target for poachers. They perform effectively with their thermal cameras; sensors have the ability to operate day and night. It enables them to monitor and research goes well on wildlife without causing any disturbance and protect forest shelter and provide insight into their habitat, behavior, and patterns.

2.4.9   Law Enforcement

Drones are also protecting the law and used for cooperation by capture footage with surveillance of large crowds and ensure public safety. Not only that they show priority the things like where mishap possibility, traffic monitoring, and also assist in monitoring criminal and illegal activities. To speak frankly, it is a fire on investigations, smugglers of migrants and illegal transportation of drugs via coastal lines and boarders are monitored like a patrolling purpose.

2.4.10   Entertainment

Drones are being developed to provide entertainment for players so that they can be used in fight clubs. Known as a cage match, two contenders and their drones are put up against each other. The destruction of any of the player's drones results in the other's win. Moreover, artificial drone intelligence is used in several ways to capture videos and photographs, for example, the dronie, which is used to take selfies.

As technology advances, drones will become more robust and advanced, accommodating longer flight times and heavier loads. The industry comes with immense opportunities for businesses, gradually becoming inevitable for them. It is, therefore, important for organizations to study the scope of drone technology in their area of business, build the required infrastructure, and test their services across it.

2.5   Drones in Enterprises: What Value Do They Add? Work Place Safety and Industry Benchmarks

Drones and drones everywhere, but how to apply them? It's not a futuristic poem. It's real. Drones in enterprises, whether in surveying and mapping or asset inspections, have become fast-growing productivity-enablers [4].

Here is a list of Questions is important and answers are provided to stay:

What is fueling this steep increase in interest for drones?
Drones have been associated with enthusiasts and hobbyists. That's changed in the last decade. Enterprises have understood the importance of going the drone-way. Industries like energy and utilities, oil and gas, mining, railway, construction and real estate, agriculture, etc. have taken to drones to improve their overall operational efficiency.

How does a drone improve industrial operational efficiency?
Drones are streamlining and optimizing processes within multiple core pillars of these industries. Think about mining or energy. Such cost-intensive industries have a lot riding on each decision they make. The input, production and operation volume often stretch to millions of units.

Site exploration and planning: Top quality surveying and mapping Right at the start, they need a stark and deep understanding of the land on which they would build or develop the plant. There are many variables to comprehend, how the land would be prepared, the apparent soil, vegetation coverage, the watershed proximity, the logistics network used to haul the materials, etc. These are all critical decisions that won't tolerate even a little error. Miscalculations would result in lasting inefficiencies for the plant. Precise surveying and mapping are extremely important. On top of operations with live and detailed asset inspections Industrial asset inspections are critical for many reasons. They depreciate over time and their value needs to be realized through consistent and highly-efficient usage [15].

2.5.1 Total Workplace Safety with Drones

This means that they can be utilized in multiple scenarios, otherwise dangerous for humans. The surveying and mapping or asset inspections often require the surveyors to access hard-to-reach areas for effective visuals. They use railings, scaffolding, cherry-pickers, rope-access, etc. for such manual inspections. These are safety hazards and should be avoided. These activities also put them in close proximity of heavy industrial fumes or dangerous equipment. The drones are an effective engagement to put all the surveyors and operators out of harm's way [5]. Drones can smoothly hover or move overhead, reach tricky areas and fly close to dangerous (toxic or high temperature) equipment or stockpiles. They give instant surveillance of many acres, regularly and whenever required. Drones have helped secure and save multiple lives across enterprises, especially with high-risk endeavors like blast planning in mining. They also act as first respondents in case of any emergency to help the operator to better direct resources and replenishment, further securing the interests of the potential victims [16].

2.5.2 Future of Drones with *Idea Forge's* Industry Benchmarks

Drones will move to be the critical turn-key engagements for total-efficiency across the upstream, midstream and downstream operations. The drone developers who keep improving their offerings will move into dominant positions, and the ones who stick to low-endurance hobby-centric drones would be pushed in a niche [5, 6].

Idea Forge has been setting the tone for high-endurance drones that have longer, faster and better flights than any other drone in their class.

This means that they...

> ➢ Cover more area under exploration or inspection;
> ➢ Fly far more stable than others giving fine and accurate readings even down to centimeters;

> Fly through adverse weather and altitude with the same consistency;
> Have extremely user-friendly and easy-to-deploy mechanisms;
> Can be used across industries and scenarios with high-accountability, durability and consistency.
> There's far more in the future of drones, however, their enterprise future is the brightest. In the next five years, drones would be almost a necessary part of all surveying, mapping and asset inspection processes.

2.6 Advantages and Disadvantages of Drones

Because of their improved safety, UAVs are used in a variety of situations. Drones use their remote-control capabilities to monitor sites, convey possible risks, and alert people to potentially dangerous situations. Refineries, pipelines, and flare stacks, to name a few. Not only that, but drone technology is also used by the military during high-risk situations. Their characteristics enable them to collect real-time data to develop and maintain a secure environment [8].

Another benefit that balances out the benefits and drawbacks of drones is the security that surrounds them. Drone operators can use an Unmanned Aircraft System (UAS) to provide safety and surveillance to private organizations, possible venues, and other expenditures with the appropriate approvals and licenses. Drones can also gather reliable data from natural catastrophes to help with safety and recovery operations [17].

For industrial specialists, acquiring appropriate information from difficult-to-reach regions is a piece of cake using UAVs. It is the best option for overcoming the limits of traditional methods in terms of worker safety, particularly in dangerous scenarios such as radiation monitoring and high-voltage line inspection. Drones also offer for a more cost-effective approach to these examinations.

Numerous concerns that crew members previously encountered, such as height, wind, weather, and radiation, have been replaced with more practical and safer choices thanks to the assistance of a Drone. Drones make inspections of tall and sophisticated structures like oil and gas refineries flare stacks, and pipelines simple and secure.

Drones are cost-effective, and they have excellent spatial resolution and reliability, giving them a viable weapon for military usage.

2.6.1   Significant Advantages

1. Cheap and cost-effective
2. High spatial resolution
3. High accuracy
4. Easier to deploy
5. No hindrances from clouds
6. A great tool for surveillance

2.6.2   Disadvantages of Drones

While drones have several advantages, they also have several drawbacks. UAVs are easily influenced and can violate the rights of a group or person. Though many people want to use drones to keep themselves secure, they may trespass on a variety of individual freedoms in the name of public safety.

Many drones have previously fired a weapon towards citizens, resulting in a significant amount of fatalities, injuries, and damage to property due to malfunctions or software errors. Other military personnel's safety is also jeopardized by drone incidents. Drones are continually being improved to reduce accidents or hazards that might endanger human health and safety.

Drones are used by many criminals as a means of identifying and monitoring their victims. The loud rotor noises are no longer a problem because they are unnoticed, allowing criminals to breach someone's privacy. Many drones equipped with thermal and night sensors detect vital parameters and efficiently target people the spy is now interested in. Because UAVs can collect reliable data, they can track routine habits and identify unusual behavior without requiring permission [12, 13].

The vulnerability of drone technology is a major downside to its development. Hackers can quickly get access to a drone's central control system and take full control of the drone. Without the original driver's understanding, the primary control system has important knowledge that hackers may leverage. Hackers can get access to personal data, corrupt or destroy files, and expose data to untrusted third parties.

As previously said, if one needs reliable, high-quality data, they must possess the required skillset. This specification would indicate that capturing, processing, and analyzing farming data would need specialized training or the use of a third-party drone service provider. Drone costs and associated resource expenses will gradually decrease as the number of operators in the sector grows [14].

2.6.3   Significant Disadvantages

1. Limited capabilities and coverage
2. Care required to use in populated areas
3. Chances of misuse due to easy operation
4. Innocent Civilian mortality due to inefficiency in recognizing targets.
5. UAV systems may possible to hack.
6. Ground Station communication fails; the vehicle may never return.
7. Refuel problems during flights.
8. If drone technology is exposed to the wrong people, Terrorist activities will increase drastically.

2.6.4   Best Uses for Drones and Its Applications

- Drone Technologies [13]
    1. Vertical Take-Off and Landing (VTOL)
    2. Radar positioning drone technology
    3. Obstacle Detection drone technology
    4. Gyro Stabilization drone technology
    5. Drone propulsion technology
    6. GPS drone technology
    7. Drone Transmission technology
    8. Live video drone technology

- Drone Technology uses and applications [13]
    1. Product Delivery
    2. Air Taxi/Drone Ambulance
    3. Disaster management
    4. Search and Rescue
    5. Aerial Photography
    6. Law Enforcement
    7. Weather Forecast
    8. Entertainment
    9. Wildlife Monitoring

2.7   Drone Technology as Career and Offered Jobs in the Current Industry

With drone technology certifications like diploma, graduation, and post-graduation, you can get plentiful job opportunities. The Association for UAVs International forecast that there will be more one lakhs new jobs by 2025 and also commercial drone technology is expected to meet more than $527.7 million by 2030, new drone job opportunities across the globe will be wide spread can prepare drone career as a future [10]. Below given type of roles in a drone industry offering with many exciting drone jobs that are being created by today. Which one is right for you based on your interest and strength?

- Precision Agricultural Surveyor
- Drone 3D Modeler Drone Farmer
- Drone Flight Instructor
- Wildlife Conservationist
- Drone Photographer and Film Maker
- Delivery/fulfillment Pilot
- Search and Rescue Worker
- Energy Inspector
- Drone Journalist
- Geographic information Systems Mapping Technician
- Police Drone Operator
- Drone-Assisted Property Manager

2.8   Societal Impact—Commercial Drones

Based on the investigation on the availability of surrounding drones, these are the identification of broad classes and issues that need further attention [8, 11].

- Safety and Security
- Ownership and Privacy
- Personal and Commercial Liability
- Regulation Attempts and Challenge

The risk of accidents both digital and physical is destined to multiply, as well as the population of civilian drones and their consumers expand

Table 2.1 Issues, major challenges, and solutions [12].

Issues	Major challenges	Possible solutions
Privacy	Detention/Access to justice	Hardware and software for device detection, and data retention/Registry owners and devices
Ownership	Accountability	Registry of owners and devices/ Assign liability for UAV owners
Security	Control/Enforcement	Creation of new infrastructure and development of proper assets: UAV trackers devices/automatic safe landing/Establishment of insurance entities etc.
Regulatory	Lack of comprehensive rules and uniformity across jurisdictions	Redefinition of "reasonable expectation of privacy"/Definition of physical aerial boundaries/ Centralization of powers
Business Models	Lack of clear guidelines to operate in compliance with the law	Promote regulations for the development UAV-related technologies

universally. How the emerging technology can be best harnessed to serve the broad interests of the society and would be depends on the ability of the future success in varied stakeholders to reconsider. However thank to internet revolution, artificial intelligence in personal computing to confluence of technical, social, regulatory and cultural trends and efforts. Table 2.1 shows the issues, major challenges and solutions [12].

2.9   Drones Research Challenges and Solutions

At present, the entire drone industry is battling on three fronts – policy, market and operational. However, what is more challenging in the current time is an operational challenge such as understanding regulations, requirement gathering, analyzing output, and project execution [20].

Firstly, it is important for organizations adopting drone technologies to understand the regulations which are currently being proposed. Regulations would impact the implementation of applications in different industries. This is the foundation of any large-scale implementation of UAS technology [8].

Secondly, the implementation of large-scale drones' technology applications would require gathering the right set of requirements across the value chain of the industry. Organizations should look at the potential of drone technology in conjunction with other emerging technology areas like 3D modeling, AR, etc. The right combination of UAS specifications along with the emerging technologies provides the optimized benefit for each use case.

Thirdly, companies should evaluate and create an operational plan to integrate drones into their operations.

There are four key dimensions that companies should evaluate – platforms for acquiring data, ways to monitor the data, frequency and accuracy of data collection and finally, the feasibility of these options [21].

The Drone market in India has a lot of potential. As seen in previous examples, there is enough scope for drone application and adoption in India for mapping and surveying. Today, there are very few players in the Indian market that can provide such services and cater to such a large market. Idea Forge is one such name that provides superior endurance and high-functioning drones with an array of products [22].

2.10 Conclusion

The importance of drone technology is quite evident from the above discussion. The wide range of technologies conveys real-time applications of drones. With more weight capacity, robust and advanced technologies, longer flight duration and maneuverability, drones can be much more useful than now. Integration of different drone technologies, and wide range of sizes and capacity of such drones, it is an inevitable asset for businesses. There is huge scope of drones in many other fields such as agriculture, garbage management and sanitation, traffic monitoring etc. Governments and businesses should build required infrastructure and create policies for Real Time applications of drones. In the future, the interconnectedness and convergence of drones will continue to increase as society embraces the next version of the Internet & mobile technology. While drones' technology is widespread, there are a number of issues and challenges for multidisciplinary researchers. There is a need to expand the collaborative research so that multitasking and parallel processing UAVs are introduced.

References

1. Shahzadi, R., Ali, M., Khan, H.Z., Naeem, M., UAV assisted 5G and beyond wireless networks: A survey. *J. Netw. Comput. Appl.*, 189, 20–31, 2021.
2. Bhoi, S.K., Jena, K.K., Panda, S.K., Long, H.V. *et al.*, An internet of things assisted unmanned aerial vehicle based artificial intelligence model for rice pest detection. *Microprocess. Microsyst.*, 80, 37–49, 103607, February 2021.
3. Kumar, R. and Agrawal, A.K., Drone GPS data analysis for flight path reconstruction: A study on DJI, parrot & yuneec make drones. *Forensic Sci. Int.*, 38, 301182, September 2021, https://doi.org/10.1016/j.fsidi.2021.301182
4. OLA Mobility (OMI) White Paper, Giving wings to a drone-powered India: Mapping enablers and opportunities, in: *Urban Mobility, UM WP*, pp. 1–27, November 2021, www.ola.institute.
5. IdeaForge, How drones boost disaster response, search and rescue, 2020. https://ideaforgetech.com/blogs/how-drones-boost-disaster-response-and-rescue
6. Invest India, The growing market for drone technologies in India, 2020, https://www.investindia.gov.in/team-india-blogs/growing-market-drone-technologies-india
7. Legins, K., Whorter, R., Sullivan, A., *Unmanned Vehicle Systems: Product Profiles and Guidance*, pp. 1–24, Unicef Supply Division, USA, Springers, July, 25–29, 2021.
8. Al-Omary, A.Y., Chapter 11 Integrating odor sensing in robotics, *Electronic Nose Technologies and Advances in Machine Olfaction 2018*, IGI Global, Pages: 20, Alauddin Yousif Al-Omary (University of Bahrain, Bahrain), 2018.
9. Norzailawati Mohd Noor, Intan Zulaikha Mastor, Alias Abdullah. *Chapter 84; 'UAV/Drone Zoning in Urban Planning: Review on Legals and Privacy*, Springer Science and Business Media LLC, 2018.
10. FICCI & EY, *Make in India for Unmanned Aircraft Systems*, 2018. https://ficci.in/spdocument/23003/make-in-india-for-uas.pdf
11. Greaves, M., Introduction to drone technology, in: *Drones on Demand*, pp. 1–15, ARPAS-UK, The UK Drone Association, UK, https://www.gov.uk/government/organisations/air-accidents-investigation-branch
12. Gopi, A.G., Rao, B., Maione, R., The societal impact of commercial drones. *Technol. Soc.*, May 2016. https://www.researchsquare.com/article/rs-7738/v1
13. Vergouw, B., Nagel, H., Bondt, G., Custers, B., Drone technology: Types, payloads, applications, frequency spectrum issues and future developments, T.M.C. Asser press and the authorsDivision eLaw, Center for Law and Digital Technologies, Organization Leiden University, Leiden, The Netherlands, 2016.

14. Vergouw, B., Nagel, H., Bondt, G., Custers, B., Chapter-2: Drone technology: Types, payloads, applications, frequency spectrum issues and future developments, in: *The Future of Drone Use, Information Technology and Law Series*, B. Custers (Ed.), vol. 27, pp. 21–47, T.M.C. Asser Press, Organization Leiden University, Leiden, The Netherlands, 2016, http://springers.com/978-94-6265-131-9.

15. Alley-Young, G., Drone (unmanned aerial vehicle), in: *Salem Press Encyclopedia*, pp. 1–3, 2015.

16. Vincenzi, D.A., Terwilliger, B.A., Ison, D.C., Unmanned aerial system (UAS) human-machine interfaces: New paradigms in command and control. *Proc. Manuf.*, 2015. https://nsuworks.nova.edu/gscis_etd/1018 https://nsuworks.nova.edu/gscis_etd/1018.

17. Mallapur, C., India tops list of drone-importing nations. *Business Standard*, 2015. https://www.business-standard.com/article/specials/india-tops-list-of-drone-importing-nations-115050400136_1.html.

18. Patrik, A., Utama, G., Gunawan, A.A.S., Chowanda, A., Suroso, J.S., Shofiyanti, R., Budiharto, W., GNSS-based navigation systems of autonomous drone for delivering items, 2019. https://journalofbigdata.springeropen.com/articles/10.1186/s40537-019-0214-3

19. Verboven, J., Chapter 1: Introduction, No Fly Drone – Drones versus the right to privacy, Thesis Master Law & Technology, June 2016.

20. Kakaes, K., What drones can do and how they can do it, in: *Drones and Aerial Observation*, pp. 9–17, http://drones.newamerica.org/primer/Chapter%201.pdf.

21. Booth, P.T., Introduction to drones as tools for research and monitoring, pp. 1–13, Otsego County Conservation Association intern, 2015. Department of Geography, SUNY Oneonta, Funding provided by the Otsego County Conservation Association, 2015.

22. Clarke, R. and Moses, L.B., The regulation of civilian drones' impacts on public safety. *Comput. Law Secur. Rev.*, 30, 3, 263–285, June 2014.

3

Drone/UAV Design Development is Important in a Wide Range of Applications: A Critical Review

M. V. Kamal[1]*, P. Dileep[1], G. Sharada[2], V. Suneetha[1] and M. Gayatri[1]

[1]Department of Computer Science & Engineering, Malla Reddy College of Engineering & Technology, Hyderabad, India
[2]Department of Information Technology, Malla Reddy College of Engineering & Technology, Hyderabad, India

Abstract

In recent years, flying robots have played an increasingly important and growing role in the mining industry, transportation, as well as civilian and military applications. Moreover, in the last decades, researchers' attention has focused on the development of new designs of drones for different applications. In this pandemic situation, drones play a significant role in delivering drugs and foods. Drones have the potential to be dependable medical delivery platforms for laboratory and microbiological samples, emergency medical equipment, vaccines, and pharmaceuticals, among other things. Drone use has been prioritized by government agencies. The next steps will include aggressive safety research initiatives, increased public awareness, industry expansion, and participation. A literature survey was carried out to understand the current state of the art and set a research directive for the advanced drones. There is also a great deal of interest in developing novel drones that can fly autonomously in such different locations and environments and perform a wide variety of missions. Besides classification, the discussion also includes the application of drones in various fields. Apart from the design and fabrication challenges of micro drones, the concept of emerging and controlling drones is also discussed in detail here. Furthermore, existing system limitations and controlling factors were revealed. The current applications of drones as well as their future potential in the industry are also discussed.

*Corresponding author: kamalmv@gmail.com

Sachi Nandan Mohanty, J.V.R. Ravindra, G. Surya Narayana, Chinmaya Ranjan Pattnaik and Y. Mohamed Sirajudeen (eds.) Drone Technology: Future Trends and Practical Applications, (53–68) © 2023 Scrivener Publishing LLC

Keywords: Drones, UAV, classifications, applications, literature review

3.1   Introduction

Old military unmanned target aircraft had a loud, rhythmic sound that was like a male bee, leading to the public word "drone" being created. Despite being widely accepted, the word has faced vehement pushback from regulators and aviation experts. In the 1980s, the term "unmanned aerial vehicle" (UAV) was used to refer to remotely controlled or autonomous, multipurpose aerial aircraft that can carry a payload [1]. This definition distinguished UAVs from other aerial systems such as gliders, balloons, cruise missiles, and ballistic vehicles. Unmanned aerial systems (UAS) is a more generally used phrase in professional circles that refers to one or more unmanned aerial vehicles equipped with a sensory array, data terminal, and electronic data link [2]. A remotely piloted aircraft system or remotely piloted vehicle (RPV) is another name for the drone (RPAS). While RPAS is a more official and generally acknowledged word, RPV has mostly been utilized in military circumstances [3]. For the sake of clarity, the term "drone" will be used throughout this article.

Drones, or unmanned aerial vehicles, have a broad definition. Given the variety of configurations available, this is understandable.

In actuality, a UAV is any aerial vehicle, whether flown remotely or autonomously, that does not depend on board human operator for flight [4]. UAVs come in a range of sizes, from substantial military drones with wingspan of about 200 feet to inch-wide micro drones that are readily available in the marketplace. Their flight ranges vary, from a few feet around the operator for commercial drones to over 16,000 miles without landing for advanced military drones. Similarly, their maximum flight altitude varies greatly, from a few feet to 60,000 feet [5]. A similar design can be found in the majority of commercially available drones today. The basic design includes a flight control microcontroller, propellers and four to eight motors, electronic speed control, a battery, a radio receiver, all of which are mounted on a light plastic or metal frame [6]. For navigation, a GPS device can be utilized, and gyroscopes and other sensors can be added to improve the drone's mid-air stability. The majority of hobbyist drones come equipped with a camera for taking aerial photos and a gimbal for improved image stabilization. Other sensors can also be attached, but increased functionality and weight come at a cost [7]. Some of the most well-known hardware manufacturers include 3DRobotics, DJI, and Parrot which sell both assembled drones and drone parts [8].

Figure 3.1 A quadcopter with a smartphone as a flight controller [16].

As new demands arise, UAV technology is rapidly developing, and UAV solutions are being proposed at a faster rate. Specific UAV uses, as well as commercial market competitiveness, affect drone characteristics [9–11, 26]. A survey of the most current uses of unmanned aerial vehicles in the cryosphere was published. In terms of flying altitudes, sensor types, data acquisition windows, overlap dimensions, viewing angles, and revisits, UAVs outperform conventional spaceborne or airborne remote sensing devices [12–15]. Small quadcopter drones have attracted a lot of attention from the general public since they are simple to fly, cost-effective, safe, and versatile. Pilot tests on the usefulness of quadcopter drones for traffic surveillance and incident monitoring were done with two quadcopter drones equipped with transmission gear and video collecting as well as a ground station. The typical quadcopter with a smartphone as a flight controller is shown in Figure 3.1 [16].

This literature proposes a new taxonomy for drones in this study that includes a broader variety of classes and gives a more thorough categorization. The remainder of this study will be organized as follows: In Section 3.2, discover how drones are classified in an unconventional way. The third section looks at and explores the various uses for drones. Finally, Section 3.4 deals with conclusions and future important of drones.

3.2   Classification of Various Categories of Air Drones

The development of a smaller air drone known as a micro air vehicle [17] has increased the demand for intelligence missions in recent decades. As a result, there is currently a big push to design and build small air drones for

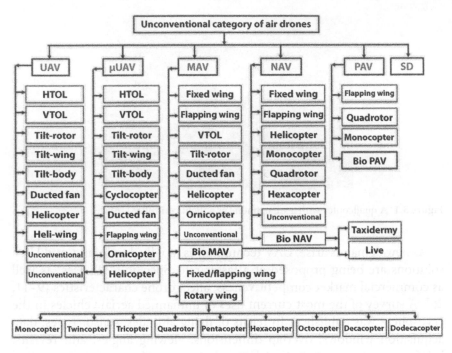

Figure 3.2 Classification of various categories of air drones [18].

Table 3.1 Various platforms of UAVs [20].

UAVs	Type	Weight (kg)	Altitude capacity (m)	Power	Endurance (min)
DJI Mavic	Rotary	0.72	5000	Battery (B)	27
DJI Mavic Air 2	Rotary	0.57	5000	B	34
Trimble UX5	Fixed	1.13	5000	B	50
DJI Phantom 4	Rotary	1.36	6000	B	28

special missions. As a result of these efforts, a variety of small drones with unique shapes and flight modes have been developed. The classification of various category of air drones are shown in Figure 3.2, where HTOL stands for Horizontal Take-Off and Landing. Production costs, wing loading, weight, speed, range, endurance, maximum altitude, and wing span are all crucial design factors that separate various drone types and improve the effectiveness of categorization systems. Additionally, drones can be categorized according to the kind of engine they have [18]. UAVs, for example, are frequently powered by gasoline engines, whereas MAVs are powered by electric motors. Drone propulsion systems vary by model. Figure 3.2 depicts the various drone models based on their configuration. Various platforms of UAVs are given in Table 3.1.

3.2.1   VTOL and HTOL UAVs

After years of research and development, HTOL drones now come in four different configurations, each defined by lift/mass balance, stability, and control. Tailplane-aft, tailless or flying, wing tailplane-forward, and tailplane-aft on booms, UAVs are some of the options [19]. The propulsion systems in the aforementioned layouts could be positioned either in front of or behind the UAV's fuselage. This kind of drone can launch and land vertically, thus there is no runway required.

3.2.2   Tilt-Body, Tilt-Rotor, and Tilt-Wingducted Fan UAVs

VTOL drones perform better than HTOL drones when hovering. They have a low cruise speed because the retreating blades stall, whereas longer-range missions often require UAVs with a higher cruise speed [21]. On the other side, taking off and landing from a vertical posture is really helpful. These limitations led to the idea of a drone that combines VTOL and HTOL capabilities being put forth [22]. Due to this, hybrid drones such as ducted fan, tilt-rotor, tilt-wing, and tilt-body UAVs are becoming a reality [23–25]. Rotors of tilt-rotor UAVs are upright during vertical flight but tilt 90 degrees forward during cruise flight. The engines of tiltwing UAVs are frequently attached to the wings and tilt with the wing.

The free wing tilt-body UAV is a new type of drone that differs from fixed wings and rotary wings. It isn't a rotary-wing aircraft or fixed-wing, nor is it a hybrid of the two. The central lifting body and the left/right wing pair are both free to rotate around the spanwise shaft, free in relation to each other and free in relation to the relative wind [26–28]. In response to external commands, the tilt-body is an unusual attachment of a boom type to a fuselage, in which the incidence angle of the boom relative to the fuselage changes. Some benefits of this kind of drone include lower sensitivity to changes in the center of gravity (CG), low speed loitering, and short take-off and landing (STOL) [29]. Drones with ducted fan thrusters have thrusters that are encased in a duct. The drone's thruster is known as 'Fan'. This fan consists of two counter-rotating sections that work together to lessen the torque that causes the body to rotate. UAVs have two counter-rotors, four control surfaces, and use ductwork as a fan in addition to vertical takeoff and landing. Despite the ease of getting on and off a cruise flight, flow separation from the duct is still a challenge [30, 31].

3.2.3 Heli-Wing and Helicopter UAVs

Unmanned helicopters for landing, vertical hovering flight, and takeoff, are currently being designed and built by researchers. Quad-rotor, single-rotor, tandem rotor, and coaxial rotor, helicopter UAVs are the four types of helicopter UAVs [30, 32]. Heli-wing other types of drones that use a rotating wing as a blade are known as unmanned aerial vehicles, or UAVs. They have the ability to fly vertically as a helicopter as well as horizontally as a fixed-wing UAV [33, 34].

3.3 Drones Acting on Various Industries

3.3.1 Military Drones

Cost, a lower risk of personnel loss, and a reduction in military financing are driving the military's migration to drones [35]. Drone pilots are unable to be shot down, tortured, or apprehended, thus this makes sense [36]. Governments place a high priority on national security, which might involve investing millions of dollars to ensure a country's safety. Drones are now being employed to assist in the detection of criminal activities and other aspects of land-border security [37, 38]. Drone technology has moved into civilian contexts as a result of military demand, particularly in the agricultural, medical, and transportation industries [39]. The military concept of using drones to deliver precision-guided munitions has been adapted or expanded to help humanitarian efforts in response to natural disasters like earthquakes and floods [40, 41].

3.3.2 Medical Drones

Drones were employed in various nations to carry automated external defibrillators (AEDs), personal protective equipment (PPE), and other supplies to emergency situations and medical facilities during the Covid-19 outbreak [42]. This assisted in reducing the spread of virus infections. In order to complete a medical diagnosis, doctors occasionally need blood samples. In these situations, the drone is preferable to the usual delivery method via road because it can complete the task more quickly [43]. Drones can help save lives in Africa, as seen by the deployment of a Zipline drone to carry blood to health facilities in Rwanda [44].

Figure 3.3 Drones to deliver medications [13].

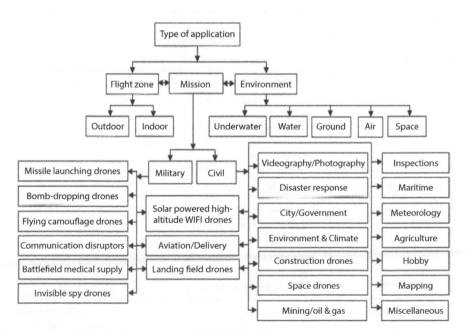

Figure 3.4 Drone applications in various fields [17].

Drones have helped cross the transportation gap in Africa amid medical situations like malaria-related anemia and pregnancy-related difficulties [45]. Despite having a fragile road infrastructure, drones and mobile technology enable poor countries to leapfrog affluent countries in terms of healthcare delivery to rural areas [46]. Drones are illustrated in Figure 3.3 delivering pharmaceuticals and Figure 3.4 depicts the use of drones in many fields. A study outlining UAV use in several medical fields and sectors summarized in Table 3.2.

Table 3.2 A literature in the application of UAV in different medical sectors.

Various application in medical sector	Objective	Sources
Environmental military missions, and healthcare	The intention is to help the reader transmit the proper techniques and procedures for both aquatic and aerial requests.	[47]
Economic and Energy sources	The authors present the current state of the art in terms of hybrid UAV energy management technique.	[48]
Consumer drone delivery	From the perspective of view of customers, the article examines consumer preferences for drone delivery.	[49]
Cardiac Arrest	Develop a method for locating a network of medical drones using a geographic strategy.	[50]
Network grid	A complex network theory is used to assess UAV resistance.	[51]
Cardiac arrest	A drone grid's optimization	[52]
Rescue	The aim is to look into the employment of motorized search and rescue workers in a hilly environment, as well as the prospective use of unmanned aerial vehicles (UAVs) in identifying and detecting casualties.	[53]

(Continued)

Table 3.2 A literature in the application of UAV in different medical sectors. (*Continued*)

Various application in medical sector	Objective	Sources
Blood transportation	The goal is to use unmanned aerial vehicles to optimize blood product transportation.	[54]
Blood sample transportation	The study's goal is to provide an overview of how blood compartment administration has progressed and to assess the experience with airborne blood transfusions.	[55]
Telesurgery	Using a UAV as a grid topology, the authors stretch a potential clinical robot.	[56]
Different sectors	The goal is to look at how drones are used in healthcare right now.	[57]
Response to natural disasters and provision of humanitarian aid	The goal is to determine how UAVs can assist fighters in the event of a natural disaster such as an earthquake, tsunami, flooding, or other natural disaster.	[58]
Blood transportation	The goal is to sketch out the need, risks, and potential of using small unmanned aerial vehicles (UAVs) to transfer blood and pharmaceuticals to hospitals.	[59]
Emergency medical	Out-of-Hospital Defibrillator Delivery Using a Drone	[60]
Infectious diseases, microbiology	Drones in healthcare-review	[61]
Cardiac arrest	Examine the viability of using a drone system to respond to cardiac arrests outside of the hospital.	[62]

3.3.3   Agricultural Drones

An exact estimate of loss can be provided using drone images. Introduce a drone as a viable tool for more precisely assessing consequently, compensation costs and wildlife damage to crops [63]. Farmers can make informed decisions on how to use farm inputs thanks to agricultural drones, which provide real-time data [64]. By flying near to the crops, micro-drones can improve and enrich the data collected, allowing for better spatiotemporal data collecting by Anthony *et al.* [65]. With drones helping with livestock management, crop monitoring, crop spraying, irrigation mapping, planting, and other aspects of the farming business, drone technology is currently assisting agricultural firms in meeting the evolving and expanding demands of the future. Drones are especially well-suited for use in agriculture because to a number of factors, including their ability to help farmers maximize their output by spotting issues before they become serious and manage crops with the aid of specialized cameras that spot pests and water shortages [66].

3.4   Conclusions and Future Scope

The different field analyses of drones listed above confirm the tremendous potential of medical, civil, agricultural, and military drones in their diverse sectors. Drones are frequently used to collect real-time images and sensor data from farm fields, helping farmers to make better input selections. Aerial drones are technical instruments that can help medical staff do their tasks more efficiently and successfully, saving more lives in the process. Drones can be used by the military while keeping civilian lives in mind. Research and development are critical in the advancement of drone technology, as it can help mitigate or reduce the risks and weaknesses of deploying drones in the specified domains. Enabling legislation are required around the world in order to fully exploit the promise of drone technology. Existing restrictions might be made available to users and potential users, which could help to prevent unauthorized use. More research is needed on integrating drones into including flight durations and extending payload, current transportation systems and supply chains.

Drones and other autonomous vehicles provide enormous potential for the healthcare business, and countries should be able to capitalize on it. They may provide a solution to the problems that have emerged as a result of the presence of coronavirus in aerial disinfection, spraying, and surveillance, consumer drone delivery, lessons learned in the fight against this virus and temperature detection using thermal cameras. Drones will be able to optimize the way of eliminating pollution with a very high percentage due to greater flight flexibility. Unmanned aerial vehicles (UAVs) are delivering crucial medicines to rural and remote areas all over the world. When road transit is threatened or hampered by disease, they are frequently the best option for getting a product to its destination as quickly as possible.

Recent research and studies in the topic offlapping wing vehicles, flying drones, and including fixed were gathered and discussed extensively. At initially, a new classification for these drones was proposed. Unmanned air vehicles, smart dust nano air vehicles, pico air vehicles, and micro air vehicles are all included in this classification. These flying drones can perform a variety of civilian and military purposes. Search and rescue, space exploration delivery, environmental protection, mailing, and were among the various missions considered. For all types of drones, the design methodologies employed and their challenges were also consolidated. The design difficulties' potential solutions were offered and considered.

References

1. Sreenath, S., Malik, H., Husnu, N., Kalaichelavan, K., Assessment and use of unmanned aerial vehicle for civil structural health monitoring. *Proc. Comput. Sci.*, 170, 656–663, 2020.
2. Mozaffari, M., Saad, W., Bennis, M., Debbah, M., Mobile unmanned aerial vehicles (UAVs) for energy-efficient Internet of Things communications. *IEEE Trans. Wirel. Commun.*, 16, 11, 7574–7589, 2017.
3. Kreps, S.E. and Wallace, G.P., International law, military effectiveness, and public support for drone strikes. *J. Peace Res.*, 53, 6, 830–844, 2016.
4. Tzelepi, M. and Tefas, A., Graph embedded convolutional neural networks in human crowd detection for drone flight safety. *IEEE Trans. Emerg. Top. Comput. Intell.*, 5, 2, 191–204, 2019.
5. Bergen, P. and Tiedemann, K., *The Year of the Drone*, p. 24, New America Foundation, New America Foundation-2010 (vcnv.org), 2010.
6. Hanford, S., Long, L., Horn, J., A small semi-autonomous rotary-wing unmanned air vehicle (UAV), in: *Infotech@ Aerospace*, p. 7077, 2005.
7. Syifa, M., Park, S.J., Lee, C.W., Detection of the pine wilt disease tree candidates for drone remote sensing using artificial intelligence techniques. *Engineering*, 6, 8, 919–926, 2020.
8. McNeil, B. and Snow, C., The truth about drones in mapping and surveying. *Skylogic Res.*, 200, 1–6, 2016.

9. Kanellakis, C. and Nikolakopoulos, G., Survey on computer vision for UAVs: Current developments and trends. *J. Intell. Robot. Syst.*, 87, 1, 141–168, 2017.

10. Otto, A., Agatz, N., Campbell, J., Golden, B., Pesch, E., Optimization approaches for civil applications of unmanned aerial vehicles (UAVs) or aerial drones: A survey. *Networks*, 72, 4, 411–458, 2018.

11. Dupont, Q.F., Chua, D.K., Tashrif, A., Abbott, E.L., Potential applications of UAV along the construction's value chain. *Proc. Eng.*, 182, 165–173, 2017.

12. Gaffey, C. and Bhardwaj, A., Applications of unmanned aerial vehicles in cryosphere: Latest advances and prospects. *Remote Sens.*, 12, 6, 948, 2020.

13. Kangunde, V., Jamisola, R.S., Theophilus, E.K., A review on drones controlled in real-time. *Int. J. Dyn. Control*, 9, 4, 1832–1846, 2021.

14. Näsi, R., Honkavaara, E., Blomqvist, M., Lyytikäinen-Saarenmaa, P., Hakala, T., Viljanen, N., Kantola, T., Holopainen, M., Remote sensing of bark beetle damage in urban forests at individual tree level using a novel hyperspectral camera from UAV and aircraft. *Urban For. Urban Green.*, 30, 72–83, 2018.

15. Zhang, J., Yang, C., Song, H., Hoffmann, W.C., Zhang, D., Zhang, G., Evaluation of an airborne remote sensing platform consisting of two consumer-grade cameras for crop identification. *Remote Sens.*, 8, 3, 257, 2016.

16. Astudillo, A., Muñoz, P., Álvarez, F., Rosero, E., Altitude and attitude cascade controller for a smartphone-based quadcopter, in: *2017 International Conference on Unmanned Aircraft Systems (ICUAS)*, Miami, FL, USA, IEEE, pp. 1447–1454, 2017.

17. Hassanalian, M. and Abdelkefi, A., Design, manufacturing, and flight testing of a fixed wing micro air vehicle with Zimmerman planform. *Meccanica*, 52, 6, 1265–1282, 2017.

18. Arjomandi, M., Agostino, S., Mammone, M., Nelson, M., Zhou, T., *Classification of Unmanned Aerial Vehicles*, pp. 1–48, Report for Mechanical Engineering class, University of Adelaide, Adelaide, Australia, 2006.

19. Sanchez, G., Escamilla, L., Hassanalian, M., Throneberry, G., Abdelkefi, A., Performance analysis and actuation mechanism selection of Albatross-inspired wing shape for tilt-wing drones, in: *AIAA Scitech 2019 Forum*, p. 2094, 2019.

20. Euchi, J., Do drones have a realistic place in a pandemic fight for delivering medical supplies in healthcare systems problems? *Chin. J. Aeronaut.*, 34, 2, 182–190, 2021.

21. Vijayanandh, R., Ramesh, M., Venkatesan, K., Kumar, G.R., Kumar, M.S., Rajkumar, R., Comparative acoustic analysis of modified unmanned aerial vehicle's propeller, in: *National Conference on IC Engines and Combustion*, Springer, Singapore, pp. 573–587, November 2019.

22. Darvishpoor, S., Roshanian, J., Raissi, A., Hassanalian, M., Configurations, flight mechanisms, and applications of unmanned aerial systems: A review. *Prog. Aerosp. Sci.*, 121, 100694, 2020.

23. Salazar, R.D., Hassanalian, M., Abdelkefi, A., Defining a conceptual design for a tilt-rotor micro air vehicle for a well-defined mission, in: *55th AIAA Aerospace Sciences Meeting*, p. 0239, 2017.

24. Sanchez, G., Salazar, R.D., Hassanalian, M., Abdelkefi, A., Sizing and performance analysis of albatross-inspired tilt-wing unmanned air vehicle, in: *2018 AIAA/ASCE/AHS/ASC Structures, Structural Dynamics, and Materials Conference*, p. 1445, 2018.

25. Elmeseiry, N., Alshaer, N., Ismail, T., A detailed survey and future directions of unmanned aerial vehicles (UAVs) with potential applications. *Aerospace*, 8, 12, 363, 2021.

26. Ucgun, H., Yuzgec, U., Bayilmis, C., A review on applications of rotary-wing unmanned aerial vehicle charging stations. *Int. J. Adv. Robot. Syst.*, 18, 3, 2021.

27. Yamaguchi, K. and Hara, S., On structural parameter optimization method for quad tilt-wing UAV based on indirect size estimation of domain of attraction. *IEEE Access*, 1678–1687, 2021.

28. Hill, A.C. and Rowan, Y.M., The black desert drone survey: New perspectives on an ancient landscape. *Remote Sens.*, 14, 3, 702, 2022.

29. Ro, K., Park, W., Kuk, T., Kamman, J., Flight testing of a free-wing tilt-body Aircraft. *In, AIAA Infotech@ Aerospace 2010*, p, 3449, 2010.

30. Austin, R., *Unmanned Aircraft Systems: UAVS Design, Development and Deployment*, John Wiley & Sons, 2011. https://www.wiley.com/en-1n/Unmanned+Aircraft+Systems%3A+UAVS+Design%2C+Development+and+Deployment-p-9780470058190

31. Ko, A., Ohanian, O., Gelhausen, P., Ducted fan UAV modeling and simulation in preliminary design, in: *AIAA Modeling and Simulation Technologies Conference and Exhibit*, p. 6375, August 2007.

32. Hoffmann, G., Huang, H., Waslander, S., Tomlin, C., Quadrotor helicopter flight dynamics and control: Theory and experiment, in: *AIAA Guidance, Navigation and Control Conference and Exhibit*, p. 6461, August 2007.

33. Singh, V., Skiles, S.M., Krager, J., Seepersad, C.C., Wood, K.L., Jensen, D., Concept generation and computational techniques applied to design for transformation, in: *IDETC/CIE in: Proceedings of the 32nd Design Automation Conference*, Philadelphia, PA, pp. 10–13, September 2006.

34. Ghazali, S.N.A.M., Anuar, H.A., Zakaria, S N A S, Yusoff, Z., Determining position of target subjects in maritime search and rescue (MSAR) operations using rotary wing Unmanned Aerial Vehicles (UAVs), in: *2016 International Conference on Information and Communication Technology (ICICTM)*, IEEE, pp. 1–4, May 2016.

35. McLean, W., *Drones Are Cheap, Soldiers Are Not: A Cost-Benefit Analysis of War*, p. 25, The Conversation, The Conversation US Inc, 2014. https://theconversation.com/drones-are-cheap-soldiers-are-not-a-cost-benefit-analysis-of-war-27924

36. Warrior, L.C., Drones and targeted killing: Costs, accountability, and US civil-military relations. *Orbis*, 59, 1, 95–110, 2015.
37. Shishkov, B., Hristozov, S., Janssen, M., Van den Hoven, J., Drones in land border missions: Benefits and accountability concerns, in: *Proceedings of the 6th International Conference on Telecommunications and Remote Sensing*, pp. 77–86, November 2017.
38. Yaacoub, J.P., Noura, H., Salman, O., Chehab, A., Security analysis of drones systems: Attacks, limitations, and recommendations. *Internet Things*, 11, 100218, 2020.
39. Vacca, A. and Onishi, H., Drones: military weapons, surveillance or mapping tools for environmental monitoring? The need for legal framework is required. *Transp. Res. Proc.*, 25, 51–62, 2017.
40. Mendoza, M.A., Alfonso, M.R., Lhuillery, S., A battle of drones: Utilizing legitimacy strategies for the transfer and diffusion of dual-use technologies. *Technol. Forecast. Soc. Change*, 166, 120539, 2021.
41. Yaacoub, J.P., Noura, H., Salman, O., Chehab, A., Security analysis of drones systems: Attacks, limitations, and recommendations. *Internet Things*, 11, 100218, 2020.
42. van Veelen, M.J., Kaufmann, M., Brugger, H., Strapazzon, G., Drone delivery of AED's and personal protective equipment in the era of SARS-CoV-2. *Resuscitation*, 152, 1–2, 2020.
43. Sachan, D., The age of drones: What might it mean for health? *Lancet*, 387, 10030, 1803–1804, 2016.
44. Ackerman, E. and Strickland, E., Medical delivery drones take flight in East Africa. *IEEE Spectr.*, 55, 1, 34–35, 2018.
45. Ling, G. and Draghic, N., Aerial drones for blood delivery. *Transfusion*, 59, S2, 1608–1611, 2019.
46. Scott, J. and Scott, C., Drone delivery models for healthcare, in: *Proceedings of the 50th Hawaii International Conference on System Sciences*, January 2017.
47. Sánchez-García, J., García-Campos, J.M., Arzamendia, M., Reina, D.G., Toral, S.L., Gregor, D., A survey on unmanned aerial and aquatic vehicle multi-hop networks: Wireless communications, evaluation tools and applications. *Comput. Commun.*, 119, 43–65, 2018.
48. Lei, T., Yang, Z., Lin, Z., Zhang, X., State of art on energy management strategy for hybrid-powered unmanned aerial vehicle. *Chin. J. Aeronaut.*, 32, 6, 1488–1503, 2019.
49. Kim, S.H., Choice model based analysis of consumer preference for drone delivery service. *J. Air Transp. Manage.*, 84, 101785, 2020.
50. Pulver, A., Wei, R., Mann, C., Locating AED enabled medical drones to enhance cardiac arrest response times. *Prehosp. Emerg. Care*, 20, 3, 378–389, 2016.
51. Xiaohong, W., Zhang, Y., Lizhi, W., Dawei, L., Guoqi, Z., Robustness evaluation method for unmanned aerial vehicle swarms based on complex network theory. *Chin. J. Aeronaut.*, 33, 1, 352–364, 2020.

52. Boutilier, J.J., Brooks, S.C., Janmohamed, A., Byers, A., Buick, J.E., Zhan, C., Schoellig, A.P., Cheskes, S., Morrison, L.J., Chan, T.C., Optimizing a drone network to deliver automated external defibrillators. *Circulation*, 135, 25, 2454–2465, 2017.

53. Karaca, Y., Cicek, M., Tatli, O., Sahin, A., Pasli, S., Beser, M.F., Turedi, S., The potential use of unmanned aircraft systems (drones) in mountain search and rescue operations. *Am. J. Emerg. Med.*, 36, 4, 583–588, 2018.

54. Amukele, T., Ness, P.M., Tobian, A.A., Boyd, J., Street, J., Drone transportation of blood products. *Transfusion*, 57, 3, 582–588, 2017.

55. Berns, K.S. and Zietlow, S.P., Blood usage in rotor-wing transport. *Air Med. J.*, 17, 3, 105–108, 1998.

56. Harnett, B.M., Doarn, C.R., Rosen, J., Hannaford, B., Broderick, T.J., Evaluation of unmanned airborne vehicles and mobile robotic telesurgery in an extreme environment. *Telemed. e-Health*, 14, 6, 539–544, 2008.

57. Amukele, T., Current state of drones in healthcare: Challenges and opportunities. *J. Appl. Lab. Med.*, 4, 2, 296–298, 2019.

58. Estrada, M.A.R. and Ndoma, A., The uses of unmanned aerial vehicles–UAV's-(or drones) in social logistic: Natural disasters response and humanitarian relief aid. *Proc. Comput. Sci.*, 149, 375–383, 2019.

59. Thiels, C.A., Aho, J.M., Zietlow, S.P., Jenkins, D.H., Use of unmanned aerial vehicles for medical product transport. *Air Med. J.*, 34, 2, 104–108, 2015.

60. Claesson, A., Bäckman, A., Ringh, M., Svensson, L., Nordberg, P., Djärv, T., Hollenberg, J., Time to delivery of an automated external defibrillator using a drone for simulated out-of-hospital cardiac arrests vs emergency medical services. *Jama*, 317, 22, 2332–2334, 2017.

61. Poljak, M. and Šterbenc, A., Use of drones in clinical microbiology and infectious diseases: Current status, challenges and barriers. *Clin. Microbiol. Infect.*, 26, 4, 425–430, 2020.

62. Claesson, A., Fredman, D., Svensson, L., Ringh, M., Hollenberg, J., Nordberg, P., Rosenqvist, M., Djarv, T., Österberg, S., Lennartsson, J., Ban, Y., Unmanned aerial vehicles (drones) in out-of-hospital-cardiac-arrest. *Scand. J. Trauma Resusc. Emerg. Med.*, 24, 1, 1–9, 2016.

63. Michez, A., Morelle, K., Lehaire, F., Widar, J., Authelet, M., Vermeulen, C., Lejeune, P., Use of unmanned aerial system to assess wildlife (sus scrofa) damage to crops (zea mays). *J. Unmanned Veh. Syst.*, 4, 4, 266–275, 2016.

64. Stehr, N.J., Drones: The newest technology for precision agriculture. *Nat. Sci. Educ.*, 44, 1, 89–91, 2015.

65. Anthony, D., Elbaum, S., Lorenz, A., Detweiler, C., On crop height estimation with UAVs, in: *2014 IEEE/RSJ International Conference on Intelligent Robots and Systems*, IEEE, pp. 4805–4812, September 2014.

66. Reinecke, M. and Prinsloo, T., The influence of drone monitoring on crop health and harvest size, in: *2017 1st International Conference on Next Generation Computing Applications (NextComp)*, IEEE, pp. 5–10, July 2017.

4

A Comprehensive Study on Design and Control of Unmanned Aerial Vehicles

P. Venkateshwar Reddy[1], P. Srinivasa Rao[1]*, M. Hrishikesh[1]
and C. Satya Kumar[2]

*[1]Department of Mechanical Engineering, Vardhaman College of Engineering,
Hyderabad, India
[2]Department of Computer Science & Engineering, Vardhaman College
of Engineering, Hyderabad, India*

Abstract

Drone construction and operation is still a hot topic of study, and we'll take a look at what's new in this subject. The discussion comes to energy usage, mobility and pace, and survival and robustness, as well as underlying physical scaling rules. The controlling of such vehicles is divided into low-level stabilization and higher-level modeling, such as trajectory planning, and we believe that integrating sensing with control and coordination is a critical issue. Finally, we go over some vehicle topologies and the exchange they imply. We examine the impact of multi-vehicle teams and multicopters with flown configurations. For Unmanned Air Vehicle (UAV) usage, fully automated activities are extremely desirable. The market for unmanned aerial vehicles (UAVs) is now one of the fastest expanding sectors in the aviation industry. UAVs have been in existence for a certain time, and there is growing trust in the skills and benefits given by autonomous operations. As a result, the deployment of these devices in combat duty will indeed be increased in the next years.

Keywords: Drones, unmanned aerial vehicles, control, design

*Corresponding author: professorpsrao@gmail.com

Sachi Nandan Mohanty, J.V.R. Ravindra, G. Surya Narayana, Chinmaya Ranjan Pattnaik
and Y. Mohamed Sirajudeen (eds.) Drone Technology: Future Trends and Practical Applications,
(69–98) © 2023 Scrivener Publishing LLC

4.1   Introduction

Drones have become popular because to significant cost savings and break-throughs in sensor technology, actuators, power storage, and processing. Remote sensing, physical contact, physical evaluation, and delivering packages are just a few of their uses. Larger Unmanned Aerial Vehicles (UAVs) may be used to transport human passengers, whether it be for fun or as flying cabs and urban transportation, referred to as Enhanced Maneuverability or Smart Urban Transportation. Drone function is challenged by (a) their typically unsteady operating aircraft, in which there is no simplified "protected" behavior inside the case of a problem; (b) the widespread restriction, which requires economic system both for sensing devices; and (c) extreme power consumption limitations, as examine the types of automatons.

Brooke-Holland [1] categorized armed UAVs into three classes during an assessment of UAVs employed by the UK armed services. There are four subcategories in Class I (a, b, c, and d). The minimal stand weight is paired with how UAVs are designed being used and in which they are anticipated to be controlled to begin the categorization process. Table 4.1 depicts this categorization.

Drones were categorized by weight, reach and durability, wing load, cruising altitudes, and engine design, according to Arjomandi *et al.* [2]. Drones were classed as super-heavy if it weighed over than 2000 kg, heavyweight if it weighed within 200 and 2000 kg, moderate if it weighed within 50 and 200 kg, light/mini if they weighed within 5 and 50 kg, and lastly

Table 4.1 Categorization of the drones based on their weights by Brooke-Holland [1].

S. no.	Class	Type	Weight range
1	Class I (a)	Nano	$W \leq 200$ g
2	Class I (b)	Micro	200 g $< W \leq 2$kg
3	Class I (c)	Mini	2 kg $< W \leq 20$kg
4	Class I (d)	Small	20 kg $< W \leq 150$kg
5	Class II	Tactical	150 kg $< W \leq 600$ kg
6	Class III	MALE/HALE/Strike	$W > 600$ kg

mini if it weighed below 5 kg [2]. Table 4.2 shows the categorization, which again is predicated on the weight of the drones.

Drones are categorized as HALE, MALE, TUAV (medium range or tactical UAV), MUAV or Mini UAV, MAV, and NAV by Gupta *et al.* [4]. Drones are divided into three categories by Cavoukian [5], notably micro and tiny UAVs, tactical Quadcopters, and major UAVs. Short distance, limited range, moderate range, wide coverage, durability, and moderate altitude long endurance (MALE) UAVs were his six divisions for tactical UAVs [5]. Drones are classed as micro, small, tactical, moderate to high height, and heavyweight by Weibel and Hansman [3]. The proposed categorization is shown in Table 4.3.

Table 4.2 Categorization of drones based on their weights by Arjomandi *et al.* [2].

S. no.	Designation	Weight range
1	Super Heavy	W > 2000 kg
2	Heavy	200 kg < W ≤ 2000 kg
3	Medium	50 kg < W ≤ 200 kg
4	Light	5 kg < W ≤ 50 kg
5	Micro	W ≤ 5 kg

Table 4.3 Categorization of drones based on their weights by Weibel and Hansman [3].

S. no.	Designation	Weight range
1	Micro	W < 2 lb
2	Mini	2 lb < W ≤ 30 lb
3	Tactical	30 lb < W ≤ 1000 lb
4	Medium and High Altitude	1000 lb < W ≤ 30,000 lb
5	Heavy	W ≤ 30,000 lb

4.2 Classification of Drones

The introduction of a relatively small drone known as a mini aircraft carrier has boosted the need for covert operations in past few decades [6]. As
a result, there is presently a concerted effort to engineering and manufacturing miniature air drones for particular missions. As a consequence, a
variety of miniature drones having diverse designs and flying modes have
really been developed. A full taxonomy of all extant drones is provided
in Figure 4.1, where HTOL stands for Horizontal Take-Off and Landing.
Drones can indeed be classified based on their functional qualities. Mass,
wing length, wing loads, reach, maximum height, acceleration, durability,
and manufacturing costs are all essential design factors that identify multiple kinds of drones and help categories work more effectively. Drones
may also be categorized according on its types of engines [5]. UAVs, for
instance, frequently employ gasoline engines, but Modern aircraft utilize
electric motors. The varieties of propulsion technology utilized in drones
vary depending on their designs. The functional objective of the device,
the materials required in its production, and the difficulty and expense
of the control system are the major features which separate UAVs from

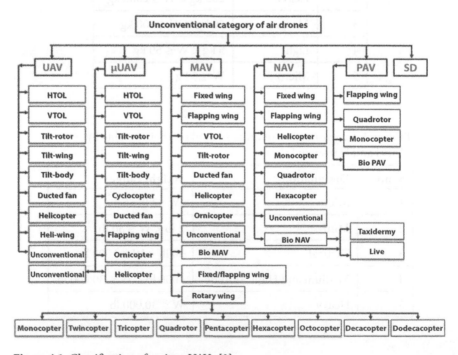

Figure 4.1 Classification of various UAVs [2].

some of the other kinds of tiny drones (such as MAVs and NAVs) [7]. UAVs come in a broad range of sizes and configurations. They might, for instance, have a maximum range as wide as a Boeing 737 or be as tiny as a radio-controlled drone [5]. Varied kinds of UAVs were developed to meet various operational needs. As a result, it's common to classify UAVs according to its operational characteristics [2]. Figure 4.1 depicts the various classifications of UAVs.

UAVs, such as drones, have always had the capacity to change the development of "internet of things" by acting as wireless aerial switches, improving connection to underground networking and delivering a variety of wireless transmission applications. Free space systems (e.g., Air-To-Ground (ATG)) [8–12] and realistic propagation systems (e.g., Okumura-Hata, and COST-231) [15, 16] are two kinds of network topologies that have been documented in the research. The confined solution for available distribution type focus incorporates line-of-sight (LoS) and non-line-of-sight (NLoS), as well as the altitude of a UAV. These values indicate a strong preference for a large covering footprint and a decent shadowing action.

A pre-defined group of variables and restrictions for distinct regions is the actual propagating type emphasis. Despite significant constraints due from their low antenna positions and narrow broadcast reach, these designs can give a high degree of precision. Both forms of transmission schemes have benefits and drawbacks in terms of factors that influence model effectiveness, such as a UAV's height, inclination angle, antenna specs, and energy usage. To attain a larger coverage, reduced power consumption, and improved LoS connection, optimization is constantly necessary [8].

Recent improvements and developments using UAVs for 5G connectivity are highlighted by researchers in [17, 18]. Many cellular services, such as wireless sensor networks, may well be enabled by the cutting-edge technology connected with 5G. Using UAVs at 28 and 60 GHz to make use of the frequency range of 5G wireless networks holds a lot of potential for increased wireless communication systems for these kinds of aerial vehicles.

Nonetheless, our research suggests that perhaps the interaction among Internet of Things (IoT) and Drones is an untouched field, which encourages scholars to continue their research. Researchers [19] provides a NN self-healing approach for optimizing a UAV's location probability in a 5G network to maximize bandwidth, range, and maximal UE. The cognitive model's effectiveness is an excellent approach depending on the minimum and maximum values of the UAVs' placement certainty, as well as attaining overall energy savings, according to the findings.

One of the UAV's capabilities is aerial surveillance and detecting with various kinds of technology, camera systems, and scanners for gathering, intelligence, and steps to investigate [20]. The difficulty of sending high-quality images and data to the ground control station, and also the difficulty of development and manufacturing, necessitates a rethinking of proposed antenna and channel design. The dynamic connections involving sensing, synchronization, and transmission utilizing drones are highlighted by researchers in [21]. They also claim that perhaps the exchange among both picture quality and payload is still an unanswered question. Researchers in [22] present a new solution wherein a UAV network supports a natural calamities control system using cloud computing. To enable IoE technologies in smart cities, the recommended work provides adaptability, portability, and quick implementation characteristics.

Drone technologies for aerial surveillance and catastrophe rescue operations are among the important areas of research highlighted in [30]. One of the most difficult challenges is establishing robust communication linkages among a drone and onshore number of internet users' goods. The influence on quality and bandwidth is substantial when a drone's synchronization incorporates elevation angles and levels, as well as the influence of poor weather on on-board sensing devices. According to the researchers, on-line decision-making process for optimization likely increases the performance of drones. The study in [23] illustrates some frequent challenges in drone facial expression surveillance, such as the influence of movement, variances instance, lighting, the height of the drone, and the sensitivity of the cameras. The proximity between both the drone and the human respondents has a negative association with all these consequences. While image recognition was not particularly successful throughout nighttime due to reduced source light, the article offers respectable findings utilizing the DroneSURF drone database for proactive and reactive surveillance.

The authors examined the latest advances in drone design and control in this study. Authors are especially interested in multicopter drones, which are vehicles that employ several rigid propellers with changing speeds to provide variable impulses and torque multiplication to force the vehicle to shift orientation. Relative to traditional aeronautical models including such helicopters or fixed-wing vehicles, this kind of aircraft is successful due to its exceptionally low technological sophistication. Furthermore, they can glide, possess well-understood control qualities, and are usually rather agile.

4.3   Flight Performance Analysis

The proposed mission for this UAV is a surveillance mission. After being launched and switching to autopilot, the UAV will travel away from the launch spot and towards the surveillance target area. In order to remain undetected, the motor will be cut off before entering the target area. In this way, the UAV may glide silently while it circles around taking pictures. After finishing surveillance, the UAV will leave the target area, turn the motor back on, and return to the launch point to be collected. It is necessary to perform theoretical calculations of the flight performance characteristics of the UAV to determine if it can complete its mission. Because the motor will be running and turned off at different points of the mission, it is necessary to perform calculations for both gliding flight and powered flight.

In order to calculate the theoretical performance analysis for the UAV, the lift and drag coefficients of the vehicle must be estimated. Vortex lattice programs such as AVL and Tornado, which are described in the previous section, are able to find lift and drag for the wings and tail. However, they are unable to provide the parasitic drag on the body of the plane. The parasitic drag on a component of an airplane can be estimated with the following equation [13]:

$$C_{D_0} = \frac{KC_f S_{wet}}{S} \tag{4.1}$$

where K is the form factor, S_{wet} is the surface area of the component, C_f is coefficient of friction, and S is the wing planform area. The form factor was found experimentally, but several people have come up with equations that fit the experimental data. An online calculator was used to output the form factor for the body and wing-like components of an airplane. For the coefficient of skin friction, estimates were made by finding the coefficient of friction of flow over a flat plate matching the dimensions of the component, depending on whether laminar or turbulent flow is expected [13]:

$$(C_f)_{turb} = \frac{0.455}{[\log_{10}(Re_L)]^{2.58}} \tag{4.2a}$$

$$(C_f)_{lam} = \frac{1.328}{\sqrt{Re_L}} \tag{4.2b}$$

Since the surface of the Bixler is rough, it is safe to estimate turbulent flow for the body. The chord lengths of the wings and the tail are small enough to estimate primarily laminar flow.

The stall speed is the lowest velocity at which the aircraft can fly without dropping out of the sky. At the angle of attack with the maximum lift coefficient, the stall speed is at a minimum. The stall speed was calculated (4.3):

$$U_{stall} = \sqrt{\frac{2W}{\rho S C_{l_{max}}}} \tag{4.3}$$

where U is velocity, W is weight, ρ is air density, S is wing planform area, and C_l is lift coefficient.

Since the mission will include shutting off the motor and gliding into the target area, the maximum range and the maximum endurance for the airplane while it is gliding must be calculated. A free body diagram of an airplane in flight was drawn to show the forces acting on the plane as shown in Figure 4.2.

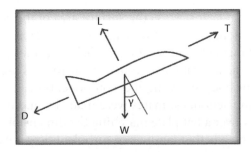

T = Thrust
L = Lift
D = Drag
W = Weight
γ = Flight path angle

Figure 4.2 Free body diagram of airplane in flight.

Assuming steady state flight, the general force equations can be written as:

$$T - D - W\sin\gamma = 0 \tag{4.4}$$

$$L - W\cos\gamma = 0 \tag{4.5}$$

$$\dot{Z} = U\sin\gamma \tag{4.6}$$

For gliding flight, thrust is equal to zero. By rearranging and dividing Equations 4.4 and 4.5, Equation 4.7 can be found for the tangent of the flight path angle.

$$\tan(-\gamma) = \frac{C_D}{C_L} \tag{4.7}$$

The range (R) of the aircraft is a function of the altitude (H) and the tangent of the flight path angle (Equation 4.8).

$$R = \frac{II}{\tan(-\gamma)} \tag{4.8}$$

The maximum range for gliding flight at a given altitude occurs where C_L/C_D is maximized. The endurance (E) of an aircraft's flight is the measure of how long the aircraft can fly. For gliding flight, endurance can be calculated using Equation 4.9. The derivation for Equation 4.9 can be found in Appendix C.

$$E = \frac{H}{\dot{Z}} = \frac{H}{\left[\dfrac{2W}{\rho S} * \dfrac{C_D^2}{\left(C_L^2 + C_D^2\right)^{3/2}} \right]^{1/2}} \tag{4.9}$$

For powered flight, the equations become more complicated due to the introduction of the power source. For this project, the UAV is powered by a lithium-polymer battery. Equations for the maximum range and the maximum endurance of a battery powered UAV were derived by Traub (Equations 4.10 and 4.11).

$$R_{max} = (Rt)^{1-n} * \left(\frac{\eta_{tot} * V * C}{\left(\frac{1}{\sqrt{\rho S}} \right) C_{D0}^{1/4} \left(2W \sqrt{k} \right)^{3/2}} \right)^n * \sqrt{\frac{2W}{\rho S}} * \sqrt{\frac{k}{C_{D0}}} * 3.6$$

(4.10)

$$E_{max} = (Rt)^{1-n} * \left(\frac{\eta_{tot} * V * C}{\left(\frac{2}{\sqrt{\rho S}} \right) C_{D0}^{1/4} \left(2W \sqrt{k/3} \right)^{3/2}} \right)^n$$

(4.11)

where $k = \dfrac{1}{\neq * AR * e}$

The range was expressed in kilometers and the endurance was in hours. Rt is the battery hour rating, expressed in hours, V is voltage, C is the battery capacity in ampere hours, AR is the aspect ratio of the wing, η_{tot} is the total propulsive efficiency, and e is the Oswald efficiency factor. The variable n is called the Peukert coefficient and typically has a value of 1.3 for lithium-polymer batteries [14].

If there is an object ahead of the vehicle that must be avoided, the max rate of climb should be calculated to determine if the obstacle can be avoided. The available battery power is defined in Equation 4.12.

$$P_b = \eta_{tot} * V * \frac{C}{Rt} \left(\frac{Rt}{t} \right)^{1/n}$$

(4.12)

The velocity at which the aircraft achieves its maximum rate of climb can be calculated using Equation 4.13. The equation for maximum rate of climb itself is given in Equation 4.14.

$$U_{max\,R/C} = \left[\frac{4}{3} * \frac{kW^2}{\rho^2 S^2 C_{D_0}} \right]^{1/4}$$

(4.13)

$$Max\ Rate\ of\ Climb = \eta_{tot} \frac{V}{W} \frac{C}{Rt} \left(\frac{Rt}{t} \right)^{1/n} - \frac{1}{2} \rho U^3 \frac{S}{W} C_{D_0} - \frac{2Wk}{\rho U S}$$

(4.14)

4.4   Dynamics and Design Objectives of Drones

This section gives a quick review of the physics that control drone flight before going through how common design goals are influenced and traded-off.

4.4.1   Drone Dynamics

The authors focus upon that near drifting characteristics of multi-copter drones and portray them only at a greater extent. Asynchronous motors operate rigid blades in standard multicopters. Rotor blades normally have two blade edges, although propellers including at least three sharp corners are also possible. At almost the same rotor speed, a larger number of rotating blades often accounts for much more significant thrusts, but the correlation between control utilization, disturbance levels, resonance, and other factors can be perplexing, see for example [24]. The rotating propeller also provides both and torque when it ejects airflow. The elevation, which corresponds to the center of rotation, overwhelms the pressure. Due to the obvious rotating stability of the blade, the force and torque elements in the construction of such hinge are usually much smaller than all those relating to a center of revolution - all of those are usually visible only while the motor is interpreting through the air.

The pressure is everywhere throughout estimated as relating to the quadratic of the propeller's rotational acceleration, in which the proportional compatible captures characteristics of the blade (such as proportions) and the environment. A thrust parallel to the plane of spin opposes the movement of the blade, which is created by a combination of aerodynamic drag here on rotor blades, which opposes its mobility. Although it is commonly described as proportionate towards the thrust forces exerted by the rotor, such response torque can also be estimated as quadratic inside the rotary angular speed.

There would be torque elements in the propeller's geometry of spin ning for such a translation blade, so these elements are generally insignificant in comparison to the instant created either by thrust operating at such a distance from the center of gravity. There are much more sophisticated propellers thrust and torque theories, such as [25], which is utilized for significant control in flexible movement, or [26], that are used for blades with forward plane. A regular activity will have an even quantity of rotors having parallel axes of motion that are equally balanced into clockwise and counter-clockwise motions. The vehicle's translating velocity is produced by directing the general thrust position such that the vectors outcome of the vehicle's propulsion, mass, and aerodynamics torque produces the necessary translational velocity. The vehicle's translational velocity is really only determined by the sum of the motor torque instead of its independent variables, because the blade thrust force seems to be all horizontal. The translation of a multi-copter approaching hovering, in wind-free situations, can indeed be easily characterized as that of the product of the mass as well as a unified "total thrust" vectors, substantially simplifying construction and operation.

Torque is generated by changes in motor torque to turn the object. The thrust vector rotates due to the torque created by the thrust operating at quite a range out from vehicle's center mass (i.e., the roll and pitch motion). Variations inside the reactions torques of both the blades cause spinning around the flow direction; note that even at hovering, the symmetrical arrangement of blades provides net zero response torque. Because of the proportional quantity of blades, the total angular velocity of both the propeller blades is 0 in normal activities.

It is worth noting that changing the propeller speeds would also change the accelerating velocity of the drone's main body due to conservation of linear momentum; however, such effect is often overlooked in comparison towards the dynamic feature. The actual propulsion intensity, as well as the three independent variables of torque, is being used to operate a multi-copter inside this fashion. As a result, a multi-copter capable of drifting should have at least four propellers [27, 28]. The majority of UAVs (and those with at least six rotors) are underactuated inside this way, while just four eventually started for each of its six levels of possibility (however special cases are talked about later). A more precise description of quadcopter mechanics can be found, for example, in research article [29].

Non-rotating raising elements, as well as perhaps associated flight controls such as wings, elevators, and thrusters, are provided on fixed-wing and powered vehicles. The forces and torques produced by some of these are generally shown as quadratic in motion in confined situations; however, whose characterization becomes greatly complicated whenever they are subjected to extreme conditions like that as slowing down or approaching extremely large distances. The book gives a good overview of specific air system modeling and kinematics [30].

4.4.2   Design Objectives and Scaling Laws

Designers look at some of the significant aspects that drive drone development and discuss some key scalable principles which control its interaction. Researchers look into power consumption (that detonates range and persistence), flexibility and velocity, survival and force, and cost/complexity in detail. Drone concepts are deeply included, such is typical of aircraft applications, and frequently handle either divide the discrepancy among competing goals. A major concern with just about any flying aircraft would be its overall mass, which is often the result of a design that focuses on a vehicle's capability in multiple design objectives at the very same point.

4.4.3 Energy Utilization

Flight time is the most important design goal for surveillance operations, whereas reach is important for transporting operations. For a constant thermodynamic efficiency development that is not constrained by the force utilization of a vehicle. A drone's lift is generated directly only by propeller blades while flying at cruising velocities, as well as its force utilization can indeed be estimated with electrohydraulic classical theory—a tarted up propeller that's not trying to interpret, and collaborating inside an inviscid, incompressible viscous fluid would then devour induced drag that really is conversely comparative towards the propeller's range, and correlating towards the force to the engine power of 1.5 [30]. Throughout this sense, if all other factors are equal, a vehicle equipped by larger rotors is projected to also have better transport endurance. Similarly, an aircraft that requires less thrust (for instance, due to its lower by it and big mass) will have much more endurance. Since the set aside energy fills immediately with in weight used for power capacity, the amazing growth in power phases as the velocity phases implies that introducing additional battery capability to something like a system actually reduces the device's flying time [31].

4.4.4 Agility and Speed

The key objective in operations like drone hustle is velocity and maneuverability [32–34]. The magnitude of weight of the vehicle, torques, and drag velocity is 0 throughout steady-state movement. All other factors being equal, a vehicle's maximal flat rate can be increased by reducing its mass, increasing the available thrust, or reducing the aerodynamic drag potential.

Because the thrust vector dominates the motion of the vehicle of a multicopter, the ability to easily change direction is critical for maneuverability. The maximum maneuverability derives from putting the blades as near towards the vehicle's center of mass as practicable for something like an aircraft having currently defined and total weight. The quadratic connection involving radius and mass moment of inertia counteracts the constant rise in thrust necessary for trajectory tracking as even the blades are moved away from either the vehicles center of mass. Because the powerful rotors positioned at a considerable distance from the vehicle's center often predominate the entire mass moment of inertia, the vehicle's attitudes maneuverability grows negatively proportional toward its total dimensions. As a result, rotors on multi-copters are often situated as close to the vehicle center as practicable. There is an obvious trade-off between effectiveness and maneuverability since fewer blades can indeed be grouped towards a more compact design. Figure 4.3 diagrammatically depicts the trade-off among agility and durability.

It briefly restates scalability argument to look at vehicle maneuverability as that of the quantity of all comparison to the rest [35]. Quite an analysis necessitates assumptions about how achievable thrust force develops increasing rotor size; this is a tough endeavor, especially because engine and battery scalability are notoriously tough to pin down. In the research presents two separate estimations for propeller airflow scalability, the first assuming that perhaps the Froude number is constant, as well as the second assuming that perhaps the propeller blade continuity equation is constant [35]. These two-substitute series of expectations lead to the conclusion that a multi-copter's linear velocity increase is either free of scale, or scales relatively to the vehicle's straight scale. The acceleration speed increase, be that as it may, scales either conversely relative to the straight scale, or the square of the linear scale, contingent upon whether Froude or Mach scaling is utilized. Consequently, more modest and more minimized multi-copters are liked in applications where agility is significant.

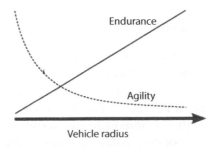

Figure 4.3 Approximate trade-off efficiency and agility with vehicle linear size (at constant mass) for a multi-copter [33].

4.4.5   Survivability and Robustness

Working in difficult settings, especially while working close to obstacles, necessitates either really precise control or perhaps the ability to withstand hits. Within the last alternative, the UAV should really be able to complete their primary mission with minimal disturbance, allowing for simpler control processes, less precise sensors, and so forth. The additional functionality that is supposed to get rid of annoyances tends to be accompanied by additional weight, which has its own set of problems. Because the concept must cover the aircraft, the added mass will generally be a considerable amount as from vehicle's central focus of weight, resulting in a greatly increased moment of inertia, which also will reduce the exact speed gains which can be achieved.

4.4.6   Low-Level Control and Stabilization

A typical approach to designing a drone control scheme is seen in the top left of Figure 4.4, for which a path generator sets benchmark levels before a reduced regulator generates actuation instructions. A distinct state prediction creates an approximation based on sensor data. The organizer often operates at a lower intensity than just the relatively little effort and estimates. This technique evaluates every design element in comparative independence, reducing both implementation complexity and computing expenditure, and perhaps keeping in mind less challenging optimal solution conflicts. Nonetheless, as the computing effort expands, tighter

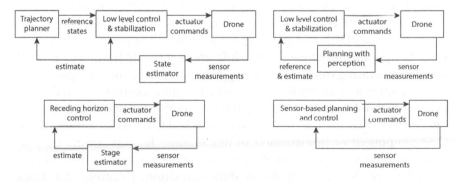

Figure 4.4 Various control system architectures [36].

component connectivity allows the developer to achieve increasingly difficult execution criteria.

The use of receding perspective controlling, in which the path prediction is repeated at a very particular intensity and no new approach is
required to follow authority, would be a first step along this route. Working
in difficult situations necessitates the use of such approaches. Normally,
state evaluation and planning are kept distinct, and the command does
not explicitly indicate the detection. The trajectory strategist and condition
estimator, on the other hand, can be combined, with the organizer taking
into account prediction reliability and sensor approach. The organizer can
operate at a lower frequency using a separate minimal influence.

The last phase is to create a single centralized platform in which the path
designer takes into account sensory restrictions and unpredictability. Even
though tighter resolution may result in improved efficiency, it is inherently
more technology and program particular.

Under typical circumstances, there really is a substantial amount of
research on control procedures for conventional multi-copters. Without
the intricate (and frequently deeply observable) aerodynamic linkages
that are typically assumed for flying devices, their motions are easily captured by reasonably uncomplicated (but still unpredictable) requirements.
It is clear from the discussion that a multi-translational copter's mobility
is dominated by its orientation, particularly the vision of the propulsion
field. If the directions and total force can really be adjusted quickly enough,
the vehicle's increased speed may be considered as either a contributor to
a higher-level temporal controller, resulting in a guessed dual integration
scheme.

A first decision while designing an attitude regulator is the portrayal
utilized for the disposition, with well-known decisions including Euler
symmetric parameters/quaternions [37], the rotation matrix itself [38], or
rotation vectors [39]. In a first request analysis, all approaches ordinarily
yield comparative outcomes, with differences possibly becoming evident
while recuperating from enormous aggravations/direction changes.

The common assumptions of wonderful state estimations and exact
existing technologies become constraining as these technologies are put
into increasingly intricate settings and are asked to do more difficult tasks.
One component of investigation in this manner focuses mostly on processes with deficiently obtained dynamics, while another focuses on portraying more significant faults in state evaluation. Controls that adapt
(either ahead of time or during action) become appealing in situations
where exact modeling is problematic.

Sustaining unpredictable payloads becomes a scenario in which training and variance may play a key role, hence the process factors used to analyze the dynamics are undetermined and may vary over time. In one of the research works, a Gaussian approach is used to approximate system dynamics, with adaptation occurring if the simulation error exceeds a certain threshold [40]. In another research work, an adaptive control strategy based on Lyapunov analysis is proposed to compensate for an uncertain payload, and adaptive control offers another one configuration of instruments to progress towards certain systems [41].

To overcome the instability problem, strong control concept is most often used. By organizing or lumping parametric uncertainties, the robust proposed controller assures that same degree of control implementation under various natural situations, and maintains the system's stability within the defined uncertainty range. As a result, when powerful regulation is given to multi-copter controller, parametric uncertainties such as infinitely mass moment of inertia instability, or unpredictable external upsetting impacts such as breeze or blow, may be overcome. When the total mass of the platform is determined, a robust control method fulfills the goal presentation in any case [42].

The interaction between both the drone's low-level control and state evaluation is another point of interest. Steering controls that need a higher angular velocity, for example, are more likely to generate uncertain movements, resulting in much more unfavorable visual sensor follow-up. As a result, it's critical to set up low-level control processes that incorporate the identifying methodology's requirements. For example, in one of the research works, a robust governor is constructed for a multi-copter employing VIO for status evaluation, where another robust governor is shown to provide more advanced implementation in hostile lighting settings at the price of a mild method of acting in suitably sunny conditions [43].

Because drones are sensitive to ambient air conditions, there is a lot of work put into recognizing the wind field surrounding a drone and compensating for it. A quadcopter is rigged with a locally accessible wind sensing element in order to assess wind fields in an urban atmosphere [44]. Similarly, a quadcopter with additional sensors evaluates the wind velocity vector, the vehicle's drag force, and external forces such as collisions [45]. Rather than using models, deep supporting learning may be supplied by relying on wide data, such as the approach for dealing with cyber physical attacks, which avoids the traditional approach of expressing fault location and discovery [46].

4.5   Design Methods and Challenges

Drones are designed in three phases, depending on their performance category, kind, size, and purpose [47–50]. Reasonable development, starting design, and definitive design are the three stages. Each step necessitates increasingly sophisticated measurement, aerodynamic, aero elastic, fundamental, motion, stability, controlling, electronic, and manufacturing analyses [51–53]. It's worth noting that, despite advancements in drone technology, they still have a few flaws in their design.

Assessing whatever provides about its optimum potential advantages of different features and workloads is among the most important tasks in the designing connection of a wide variety of drones [51, 53]. Drone sizing is often divided into five stages: (1) mission definition, (2) flying mode selection depending on type, (3) wing shape (planform) and aspect ratio determination, (4) constraints analysis, and (5) weight assessment [51–53]. The examination of the route is guided in the context of the mission, resulting in the guarantee of the flying duration, trip speed, and rationing speed. Following that, the confirmation of flying modes, the condition of the wings, and its prospective proportions are addressed in accordance with the mission type. Then, in order to determine the correct wing layering and propulsion building of the drone, an essential analysis is performed under which the flight's translational and rotational circumstances are replicated. Various weight estimate strategies can be used in addition to the in advance of cited improvements. As a consequence of the interaction, the math and features of drones are guaranteed, as well as the computing of a few aerodynamic characteristics for every type [53].

The sizing procedure should be followed as closely as possible [54]. A schematic view of the costs of designing and manufacturing several types of drones is illustrated in Figure 4.5 [55]. The feasible and trial provides that develop when scaling a drone, such as enlarged or reduced force thickness of propulsion systems, technological sheets, manufacturing methods, and so forth, should bring about such pattern shown in Figure 4.5. Drones of all sizes (UAV, MAV, NAV, and PAV) are not merely scaled-down versions of larger designs [56]. Because all of the characteristics of larger designs must be contained in a small container, the difficulty and sophistication of its design and fabrication increases dramatically. Despite the fact that engineers have endeavored to create insect-sized drones, work on such drones has slowed recently due to the physical and creative challenges posed by their smaller size [57, 58]. The low Reynolds number, which derives from its slower speeds and small size, is a serious difficulty with any of these

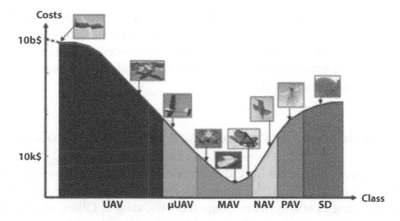

Figure 4.5 A schematic view of the costs for design and fabrication of different types of drones [55].

types of drones [53, 59]. The majority of the time, travelling across this flow path is more difficult. More along the lines, researchers began to focus on insect travel [60–62].

4.5.1 Proposed Solutions for Design Challenges

To defeat the referenced difficulties for various kinds of miniature drones, engineers and designers of drones should to consider different boundaries in the design interaction which can bring about creating aerodynamic drones. As talked about in the past areas, each sort of drones and their design strategies enjoy benefits and hindrances. Thusly, by utilizing hypothetical, factual, reconsidered allometrically, and bio-motivation strategies, a complete procedure can be proposed which tracks down answers for the disadvantages of past techniques.

Taking inspiration from nature, several types of drones might be exhibited [63]. The ability to adapt and adjust the configuration may be regarded another subject of investigation in ebb and flow design theories for drones. Given the fact that only a few ways are now being explored to include transducer drone development, it can be further studied for something like the design of lightweight, fast deployment, functionally operated, and small production wings for autonomous and small flying vehicles [64]. It's worth noting that natural inspiration can help a few different innovations get acquainted with design.

In rundown, in the design interaction of drones, two sections should be noted, the first is drones' setup and the subsequent one is their design strategy. As of late, there are a few efforts to design drones with unpredictable setups which nearly are enlivened from nature, like birds, bugs, marine creatures, and so forth [65].

4.6   Guidance, Navigation, and Control of Drones

Several research projects have concentrated on drone guidance, navigation, and control (GNC) throughout the last 20 years, leading to a variety of methodologies and systems. Some scholars, such as Ollero and Merino [66] for navigation systems, Chao et al. [67] for autopilots, Goerzenet et al. [68] for path optimization algorithms, and Valavanis [69] for UAVs in total, have attempted to examine various GNC methods and subgroups [70]. Kendoul [70] has also just completed a detailed survey study that organizes the wide range of GNC approaches. He gave an overview of GNC technologies that would help drones become more autonomous. Control, navigation, and guiding are the three primary categories in which the methods have been documented. Methods are categorized there at highest possible level for every category depending on the amount of independence they give, and then by the computational technique utilized, which is also in general instances closely related with both the type of sensors used [70]. Different sorts of GNC technologies are presented in Figure 4.6 depending on Kendoul's [70] research.

Drone guidance, navigation, and control (GNC) is typically accomplished using three methods: radio controller, visual base, and autonomous [71]. A communications system is amongst the most frequent methods to manipulate and move drones. Drones are controlled using this manner using a wireless program that integrates a transmitter and a receiver. Directions are sent to the drone's electronic systems using electromagnetic radiation in this guidance system [71]. Remote Control (RC) equipment comprises mostly of a transmission medium with many radio channels. The operator can send commands to the drone through either of these methods [72]. The transmitter range in remote control systems varies, but it normally covers an area of roughly 5 km.

Extra connections are being used to operate the lens. The receiver is often used in such technology to convey instructions to the stepper motors and frequency regulators [71]. A sensor is equipped upon that drone for video-based GPS devices, which is used to collect recordings and images while flying through locations and relay them towards the ground control

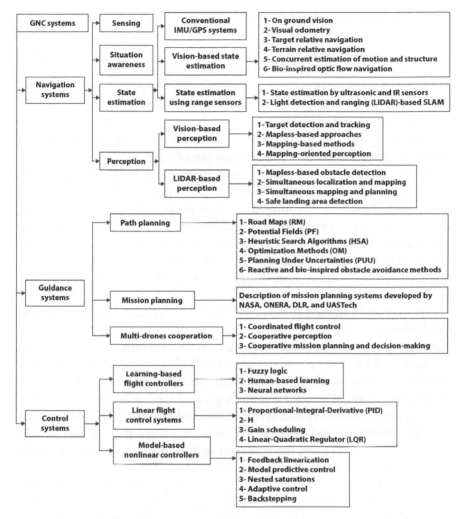

Figure 4.6 Classification of GNC systems developed for drones based on Kendoul [70].

station via video transmitters. The key dynamics of a camera system are small size, lightweight, and great perceptibility and brightness [71]. The photos transmitted from the video transmitters as well as the images received via the receiver antenna are displayed on a display at the ground control station in a visual database engine. The produced waves can be used to evaluate radio cables. Supplements are also used in conjunction with receiver antenna to reduce the time needed to obtain images [73]. Different types of sensors, shade, heated, or thermal imaging cameras are being used to collect data upon that drones' overall climate. Small drones frequently employ shading lenses, which are only useful during the day and cannot provide massive range or depth to the captured climate. PC vision plays an important role in the computerized system of a video basis guidance system. PC vision methods are used in these devices to extract the desired data. These captured images are used for guidance, correction, and data collection [74]. In practice, video emitters can send messages within a certain distance, but information can't be captured over long distances on several flights. The business types of transmitters only function within a specific range. When UAVs are already out of sight, they create a no-land man's and force drones to reduce their flying range [71]. The navigation system is by far the most effective technique of guiding, investigating, and controlling the drone. An autopilot is a set of software and hardware that drives drones to carry out respective flight tasks in a predictable manner. For example, heading and pace can indeed be expressed in different parts of the flight by describing flight design ideas, and the drone will automatically conform to this aircraft design and seek to carry out its main aim with minor errors [71].

Micro pilot [75], Piccolo and Paparazzi [76] are examples of autopilots available today. Micro pilot automated systems can carry a weight of 28 g, have dimensions of 4 cm by 10 cm, and operate 24 servos while flying reach a height of 12 km and a perimeter of 50 km [75]. In an autopilot system, the aircraft should be transferred to the structure during takeoff, and the drone should have been in contact also with ground control station at all times, sending data such as altitude, acceleration, and so forth. Various directives may be supplied to the drone via RF modems since the very beginning. After receiving directives, the autopilot delivers them to the servo, which causes the drone to perform the perfect reaction [71].

Despite these strategies, experts offered strategies for coping with drone exploration that might be used in the future for small drones. Bublitz [77] used Smart Glasses to use head movements to operate a quadcopter drone. Head motions can be captured by Smart Glasses, which can then be converted into flight commands and sent to the drones. As a result, when this technology is used, the quadcopter is simply confined by the commands recognized by the head mounted device, whereas the human pilot has been in charge of the directions and navigating responsibilities. This method may be appropriate for small drones with limited flying capabilities that can do drifting flight, such as rotating wings MAVs.

The establishing mechanism is amongst the most important components of drone location tracking. Drones may be located using a variety of methods, including the Global Positioning System (GPS) and the Inertial Navigation System (INS) [78]. GPS is commonly used in prattle to determine location, speed, and altitude. To provide an accurate location for the drone, GPS ought to be in contact including at least four satellites throughout all times [71]. Outer commotion or blockage has little effect on GPS signals [56]. As a result, for drones which are only equipped with GPS, it has been observed that some of the drones may lose its GPS connection for a short time. Drones must always be grounded under these conditions, and its primary mission is ended prematurely due to safety considerations. To avoid this problem, it is necessary to develop a system that can calculate the region of drones whenever aircraft interrupt its GPS connection for a small period of time [79]. In these situations, an inertial guidance program is the main option. The INS includes tiny and medium-sized motion sensors, which are used to calculate the drones' location and direction. Nowadays, GPS is frequently used with the INS to avoid positioning errors [80, 81]. These two types of indicators are combined to provide precise navigating data. The Kalman filter is regarded as the standard algorithm for fusing estimates [82]. Overall, the Extended Kalman Filter (EKF) is used to assess the location of drones that have momentarily lost its GPS connection [79].

Another technique for investigating and guiding small drones that might have been proposed and evaluated would be to use telecommunication and the internet to transmit drones' instructions. This approach can be used to overcome the reach limitations of previous strategies. This technique may be a good strategy for guiding drones, such as MAVs, as evidenced by the growing spread of telecommunication services around the globe and the low altitude journey of micro drones. Despite its low cost, this technology can have a significant impact when used in conjunction with other management measures. Drones provided by this technology might be useful for knowledge exercises [71].

Unmanned Aerial Vehicles (UAVs) are growing in popularity as instruments in a variety of industries, both civil and military. Numerous UAV technologies are expensive, and every operation requires a substantial quantity of people and equipment on the base. The use of an unmanned aerial vehicle (UAV) entails danger and the use of crucial applications. Destruction to the aircraft, especially throughout final approach, as well as harm towards the control system and the mission's cargo, are all dangers. Any harm to the flying equipment is obviously undesired, however when the UAV is placed in a remote region, any benefit in launching UAVs is entirely nullified. UAVs become such a burden if damages on arrival preclude them from undertaking additional operations after the first flying. Landing hazards can be decreased by using skilled personnel who are familiar with aviation and landing tactics. Nevertheless, an expert pilot might not even be available when needed, and keeping in a distant and possibly dangerous position may well be extremely expensive. Airline pilots are believed to make their living during the first 3 minutes and the last 3 minutes of each trip. The substantial increase in closeness towards the earth at these periods is one probable and pretty evident cause for this perception. Even being autonomous, UAVs are not typically considered expendable, especially in home remodeling uses. During the evaluation of several UAV aircraft structures, it was discovered that even a huge amount of time, money, and effort went through into construction and design of a UAV aircraft and its payloads, yet it simply took minutes to demolish it all. Even the most professional pilots have terrible days, whether it's because the weather isn't cooperating or because there are too many things going on at once. This research took into account the necessity for support in such crucial situations. It is important to reduce the complexity of UAV installation in order to improve its appearance, particularly to reduce the danger of damage during the principal components of UAV recovery.

4.7   Conclusion

Current investigations and research in the subject of unmanned aircraft, including stationary and flapped winged vehicles, were merged and extensively discussed. It was initially recommended that these drones be classified in a different way. This classification includes autonomous air vehicles, micro air vehicles, nano air vehicles, Pico air vehicles, and intelligent dirt, among other drone types. These flying drones may be used for a variety of civilian and military operations. Rescue operations, environmental monitoring, shipping and transportation, and scientific study were among the tasks considered. For a wide spectrum of drones, pre-existing design techniques and their challenges were also combined. Potential solutions to the design issues were offered and discussed. Furthermore, pre-existing assembling procedures and challenges, motion control and sensors, power supply and endurance, control and guidance of drones were examined, with ground-breaking solutions proposed to overcome present limitations. The importance of swarm flying and drone dispersion were also emphasized. The problems facing UAV operators include integrating the UAV fleet with both the pilot force whilst following the established and proven operating procedures currently in place while sacrificing the flexibility that UAVs provide. In the Automated Landing setting, this is especially difficult. This difficulty poses two fundamental questions about the revival of UAV capabilities. The first is to find the most effective maneuvering method for guiding a UAV through the take - off and landing phase, taking into account existing systems, practices, and possible military navigation system objectives, as well as the objective of effectively blended the UAV fleet with both the remotely controlled fleet while achieving maximum personal freedom. The second step is to figure out the best strategy to operate a UAV throughout the recovery's take - off and landing phases.

References

1. Brooke-Holland, L., *Unmanned Aerial Vehicles (Drones): An Introduction*, House of Commons Library, London, UK, 2012.
2. Arjomandi, M., Agostino, S., Mammone, M., Nelson, M., Zhou, T., *Classification of Unmanned Aerial Vehicles*, pp. 1–48, Report for Mechanical Engineering Class, University of Adelaide, Adelaide, Australia, 2006.
3. Weibel, R. and Hansman, R.J., Safety considerations for operation of different classes of UAVs in the NAS, in: *AIAA 3rd Unmanned Unlimited Technical Conference, Workshop and Exhibit*, p. 6421, September 2004.
4. Gupta, S.G., Ghonge, D., Jawandhiya, P.M., Review of unmanned aircraft system (UAS). *IJARCET*, 2, 1646–1658, 2013.
5. Cavoukian, A., *Privacy and Drones: Unmanned Aerial Vehicles*, pp. 1–30, Information and Privacy Commissioner of Ontario, Canada, Ontario, 2012.
6. Hassanalian, M. and Abdelkefi, A., Design, manufacturing, and flight testing of a fixed wing micro air vehicle with Zimmerman planform. *Meccanica*, 52, 6, 1265–1282, 2017.
7. Cai, G., Dias, J., Seneviratne, L., A survey of small-scale unmanned aerial vehicles: Recent advances and future development trends. *Unmanned Syst.*, 2, 02, 175–199, 2014.
8. Almalki, F.A. and Angelides, M.C., Evolution of an optimal propagation model for the last mile with low altitude platforms using machine learning. *Elsevier Comput. Commun. J.*, 142, 9–33, 2019.
9. Almalki, F.A. and Angelides, M.C., Deployment of an aerial platform system for rapid restoration of communications links after a disaster: A machine learning approach. *Computing*, 102, 4, 829–864, 2020.
10. Khawaja, W., Guvenc, I., Matolak, D.W., Fiebig, U.C., Schneckenburger, N., A survey of air-to-ground propagation channel modeling for unmanned aerial vehicles. *IEEE Commun. Surv. Tutor.*, 21, 3, 2361–2391, 2019.
11. Almalki, F.A. and Angelides, M.C., Propagation modelling and performance assessment of aerial platforms deployed during emergencies, in: *2017 12th International Conference for Internet Technology and Secured Transactions (ICITST)*, IEEE, pp. 238–243, December 2017.

12. Almalki, F.A.E., *Optimisation of a Propagation Model for Last Mile Connectivity with Low Altitude Platforms Using Machine Learning*, Doctoral Dissertation, Brunel University London, 2017.

13. Bertin, J.J. and Cummins, R.M., Characteristic parameters for airfoil and wing aerodynamics, in: *Aerodynamics for Engineers*, pp. 236–304, Prentice-Hall, Englewood Cliffs, NJ, 1979.

14. Traub, L.W., Range and endurance estimates for battery-powered aircraft. *J. Aircr.*, 48, 2, 703–707, 2011.

15. Alsamhi, S.H., Ansari, M.S., Ma, O., Almalki, F., Gupta, S.K., Tethered balloon technology in design solutions for rescue and relief team emergency communication services. *Disaster Med. Public Health Prep.*, 13, 2, 203–210, 2019.

16. Alsamhi, S., Almalki, F.A., Gapta, S., Ma, M.A.O., Angelides, M., Tethered balloon technology for emergency communication and disaster relief deployment. *Springer Telecommun. Syst.*, 13, 1–10, 2019.

17. Li, B., Fei, Z., Zhang, Y., UAV communications for 5G and beyond: Recent advances and future trends. *IEEE Internet Things J.*, 6, 2, 2241–2263, 2018.

18. Almalki, F.A., Soufiene, B.O., Alsamhi, S.H., Sakli, H., A low-cost platform for environmental smart farming monitoring system based on IoT and UAVs. *Sustainability*, 13, 11, 5908, 2021.

19. Sharma, V., Jayakody, D.N.K., Srinivasan, K., On the positioning likelihood of UAVs in 5G networks. *Phys. Commun.*, 31, 1–9, 2018.

20. Alsamhi, S.H., Almalki, F., Ma, O., Ansari, M.S., Lee, B., Predictive estimation of optimal signal strength from drones over IoT frameworks in smart cities. *IEEE Trans. Mob. Comput.*, 22, 1, 402–416, 1 Jan. 2023.

21. Yanmaz, E., Yahyanejad, S., Rinner, B., Hellwagner, H., Bettstetter, C., Drone networks: Communications, coordination, and sensing. *Ad Hoc Netw.*, 68, 1–15, 2018.

22. Al-Khafajiy, M., Baker, T., Hussien, A., Cotgrave, A., UAV and fog computing for IoE-based systems: A case study on environment disasters prediction and recovery plans, in: *Unmanned Aerial Vehicles in Smart Cities*, pp. 133–152, Springer, Cham, 2020.

23. Kalra, I., Singh, M., Nagpal, S., Singh, R., Vatsa, M., Sujit, P.B., Dronesurf: Benchmark dataset for drone-based face recognition, in: *2019 14th IEEE International Conference on Automatic Face & Gesture Recognition (FG 2019)*, IEEE, pp. 1–7, May 2019.

24. Traub, L.W., Propeller characterization for distributed propulsion. *J. Aerosp. Eng.*, 34, 3, 04021020, 2021.

25. Faessler, M., Falanga, D., Scaramuzza, D., Thrust mixing, saturation, and body-rate control for accurate aggressive quadrotor flight. *IEEE Robot. Autom. Lett.*, 2, 2, 476–482, 2016.

26. Gill, R. and D'Andrea, R., Computationally efficient force and moment models for propellers in UAV forward flight applications. *Drones*, 3, 4, 77, 2019.

27. Mueller, M.W. and D'Andrea, R., Relaxed hover solutions for multicopters: Application to algorithmic redundancy and novel vehicles. *Int. J. Robot. Res.*, 35, 8, 873–889, 2016.

28. Zhang, W., Mueller, M.W., D'Andrea, R., Design, modeling and control of a flying vehicle with a single moving part that can be positioned anywhere in space. *Mechatronics*, 61, 117–130, 2019.

29. Mahony, R., Kumar, V., Corke, P., Multirotor aerial vehicles: Modeling, estimation, and control of quadrotor. *IEEE Robot. Autom. Mag.*, 19, 3, 20–32, 2012.

30. Nangia, R.K., *Aerodynamics, Aeronautics and Flight Mechanics*, Second Edition, B.W. McCormick (Ed.), p. 652pp. Illustrated.£ 22.50, John Wiley & Sons, Baffins Lane, Chichester, West Sussex P019 1UD, 1995, Baffins Lane, Chichester, West Sussex P019 1UD, *Aeronaut. J.*, 100, 991, 36–36, 1996.

31. Jain, K.P., Tang, J., Sreenath, K., Mueller, M.W., Staging energy sources to extend flight time of a multirotor UAV, in: *2020 IEEE/RSJ International Conference on Intelligent Robots and Systems (IROS)*, IEEE, pp. 1132–1139, November 2020.

32. Delmerico, J., Cieslewski, T., Rebecq, H., Faessler, M., Scaramuzza, D., Are we ready for autonomous drone racing? The UZH-FPV drone racing dataset, in: *2019 International Conference on Robotics and Automation (ICRA)*, IEEE, pp. 6713–6719, May 2019.

33. Kaufmann, E., Loquercio, A., Ranftl, R., Dosovitskiy, A., Koltun, V., Scaramuzza, D., Deep drone racing: Learning agile flight in dynamic environments, in: *Conference on Robot Learning*, PMLR, pp. 133–145, October 2018.

34. Pfeiffer, C. and Scaramuzza, D., Human-piloted drone racing: Visual processing and control. *IEEE Robot. Autom. Lett.*, 6, 2, 3467–3474, 2021.

35. Kushleyev, A., Mellinger, D., Powers, C., Kumar, V., Towards a swarm of agile micro quadrotors. *Auton. Robots*, 35, 4, 287–300, 2013.

36. Mueller, M.W., Lee, S.J., D'Andrea, R., Design and control of drones. *Annu. Rev. Control Robot. Auton. Syst.*, 5, 161–177, 2022.

37. Brescianini, D. and D'Andrea, R., Tilt-prioritized quadrocopter attitude control. *IEEE Trans. Control Syst. Technol.*, 28, 2, 376–387, 2018.

38. Lee, T., Leok, M., McClamroch, N.H., Geometric tracking control of a quadrotor UAV on SE (3), in: *49th IEEE Conference on Decision and Control (CDC)*, IEEE, pp. 5420–5425, December 2010.

39. Yu, Y., Yang, S., Wang, M., Li, C., Li, Z., High performance full attitude control of a quadrotor on SO (3), in: *2015 IEEE International Conference on Robotics and Automation (ICRA)*, IEEE, pp. 1698–1703, May 2015.

40. Yel, E. and Bezzo, N., Gp-based runtime planning, learning, and recovery for safe UAV operations under unforeseen disturbances, in: *2020 IEEE/RSJ International Conference on Intelligent Robots and Systems (IROS)*, IEEE, pp. 2173–2180, 2020.

41. Sankaranarayanan, V.N., Roy, S., Baldi, S., Aerial transportation of unknown payloads: Adaptive path tracking for quadrotors, in: *2020 IEEE/RSJ International Conference on Intelligent Robots and Systems (IROS)*, IEEE, pp. 7710–7715, October 2020.

42. Dhadekar, D.D., Sanghani, P.D., Mangrulkar, K.K., Talole, S.E., Robust control of quadrotor using uncertainty and. *Intell. Robot. Syst.*, 101, 3, 1–21, 2021.

43. Jarin-Lipschitz, L., Li, R., Nguyen, T., Kumar, V., Matni, N., Robust, perception based control with quadrotors, in: *2020 IEEE/RSJ International Conference on Intelligent Robots and Systems (IROS)*, IEEE, pp. 7737–7743, October 2020.

44. Patrikar, J., Moon, B.G., Scherer, S., Wind and the city: Utilizing UAV-based *in-situ* measurements for estimating urban wind fields, in: *2020 IEEE/RSJ International Conference on Intelligent Robots and Systems (IROS)*, IEEE, pp. 1254–1260, October 2020.

45. Tagliabue, A., Paris, A., Kim, S., Kubicek, R., Bergbreiter, S., How, J.P., Touch the wind: Simultaneous airflow, drag and interaction sensing on a multirotor, in: *2020 IEEE/RSJ International Conference on Intelligent Robots and Systems (IROS)*, IEEE, pp. 1645–1652, 2020.

46. Fei, F., Tu, Z., Xu, D., Deng, X., Learn-to-recover: Retrofitting uavs with reinforcement learning-assisted flight control under cyber-physical attacks, in: *2020 IEEE International Conference on Robotics and Automation (ICRA)*, IEEE, pp. 7358–7364, May 2020.

47. Sadraey, M., A systems engineering approach to unmanned aerial vehicle design, in: *10th AIAA Aviation Technology, Integration, and Operations (ATIO) Conference*, p. 9302, September 2010.

48. Verstraete, D., Coatanea, M., Hendrick, P., Preliminary design of a joined wing hale uav, in: *International Congress of the Aeronautical Sciences, Anchorage*, Alaska, USA, pp. 14–19, September 2008.

49. Periaux, J., Gonzalez, F., Lee, D.S.C., *Evolutionary Optimization and Game Strategies for Advanced Multi-Disciplinary Design: Applications to Aeronautics and UAV Design*, vol. 75, Springer, Colorado, USA, 2015.

50. Amirreze, K., Marzieh, D., Foad, S., Fatemeh, A., A new systematic approach in UAV design analysis based on SDSM method. *J. Aeronaut. Aero. Eng.*, S1, 001, 2013.

51. Hassanalian, M., Khaki, H., Khosravi, M., A new method for design of fixed wing micro air vehicle. *Proc. Inst. Mech. Eng. G J. Aerosp. Eng.*, 229, 5, 837–850, 2015.

52. Hassanalian, M., Abdelkefi, A., Wei, M., Ziaei-Rad, S., A novel methodology for wing sizing of bio-inspired flapping wing micro air vehicles: Theory and prototype. *Acta Mech.*, 228, 3, 1097–1113, 2017.

53. Hassanalian, M. and Abdelkefi, A., Design, manufacturing, and flight testing of a fixed wing micro air vehicle with Zimmerman planform. *Meccanica*, 52, 6, 1265–1282, 2017.

54. Amirreze, K., Marzieh, D., Foad, S., Fatemeh, A., A new systematic approach in UAV design analysis based on SDSM method. *J. Aeronaut. Aero. Eng.*, S1, 001, 2013.

55. Gertler, J., *US Unmanned Aerial Systems*, Library of Congress Washington DC Congressional Research Service, US, January 2012.

56. Austin, R., *Unmanned Aircraft Systems: UAVS Design, Development and Deployment*, John Wiley & Sons, US, 2011.

57. Petricca, L., Ohlckers, P., Grinde, C., Micro-and nano-air vehicles: State of the art. *Int. J. Aerosp. Eng.*, 2011, 2011.

58. Serokhvostov, S.V., Ways and technologies required for MAV miniaturization, in: *Proceedings of the European Micro Air Vehicle Conference (EMAV'08)*, July 2008.

59. Shyy, W., Lian, Y., Tang, J., Liu, H., Trizila, P., Stanford, B., Ifju, P., Computational aerodynamics of low Reynolds number plunging, pitching and flexible wings for MAV applications. *Acta Mech. Sin.*, 24, 4, 351–373, 2008.

60. Harbig, R.R., Sheridan, J., Thompson, M.C., Reynolds number and aspect ratio effects on the leading-edge vortex for rotating insect wing plan forms. *J. Fluid Mech.*, 717, 166–192, 2013.

61. Nguyen, T.T., Sundar, D.S., Yeo, K.S., Lim, T.T., Modeling and analysis of insect-like flexible wings at low Reynolds number. *J. Fluids Struct.*, 62, 294–317, 2016.

62. Taira, K. and Colonius, T., II, Three-dimensional flows around low-aspect-ratio flat-plate wings at low Reynolds numbers. *J. Fluid Mech.*, 623, 187–207, 2009.

63. Hassanalian, M., Abdelmoula, H., Ayed, S.B., Abdelkefi, A., Thermal impact of migrating birds' wing color on their flight performance: Possibility of new generation of biologically inspired drones. *J. Therm. Biol.*, 66, 27–32, 2017.

64. Quintana, A., Hassanalian, M., Abdelkefi, A., Conceptual design and performance improvement of growing micro unmanned air vehicle, in: *AIAA Science and Technology Forum and Exposition*, Grapevine, Texas, pp. 9–13, January 2017.

65. Ibrahim, M.M.S., Shanmugaraja, P., Vini, M.M.T., The roles, benefits and design challenges of multi versatile unmanned drones in flying ad-hoc network, in: *Contemporary Research in Electronics, Computing and Mechanical Sciences*, vol. 1, 2018.

66. Ollero, A. and Merino, L., Control and perception techniques for aerial robotics. *Ann. Rev. Control*, 28, 2, 167–178, 2004.

67. Chao, H., Cao, Y., Chen, Y., Autopilots for small unmanned aerial vehicles: A survey. *Int. J. Control Autom. Syst.*, 8, 1, 36–44, 2010.

68. Goerzen, C., Kong, Z., Mettler, B., A survey of motion planning algorithms from the perspective of autonomous UAV guidance. *J. Intell. Robot. Syst.*, 57, 1, 65–100, 2010.

69. Valavanis, K.P., *Advances in Unmanned Aerial Vehicles: State of the Art and the Road to Autonomy. Intelligent Systems, Control and Automation: Science and Engineering*, p. 33, Springer, U.S.A. 2007.
70. Kendoul, F., Survey of advances in guidance, navigation, and control of unmanned rotorcraft systems. *J. Field Robot.*, 29, 2, 315–378, 2012.
71. Hassanalian, M., Radmanesh, M., Ziaei-Rad, S., Sending instructions and receiving the data from MAVs using telecommunication networks, in: *Proceeding of International Micro Air Vehicle Conference (IMAV2012)*, Braunschweig, Germany, pp. 3–6, July 2012.
72. Gerdes, J.W., *Design, Analysis, and Testing of a Flapping Wing Miniature Air Vehicle*, University of Maryland, College Park, 2010.
73. Kurdila, A., Nechyba, M., Prazenica, R., Dahmen, W., Binev, P., DeVore, R., Sharpley, R., Vision-based control of micro-air-vehicles: Progress and problems in estimation, in: *2004 43rd IEEE Conference on Decision and Control (CDC) (IEEE Cat. No. 04CH37601)*, vol. 2, IEEE, pp. 1635–1642, December 2004.
74. Máthé, K. and Buşoniu, L., Vision and control for UAVs: A survey of general methods and of inexpensive platforms for infrastructure inspection. *Sensors*, 15, 7, 14887–14916, 2015.
75. Trites, S., Miniature autopilots for unmanned aerial vehicles. *Micro Pilot.* URL: http://www. micropilot.com.
76. Coleman, C., Funk, J., Salvati, J., Whipple, C., Padir, T., Wyglinski, A., *Design of an Autonomous Platform for Search and Rescue UAV Networks*, Worcester Polytechnic Institute, Project Number: WND1, Massachusetts, U.S., April 26th, 2012.
77. Oljača, M.V., Pajić, M., Gligorević, K., Dražić, M., Zlatanović, I., Dimitrijević, A., Balać, N., Design, classification, perspectives and possible applications drones in agriculture of Serbia. *Poljoprivrednatehnika*, 43, 4, 29–56, 2018.
78. Pacholski, N., Szabo, C., Falkner, K., *Extending the Sensor Edge Smart Drone Positioning System*, The University of Adelaide, South Australia, 2013.
79. Mao, G., Drake, S., Anderson, B.D., Design of an extended kalman filter for uav localization, in: *2007 Information, Decision and Control*, IEEE, pp. 224–229, February 2007.
80. Brown, R.G. and Hwang, P.Y., *Introduction to Random Signals and Applied Kalman Filtering: With MATLAB Exercises and Solutions*, Wiley, U.S., 1997.
81. Theilmann, C.A., *Integrating Autonomous Drones into the National Aerospace System*, Master of Science, University of Pennsylvania, 2015.
82. Nemra, A. and Aouf, N., Robust INS/GPS sensor fusion for UAV localization using SDRE nonlinear filtering. *IEEE Sens. J.*, 10, 4, 789–798, 2010.

5

Some Studies of the Latest Artificial Intelligence Applications of Drones are Explored in Detail with Application Phenomena

G. Vaitheeswaran[1], B. Sundaravadivazhagan[2]* and Karthikeyan[3]

[1]Stack Innovations, Trichy, India
[2]Department of Information and Technology, University of Technology and Applied Science-AI-Mussanah, Oman
[3]Department of Artificial Intelligence & Machine Learning (CSE), Varadhaman College of Engineering, Hyderabad, India

Abstract

Generally, drones are known as Unmanned Aerial Vehicle (UAV). Humans or robots can control the UAV remotely. Drone technology was introduced in twentieth century for military purposes and it came to common people hands by twenty-first century. Artificial Intelligence (AI) helps the machine to think artificially and it makes them to behave like humans. In this digital era, most of electronic devices are converted into smart devices with help of the AI technology. This chapter explores the role of AI in drone technologies. Drone technologies are used by the following organizations: military, industries, and logistics. AI applications of drones are discussed in this chapter with detailed uses cases. AI technology helps the drone to collect data and predict the outcome instantly. The smart drone greatly helps people to execute the hard work they are reluctant to do and to produce more efficient results. This chapter provides an overview of the issues and challenges of drone technologies using AI applications for researchers and industry persons.

Keywords: Unmanned aerial vehicle (UAV), artificial intelligence (AI), drone technologies

*Corresponding author: bsundaravadivazhagan@gmail.com

Sachi Nandan Mohanty, J.V.R. Ravindra, G. Surya Narayana, Chinmaya Ranjan Pattnaik
and Y. Mohamed Sirajudeen (eds.) Drone Technology: Future Trends and Practical Applications,
(99–116) © 2023 Scrivener Publishing LLC

5.1   Introduction

Drones are commonly referred as flying robots or Unmanned Autonomous Vehicles (UAV), but at present most of them are controlled by human pilots. UAV perform a range of benefits in the field of industries, from agriculture to real estate and from defense system to e-logistics delivery system. Industrialists and researchers are developing intelligent drones in such a manner that can read, calculate, analyze, and predict data in order to reduce human interference and provide useful information. No humans are required, as the drones would rely on built-in machine learning mechanisms to operate. Since 2016, drone technology has been rapidly popularized among consumers with the market size growing from two billion USD in 2016 to twenty-two billion USD by 2020. Technavio, a leading global technology research and advisory company reported[1] in the year 2021 that the drone market will be increased from 2021 to 2025 with 21 billion USD. The Figure 5.1[2] states the global drone market. It can be clearly identified that there is an increase in the use of drones that acts as a key factor for the growth of the market. Legitimizing UAV operations for commercialized applications will be important in stimulating market growth during the forecast period.

Several companies have embraced the drone devices and utilizes in various domains containing cable assessment, monitoring (surveillance), civil planning, farming and public safety. Nowadays, most of the research is done on this drone technology, which is mostly used for domain-based data collection and analysis of those domains. However, the direct development of navigation systems to provide advanced computerization of drone operations has become a realistic objective. Deep neural networks (DNNs) and the related research area play an important role in the purpose mentioned above.

With the emergence of advancement in electronic and application technologies, Deep Learning (DL) has become more adaptable with the most recent advances in hardware and software skills. There is no deficiency of credentials associated to its application for drone autonomy. Even though, the solutions of domain-knowledge system utilize the accurate GPS, Lidar, computer vision to produce a system for autonomous navigation. The above discussed statements are not conclusive, are expensive to use. The network

---

[1]https://www.prnewswire.com/news-releases/usd-21-bn-growth-in-drone-market-from-2020-to-2025-driven-by-increasing-applications-of-drones--technavio-301410682.html
[2]https://www.prnewswire.com/news-releases/usd-21-bn-growth-in-drone-market-from-2020-to-2025-driven-by-increasing-applications-of-drones--technavio-301410682.html

Figure 5.1 Global drone market.

connected drone system requires right network access to provide continuous operations.

This chapter discusses navigation tasks that use Deep Learning (DL) models for implementing navigation works to drone autonomy. As with the implementation of deep learning in self-driving vehicles, the capacity of deep learning models skilled to supply well-built understanding of visual and other sensor data in drones is crucial to the capacity of drones to attain complete autonomous navigation.

The following are the core contributions of this chapter:

- Provides general features of drones
- Deliberates the role of artificial intelligence in drone technology
- Analyses case studies of drones in various sectors

In recent days, research projects that focus on developing new navigation methods, with or without the collaboration of industry collaborators, create a definition of what the state of the art is - solutions that are not currently practiced in the industry, but solutions and processes that are actively explored for the future growth.

5.2   Evolution of the Drone

The drone was originally developed for military use in early 19th centuries. Later with the uncontrolled development of computer and internet

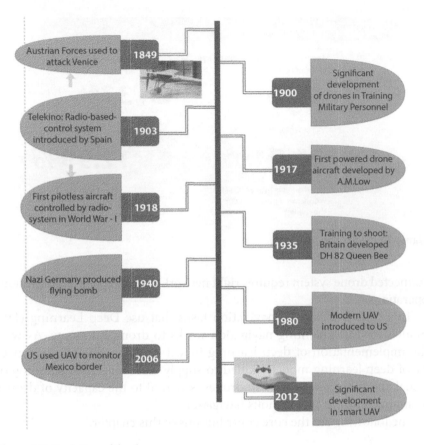

Figure 5.2 Evolution of the drone.

technology the drone was also used on commercial applications. Nowadays, the drone is inevitably thing among the people. This section discusses the evolution of drones from military to commercial. Figure 5.2 clearly shows the evolution of drones.

5.2.1   Military Drones

Drones sailed for Italy in 1849, while Venice opposed for independence from Austria. The soldiers of Austria assaulted Venice by balloons that filled with hot air and hydrogen.

In World War, the unmanned radio-forced aircraft was utilized. Likewise in 1918, the U.S. military created a test drone Kettering buck that is unmanned "flying bomb". This drone had never been utilized in battle.

In 1935, most commonly utilized drone was emerged like a complete-scale reprint of the de Haviland DH82B "Queen Bee" bi-aircraft. This was filled with controls operated by radio and servo procedures in the rear place. This type of aircraft can usually operate from the front place. However, it is usually flown unmanned by weaponry gunners who are in teaching to shoot.

The word drone comes from this premature utilize, a play titled "Queen Bee". UAV technology caught the attention of the military. Although, it was frequently unpredictable and expensive. After concerns arose about the downing of emissary planes, the military reconsidered the topic of unmanned aerial vehicles. Following concerns over the downing of spy planes arose, the army reconsidered the theme of unmanned airborne transports. Army drones soon abandoned the leaflets and adopted roles that acted as spy processors.

5.2.2 Commercial Drones

According to the company Precept, a report by military research and development provides that, from the past one hundred and fifty years, in the year 2006, the drones are used in the non-military enterprises. Federal Aviation Administration (FAA) had granted license for the first commercial drone in the same year.

In developed countries such as the United States and the European Union, the government is rapidly adopting drone technology to provide relief in times of disaster and to monitor the country's borders. Large companies began to use the drone for commercial purposes such as oil-pipeline monitoring, and surveillance.

In late 2013, Amazon became the first major company to announce the use of drones for product distribution and it brings public awareness of drone technologies. Due to government regulations and vulnerable measures, commercial drones faced many struggles. Technological advances have made the drone industry grow faster for personal use. Recently developed drones are based on quadcopters. They are mainly used for entertainment purposes. In the field of photography, the drone business has reached all over the world. Unlike military drones, recreational drones are inexpensive and are available from one thousand dollars. The US government has allowed medical companies to use drones for drug distribution. The global research team predicts that the drone business in the medical field will reach nine thousand and forty-seven million dollars by 2027.

In the year 2018, China and Israel started to invest in research into the use of drones for pilot-less vehicles, photography, and indoor applications. According to Percepto, the commercial drones will reach around 122 million by the year 2023. In the commercial sector, agricultural and industrial are the first two major organizations started to use drones. In agriculture, drones are being used to manage and research crops. Due to the cost margins in the agricultural market, other industries such as renewable energy, mining, logistics, ports and harbors are doing better than that in terms of adoption. The value of surveys for limited space (e.g., oil and gas, power generation, mining, chemicals and marine ships) is projected to be $ 795.12 million in 2019 and $ 1,936.32 million in 2027.

This growth was partly due to the approval of air traffic controllers in the United States for fully automated commercial drones from 2021 onwards. That is, no authorized company can operate drones onsite without control or surveillance.

5.3   Drone Features

Generally, drones are officially recognized as Unmanned Aerial Vehicles (UAVs). Basically, a drone is an airborne robot that can fly remotely or automatically using software control aeronautics on its embedded systems. This work is integrated with the internal sensors and the Global Positioning System (GPS). The drones' applications are divided into four types based on its types, as follows:

- See
- Sense
- Move
- Transform

Mostly, the UAVs were linked with the military. In which, anti-plane objective training was utilized at first used like knowledge gathering and especially controversial military hardware bases. There were several civilian roles of Drones which used recently that are following:

- Explore and release
- Scrutinizing thefts
- Traffic observing
- Climate observing
- Fire combating

- Individual use
- Drone-based picture making
- Videography
- Farming
- Delivery services

5.4   AI Meets Drones

Artificial intelligence (AI) is growing rapidly due to technology and development Innovative applications. From virtual assistants to self-driving cars, AI is used all the fields. However, there are still big computational challenges Creating Advanced AI.

In AI, computer vision trains the computer to identify, examine, and monitor the objects that are capture from the visual data (i.e., images and videos). Generally, the computer visions are powered by the system authentication. The computer visions are trained by millions of data with labeled inputs. The advancement in machine learning (ML) and deep learning (DL) has made great strides in recent years. In the future it overcomes the difficulties of human effort on certain objects.

There are good image processing solutions and they work just like the Machine Learning and Deep Learning techniques for some missions. Forever consider that any Machine Learning approach needs large data sets and a bunch of teaching. So, if there is only a restricted number of an image, the picture processing software would be the greatest clarification. ML or DL techniques have better picture processing software results when large datasets are available and various tasks need to be performed. In other words, the more tasks and complexity in image processing software, the more ML/DL approaches can resolve them like the range of data increase.

Several applications of Machine Learning and Deep Learning techniques are presently identified in the research and maintenance domains. Companies like Sky-Futures[3], Scopito[4] and DRONE VOLT[5] utilizes dissimilar ML and/or DL techniques for different research missions. For instance, insulators for electrical connections are robotically found in pictures and ensured for contradictions or techniques are utilized to discover decay on metal planes. On the report of Sky-Futures, their recognition rate is

---

[3]https://www.sky-futures.com/drone-inspections-surveys/
[4]http://scopito.com/
[5]https://www.dronevolt.com/en/

80–90%. An additional instance is Ardenna, an American software industry presently improving from computer visualization software to detect 30 damages simultaneously from BNSF, the railway company for ML-based train inspection. Several AI techniques are already being utilized for data analysis in the fields of energy, farming, actual estate, production, and forestry. The list of data analysis applications which depends on Artificial Intelligence already appears nonstop and the instances stated now are just a little part of already existed on the market.

The artificial intelligence-based drone market is huge and many existing companies, new start-ups will give the opportunity to make the idea come true. Many industrialists and researchers have proposed user-to-user solutions based on drones AI.

Moreover, the best solution can be proposed based on the knowledge in computer visualization and DL, which suitable for the requirements of the client. The following are some important aspects that have to be considered to build AI for drones: embedded, ground station based and cloud solutions. Figure 5.3 provides the big picture of AI and drones.

Current research provides the ability for artificial intelligence machines to communicate intelligently. Additionally, the combination among drones

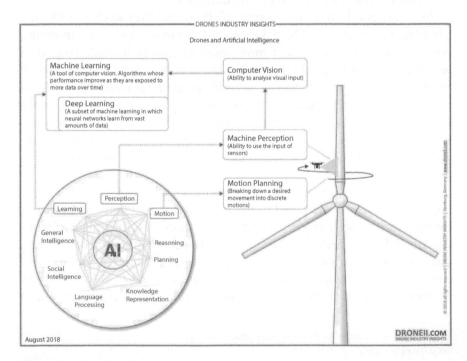

Figure 5.3 Drones and AI.

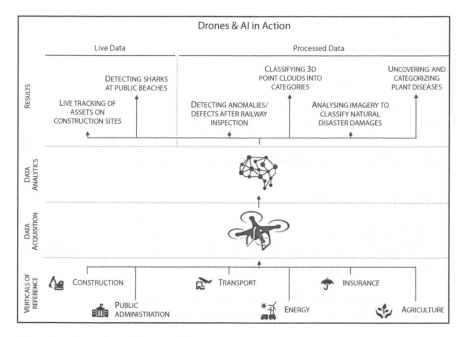

Figure 5.4 Examples of drone AI.[6]

and AI represents a reply to several requires of airborne images. Also, it gives a new caption in the potential of aerospace technology for various fields such as energy, construction, defense, and farming. Figure 5.4 represents the actions of drone AI. Thanks to the expertise in artificial intelligence for offering solutions based on drone, computer vision, and DNN.

AI enabled drones are using in the following fields:

- Entity recognition, counting, separation and tracing
- Recognition and monitoring of the person or animal
- Number of crowds
- Heat detection
- Identify the usage of face shields in community and specialized places
- Detecting the usage of safety tools (like helmets, glasses, etc.)
- Face recognition
- Fire and smoke discovery
- Authorize shield reading

---

[6]https://medium.datadriveninvestor.com/ai-powered-drones-a-technological-benison-c734f6135c2c accessed on April-04-2022

- Detection of scratch on surfaces
- Mask discovery

Embedded Solutions

Currently, drones can operate at high levels of autonomy by utilizing an intellectual camera with an NVIDIA graphics card. This facilitates the superior AI applications on the Edge. An optimal resolution is needed for those applications which demands real-time solutions.

Station-Based Solutions

Depending on the purpose and the work surroundings, several companies provide two resolutions for running Artificial Intelligence depended domains on the position of during or after flight:

- If Online: the video is processed in real-time.
- If Offline: the video is processed after flight.

Cloud-Based Solutions

In general, the drone servicers provide new appearances for cloud depended AI applications with high competence GPU. This result produces the capacity to remotely process the customer data on GPU servers.

Deep Learning (DL) Models

The more specific DL patterns often appear in the research area of drone navigation decision making are discussed as follows.

First, "VGG-16" [1] is a CNN film Trained assortment in the "ImageNet" database of 14 million images [2] Fits with 1000 of labels. VGG-16 supports a wide range of image assortments or can serve as a platform for changing learning by fine-tuning using target-specific images Drone environment. Most research is about accepting it or finding the meaning Model "YoloV3" [3] is used as a platform to avoid conflict or object in the research pool Diagnosis/Difference. "ResNet" architecture [4] is derived from CNN-based paper discussing application through optimization of the "AlexNet" framework [5], the remaining layer "shortcuts" that can approximate the function of the entire neural layer.

Similar to VGG-16, ResNet is trained in ImageNet database. Benefit of Resnet The shortcut configuration is a significant reduction in overhead processing, as a result Efficient models with low response times but maintain comparable accuracy. This is favorable for drone operations that require low CPU overhead. "DroNet" much Autonomous drone specified for navigation area and uses manually labeled car Bicycle displays as training data for navigation in urban environment. Publications for this the drone from an image is specific to the purposes of drone navigation A steering angle will guide the drone while avoiding obstacles and collisions Probably, the UAV will identify hazardous situations and allow them to react immediately. For an autonomous drone network built for one purpose, drone work [6] was highly cited and Used as a base network for many documents in the research team.

5.5   Use Cases

5.5.1   Army

Drones may be equipped with various surveillance devices to collect HD video and still photos at all times of the day and night. It may be equipped with equipment that allows you to listen to phone calls, track GPS movements, and collect license plate data. Drone surveillance is the practice of capturing still photos and video by unmanned aerial vehicles (UAVs) and targeting specific targets, individuals, companies, or locations. Drone tracking allows you to collect data about when a target is undetected from a range or altitude. It monitors and allows covert operations. All of these restrictions simply cannot be overcome, but can be efficiently modified using artificial intelligence (AI)-driven drones. These drones combine multiple altitude views with high-resolution visual and thermal imaging. Multi-zoom capabilities and weather-sensitive stability mean that the drone will enhance all ISR functions.

IdeaForge[7], Big Bang Boom[8] are few Indian-based drone manufacturing companies that uses AI for unparalleled surveillance-support. Troops can be easily positioned by the drone from a container, anywhere at any time. This adds an excellent tactical capability in which troops can detect the drone's forward path or target location. Direct monitoring helps to protect their movement and/or increase the probability of campaign success.

There are a number of AI-factors that make drones the best surveillance partner for security services.

---

[7]https://www.ideaforge.co.in/
[8]https://bigbangboom.com/

Automatic Drone Navigation

Players can set up these drones anywhere, anytime. The drone fits in a small bag with a complete ground control station (GCS). Drones can be used from any known or obscure point, even at punitive heights. Once used, the drone can move on an automatic pilot with the help of its built-in AI. It will automatically move from the lane (desired checkpoint) to the lane. If instructed, the drone can perform automated patrols of the region, providing direct (and secure) high-resolution visual/thermal feedback to the home site.

Automatic Repair of Airways—Automatic Detection of Obstacles and Collisions

But what about obstacles such as rocks, trees, or man-made structures in the drone's flight path? The AI-driven automation pilot detects these obstacles in real time and calculates potential collision probabilities based on altitude, trajectory, and speed. The drone adjusts its trajectory to avoid this collision. Automated pilot and automatic collision avoidance make these drones versatile surveillance 'security units'. Such self-guided intelligence helps to reduce this workload of ground soldiers. These players can focus on important strategy and decision making.

Moving-Target Indicator

One of the key components of an ISR is camouflage motion-detection. For example, areas around the "Line of Control" (LoC) or the "Line of Actual Control" (LAC) are always at risk of a rogue enemy infiltration. These enemy troops use Lay-off-the-Land to move their units and equipment camouflage. Drones flying at this altitude, with their deep view and thermal imaging, use AI to automatically detect any irregularities. This means that the drone in an automated pilot, no matter how well or efficiently concealed, will immediately pick up any irregular behavior or movement. Such advanced and domestic drone-technology enables ISR and counter-intelligence to be upgraded in real time.

Moving-Target Tracking

As mentioned above, these drones are designed for strong weather and harsh conditions. They can operate automatic aircraft, automatically adjust their trajectories to avoid collisions, and automatically detect hidden movement or activity. Now, when shifting gears from drone tracking to action, the drone can mark almost any moving target and follow the target's movement (tracking). The target may be a person fleeing troops or a vehicle with enemy arsenal.

This AI-enabled and drone-led moving-target surveillance is a great boon to active campaign intelligence. This increases the overall success rate and impact of these military campaigns.

AI-Enabled Decision-Making Drone-Driven Security Environment

Artificial intelligence is needed to turn drones into active military units. These drone-units not only act as resource amplifiers, but also protect the lives of ground soldiers. Beyond drone units, AI-enabled ISR initiatives help to create a larger strategic picture for decision-making at the regional and national levels. The Army is keen to prepare each battalion for unmanned aircraft for greater situational awareness and faster reaction times. The intelligence collected by the drone from each battalion, while cooperating with a central command center, intercepts all information, creating a complete and detailed map of direct operations on the front. This information indicates the difference between winning a border contest and failing to prevent enemy infiltration.

The AI technology that is embedded with drones will ensure that live footage of the drone (along with other bundled components such as CCTV and motion-sensor feeds) is delivered securely and instantly to any certified receiving station in the world. This means that senior officials at the Regional Command Centre or the Minister of Defence in Delhi can view the live monitoring feed at any time.

The use of such drone-technology, through AI-operation, gives the country an edge in current and future security decisions. These applications are rapidly being monitored with the aim of rapidly modernizing the entire army. AI-enabled drone-technology that makes such an impact and when made 100% locally, it multiplies their benefits many times over. It has been central to the drone-based start-ups vision and mission for over a decade, and will define it as a way forward for the future.

5.5.2   Weather Forecast

The environment is changing. The same is true of natural disasters. Drones are less accurate than satellite images in anticipation of severe weather events. Nevertheless, they can provide vital assistance in the event of a tragedy. Government agencies and insurers are increasingly aware of the potential for hiring people to assess post-disaster losses, especially in areas not designated as safe places to access. This is a significant improvement over traditional data collection methods for collecting wind samples capable of increasing the reliability of climate forecasting models.

Weather drones are specially designed drones that fly in the lowest layer of the Earth's atmosphere, the boundary layer. Sensors are fitted to collect information about temperature, humidity, and air in the atmosphere, ultimately helping to improve weather forecast models.

5.5.3 Industry

Drones are playing major role in the manufacturing industries. The advancement in the drone technologies make easy to mount the pillars, walls in the difficult places where human not able to work. This reduces the human risk and costs. The drones are helping to move the cranes and vehicles to remote area. The drones are broadly classified as indoor and outdoor drones in the manufacturing industries. The indoor drones are used for the safety purposes and for installing cables, doors, etc., in the limited spaces. The outdoor drones can be used in the complex work-fields such laser rangefinders, ultra-wideband radio signals, etc.

According to PwC, drone use in construction and mining will eventually become a $ 28.3 billion global market. Businesses in these industries use drones to easily comply with the comprehensive laws and regulations surrounding labor protection.

Construction companies need U.S. government laws to constantly inspect their sites to ensure they are safe for workers. This process can take anywhere from 10 hours to a few days - but with drone technology, studies can be completed as quickly as 15 minutes.

The advantages of drones in manufacturing industries are as follows:

- It provides safety and reduce human risks
- Cost savings
- Unlike human, drone finishes work quickly
- Efficient data collection
- Increase in productivity

According to authors [7] Omid Maghazei and Torbjorn H. Netland, the drones in manufacturing industries will reach high by the year 2025. Many manufacturing industries will use drone for four typologies such as see, sense, move, and transform.

5.5.4   Agriculture

The use of drones has many benefits in agriculture and, moreover, plays an important role in giving precision farming [8–10]. Precision farming is the science of using the latest technology to improve agricultural productivity, crop yields and produce profitability. According to analysts[9], drones in agriculture will be worth $ 5 billion by the end of 2025.

The application of drones in agriculture provides the following advantages:

    i.   It provides farm analysis accurately
    ii. It reduces time and cost
    iii. It improves crops and produce more yields
    iv. It helps to spray the fertilizers where human cannot
    v. It helps to fight climate change by reducing the vehicles pollution

According to the United Nations researchers, in the year 2050, the world population will reach 9.7 billion with the agricultural consumption set to rise by 69%. The development of drone technologies will make the evolution in farming.

The following are the common tasks that are performed by drone in farming:

- Crop monitoring
- Irrigation management
- Fertilizer management
- Climate management

A drone camera lens magnifies the yellow flower of a tomato seedling and can use these images in an artificial intelligence algorithm that accurately predicts how long it will take for the flowers to turn into ripe tomatoes and the baking and grocery preparation area. Figure 5.5 portraying the sample of drone usage in agriculture.

5.5.5   Logistics

Healthcare-drones can deliver medical supplies, including blood, vaccines, and medicines, as well as other supplies such as medicines and medical samples. It began providing personal protective equipment and COVID-19 testing in the United States and Israel during the COVID-19 eruption.

---

[9]https://yourdronereviews.com/benefits-of-drones-in-agriculture accessed on April-04-2022

Figure 5.5 Agricultural drone.

As of October 2020, Zipline has completed more than 70,000 medical deliveries.

Food: It is recommended as a means of quickly moving prepared items such as pizza, tacos, and frozen drinks. Star Simpson's Tacopter[10] Demo is a taco delivery idea that uses a mobile application to order unmanned aerial vehicles (UAV Tacos in the San Francisco area), an early prototype of food delivery drones.

Mail: Various postal businesses in Australia, Switzerland, Germany, Singapore, the United Kingdom, and Ukraine conducted drone tests to see if unmanned delivery drone services were feasible and profitable. United States of America Postal Service (USPS) uses horse-flying drones to test delivery systems. According to a new study by Martin Luther University of Haley-Wittenberg, drones have a lower energy balance when carrying luggage than conventional delivery vehicles.

Shipping Redistribution: Instead of sending small boats, shipping companies Mark and Rotterdam tried to use drones to rearrange foreign ships. It carries supplies to troops in the field and quickly distributes emergency food and ammunition. Automated drones are capable of carrying 60 lb and can fly up to 6 km. TRUAS (Tactical Resupply Unmanned Aircraft System) is running a $ 225,000 competition to deliver drones.

---

[10]https://www.wired.com/2012/03/qa-with-tacocopter/. March 23, 2012. Retrieved March 8, 2022.

5.6   Conclusion

As an industrial world, many industries in the most developed and developing countries require the use of drones. In this chapter, the applications of drones in developing countries for various industrial opportunities are discussed with major use-cases. The development of drone has several benefits for various applications due to its economic constraints. The practice of industrial drones can support in several ways that are as follows: taking decisions, improving the business activities, reducing the operation costs, speeding the task analysis and improving the surveillance.

It is very good to develop sophisticated applications of artificial intelligence to help the local industry. In fact, it is better to build drones using artificial intelligence without being misused by irresponsible parties. The researchers and industrialists need to develop drones with artificial intelligence to overcome complex challenges and solve difficult problems for humans. By using artificial intelligence drones for industry, they can work hard and turn it into something more effective and efficient.

References

1. Simonyan, K. and Zisserman, A., Very deep convolutional networks for large-scale image recognition, in: *Proceedings of the 3rd International Conference on Learning Representations, ICLR 2015—Conference Track Proceedings*, San Diego, CA, USA, pp. 1–14, May 7–9, 2015.
2. Deng, J., Dong, W., Socher, R., Li, L.J., Li, K., Li, F.-F., ImageNet: A large-scale hierarchical image database, in: *Proceedings of the 2009 IEEE Conference on Computer Vision and Pattern Recognition*, Miami, FL, USA, pp. 248–255, June 20–25, 2009.
3. Redmon, J. and Farhadi, A., YOLOv3: An Incremental Improvement, Computer Vision and Pattern Recognition, 2018, arXiv, arXiv:1804.02767.
4. He, K., Zhang, X., Ren, S., Sun, J., Deep residual learning for image recognition, in: *Proceedings of the 2016 IEEE Conference on Computer Vision and Pattern Recognition (CVPR)*, Las Vegas, NV, USA, pp. 770–778, June 27–30, 2016.
5. Krizhevsky, A., Sutskever, I., Hinton, G.E., ImageNet classification with deep convolutional neural networks. *Commun. ACM*, 60, 84–90, 2017.
6. Loquercio, A., Maqueda, A.I., Del-Blanco, C.R., Scaramuzza, D., DroNet: Learning to fly by driving. *IEEE Robot. Autom. Lett.*, 3, 1088–1095, 2018.
7. Maghazei, O. and Netland, T.H., Drones in manufacturing: Exploring opportunities for research and practice. *J. Manuf. Technol. Manage.*, 31, 6, 2019.

8. Content-based mammogram retrieval using mixed kernel PCA and curvelet transform. *Towards Automated Drone Surveillance in Railways: State-of-the-Art and Future Direction Conference Paper*, pp. 336–348, October 2016.

9. Metha, P.A., Siddhant, M.U.I., Implementation of voice activated autonomous quadcopter. *Int. J. Eng. Sci. Res. Technol.* 6, 9, 82–85. September 5, 2017.

10. Ganeshwar, R., Ramprabhu, J., Manochitra, P., Kaliappan, S., Surveillance and security management system using drone. *International Journal of Advanced Research Trends in Engineering and Technology (IJARTET)*, 5, 5, May 2018.

6

Drone Technologies: Aviation Strategies, Challenges, and Applications

Devshri Satyarthi[1]*, K.V. Arya[2] and Manish Dixit[3]

[1]*Department of Computer Science and Engineering , Atal Bihari Vajpayee-Indian Institute of Information Technology and Management, Gwalior, India*
[2]*Department of Computer Science and Engineering, Atal Bihari Vajpayee-Indian Institute of Information Technology and Management, Gwalior, India*
[3]*Department of Computer Science & Engineering, Madhav Institute of Technology and Science, Gwalior, India*

Abstract

A drone or an unmanned aerial vehicle (UAV) is utilized in a wide range of applications with different purposes, such as military, agriculture, surveillance, and photography. Nowadays, we have next-generation smart UAVs with great potential because of digital advances, including remote sensing, wireless connectivity, search and rescue, delivery of commodities, surveillance and security, and crop monitoring. Particularly in civil infrastructure, in terms of reduced risks and lower costs, UAVs have many applications, namely real-time monitoring, disaster control, traffic detection, logistics, and inspection. UAV, which is also called a drone, is capable of capturing an aerial view or a bird's eye view. In this chapter, we focused on the recent trends in drone research technology and identified issues and challenges. The chapter also provides future insights on other potential UAV uses and application challenges, with the help of existing literature reviews and research efforts that could unlock the problems and constraints described. Furthermore, we discussed a relevant comprehensive study and the specifications of different drone models and UAV simulators. We also defined the UAV applications, including their classification, requirements, and equipment. Finally, future insights on UAV applications in artificial intelligence are also given.

Keywords: Unmanned aerial vehicle (UAV), aviation, challenges, wireless connectivity

---

*Corresponding author: devshri@iiitm.ac.in, devshri03@gmail.com

Sachi Nandan Mohanty, J.V.R. Ravindra, G. Surya Narayana, Chinmaya Ranjan Pattnaik and Y. Mohamed Sirajudeen (eds.) Drone Technology: Future Trends and Practical Applications, (117–152) © 2023 Scrivener Publishing LLC

6.1   Introduction

UAV is also known as drone, it is an aircraft without any human pilots or operator, crew members and passenger on board, that mean aircraft take instruction autonomous, by remote control, or both. So, it able to fly at a regulated, sustained level and reciprocating engine, powered jet, or electric motor. Unmanned aerial vehicle is a component of an unmanned aircraft system (UAS); it adds a controller on the surface and a communications infrastructure for the unmanned aerial vehicle. Remotely piloted aircraft are UAVs that can be flown by a trained pilot using a remote control (RPA), or is also called as autopilot assistance, up to fully autonomous aircraft that have no rider for human interference. In twentieth century UAV especially developed military missions too "dull, dirty or dangerous" for human military force. Now a day UAV technology improved with low costs and also high mobility and ability to fly so it uses many nonmilitary applications such as agriculture, surveillance, security, aerial photography, civil infrastructure, some drone games etc. so coming days researcher working SAR and save operations, delivery of goods, surveillance and security, agriculture and civil infrastructure scrutiny. The availability of drones with enhanced military mission safety and lower operational costs has led to the widespread usage of unmanned vehicles and unmanned systems in a variety of contexts.

These are major three categories of Drone or UAV are follows:

1. Unmanned/Autonomous Aerial Vehicle (UAV/AAV)
2. Unmanned/Autonomous Ground Vehicle (UGV/AGV)
3. Unmanned/Autonomous Underwater Vehicle (UUV/AUV)

The primary feature of drone is flying. There are three types of drones classified into based on flying mechanism and altitude are single rotor (helicopter), fixed wing (airplane) and multicopter (multirotor). In single rotor is a traditional helicopter, it is difficult to fly because it has single lifting with two or more blades but it is strong, fast and efficient. In fixed wing cannot move easily over there wing to generate lift, so it move stay in motion but it is light weight, less energy consume and generous fight time. It uses 4D mapping, agriculture survey etc. In multicopter is easy to use, it has more than one rotor generally, four to eight rotors. So, accuracy is most important feature of multicopter. There are three types of

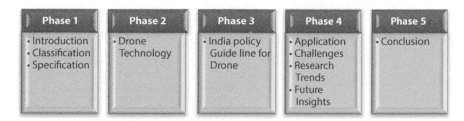

Figure 6.1 Detailed structure of the drone.

multicopter: Quadcopter, Hexacopter and Octocopter. Figure 6.1 describes the structures of drone; this chapter idea of the survey analyzes some UAV applications, challenge, and issues and identifies the hurdles. Significantly, this chapter consists of following point:

❖ Define unmanned aerial vehicle (UAV) categorization and specifications depending on their forms, lifespan, mass, payload, parts, sensors, and operations.
❖ Explain the drone technology.
❖ Discusses on guideline or rule for drone by the government on 15 July 2021.
❖ Summarize developments in research, important problems, and upcoming understandings in the UAV usage in the application area.

6.1.1 Categorization of Unmanned Aerial Vehicle (UAV)

This has a substantial attempt to improve also develop drone for particular purposes in recent years. Figure 6.2 discusses different classifications of UAV that have planned according to various parameters such as based on size, based on weight, based on range and endurance, based on altitude, based on engine type, and also based on configuration.

6.1.1.1 Classification Based on Size

The size of UAVs is one of the most crucial factors in identifying them. UAVs today exist in various sizes and are used for a range of tasks. At one extreme, they are as small as insects, while at the other, they are as big as enormous aircraft. UAVs that are so tiny they could be handled by one

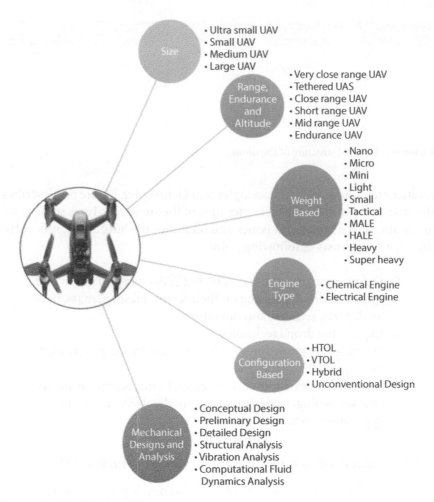

Figure 6.2 Classification of UAVs.

Table 6.1 Size-based classification.

S. no.	Class	Maximum dimension (cm)
1	Ultra-Small (NAV) 10 m	<7.5
2	Very Small (MAV)	7.5-15
3	Small	15-200
4	Medium	200-1000
5	Large	>1000

individual. They can potentially be as big as a bird or a big insect, says [1, 2], such as Ultra small UAV, Small UAV, Medium UAV, Large UAV. Table 6.1 shows the summary of size-based classification.

6.1.1.2   Classification Based on Range, Endurance, and Altitude

This classification of UAV performance based on range, endurance and altitude. One of the most crucial characteristics is range; that is working with different parameters. Range is especially depending on weight of the payload. It detects how far a UAV can fly away from its ground control station. Another crucial parameter is endurance, which measures how long a UAV can operate without refueling or charging. Depending on the job, a UAV's endurance ranges from 1 to more than 46 hours. Endurance and efficiency is affecting by volume and mass of the fuel or battery load of the UAV. Classification of UAVs based on range and endurance developed by US military. They classified into five categories are Very close range UAVs, Tethered UAS, Close range UAV, Short-range UAVs, Mid-range UAVs and Endurance UAVs. Table 6.2 summarizes the classified classification based on range, endurance, and altitude of UAV.

6.1.1.3   Classification Based on Weight

The main purpose of weight-based UAV developed for safety of missions; they correspond to the anticipated kinetic energy generated during impact. UAV classified into 10 categories based on weight; they are nano UAVs, micro UAVs, mini UAV, light small UAV, tactical UAV, MALE UAV, and

Table 6.2 Classification based on range, endurance, and altitude of UAV.

S. no.	Class	Maximum range (km)	Maximum endurance (h)
1	Very close range	5	<6
2	Tethered UAS	10	24
4	Close range UAVs	50	6
4	Short-range	150	12
5	Mid-range	650	12-46
6	Endurance	400	46

Table 6.3 Classification based on weight.

S. no.	Class	Maximum weight (kg)	Maximum range (km)
1	Nano	0.2	5
2	Micro	2	25
4	Mini	20	40
4	Light	50	70
5	Small	150	150
6	Tactical	600	150
7	MALE	1000	200
8	HALE	1000	250
9	Heavy	2000	1000
10	Super heavy	24,950	1500

HALE UAV, heavy UAV and super heavy UAV. Ravn X is considered one of the heaviest UAV in worlds, which weighs is 24,950 kg. Table 6.3 represents summary of classified classification based on weight of UAV.

6.1.1.4   Classification Based on Engine Type

According to this classification UAVs are available in a variety of shapes, dimensions, and performance levels. As the UAV's load increases, so does the size of its engine. The range and endurance of a UAV might be affected by its engine configuration. Chemical engines and electrical engines are the two basic types into which engines can be divided. The most popular forms of engines are electric and piston engines. Although piston engines are typically utilized with large UAVs that must handle massive payloads, electric engines are typically employed with light and small UAVs [3, 4].

6.1.1.5   Classification Based on Configuration

Configuration is most important part of UAVs. These configurations are depending on their missions. There are classified into four classes are horizontal take-off, landing (HTOL), vertical takeoff and landing (VTOL), hybrid UAVs, and unconventional UAVs.

6.1.1.6   Classification Based on Mechanical Design and Analysis

UAV design is an iterative procedure that involves careful preparation; there are various designs for UAVs. There are classified into seven classes are conceptual design, preliminary design, detailed design, structural Analysis, Vibration Analysis and Computational Fluid Dynamics Analysis.

6.1.2   Specification of Drones

The technical specification of a drone describes the information of specific functionality, specific uses and performance levels for the drone; its represent the set of KPIs (Key Performance Indicators). Table 6.4 shows UAV's technical specifications, operational conditions and provided by the manufacturer [5]. Intel Aero Ready to Fly Drone is a kit for developers. According to professional judgment the developers can modify the kit, but Intel hasn't been established operating limitations of the kit that haven't tested any type of configurations as compare to another basic configuration. In this kit developers are responsible for testing, ensuring the safety and operating limits of those configurations [6] Table 6.5. According to Intel® Aero Ready to Fly Drone Specifications. There are some basic key indicators gives good analysis of the drone are follows:

- ❖ **Flight autonomy**: Those find or measure the maximum flight time with the help of single battery pack and is reported in seconds/minutes.
- ❖ **Maximum ceiling altitude**: It finds maximum height the drone can reach above sea level. It unit is meters.
- ❖ **Maximum horizontal distance reachable**: It finds maximum horizontal distance reachable by the drone with respect to the pilot's position. It unit is km.
- ❖ **Maximum video transmission distance**: To show clear distance and fluid reception for video captured by the drone camera. Its standard unit is km or meters.
- ❖ **Max wind resistance**: The wind is constant or gust disturb by the drone during flight. The maximum value in m/s.
- ❖ **Max transportable payload (weight)**: The maximum transportable weight shows in grams or kg.
- ❖ **Sensors**: On-board sensors available.

Table 6.4 Specifications and operational conditions of the model md4-1000 UAVs [5].

S. no.	UAV specification	Value
1	Climb rate	7.5 m/s
2	Cruising speed	15.0 m/s
4	Peak thrust	118N
4	Vehicle mass	2.65 kg approx. (depends on configuration)
5	Recommended payload mass	0.80 kg
6	Maximum payload mass	1.25 kg
7	Maximum takeoff weight	5.55 kg
8	Dimensions	1.04 m between opposite rotor shafts
9	Flight time	Up to 45 min (depends on payload and wind)
Operational condition		
1	Temperature	-10°C to 50°
2	Humidity	Maximum 90%
4	Wind tolerance	Steady pictures up to 6 m/s
4	Flight radius	Minimum 500 m using radio control, with waypoints up to 40 km
5	Ceiling altitude	Up to 1,000 m
6	Take-off altitude	Up to 4,000 m about sea level

6.2   Drone Technology

We have discussed newest drone plenty information on latest drone technology with equipment and component are available in market, so in this phase focuses on the different types appear anti drone with the help of pros and cons of each technology. There are classified in two way monitoring equipment and countermeasures.

Table 6.5 Intel® Aero ready-to-fly drone specifications [6].

Name	Measurement
Drone dimensions [hub-to-hub (diagonal)]	460 mm
Drone height [from the base to the top of a GPS antenna]	222 mm
Propeller [length]	240 mm
Weight of drone [basic configuration without battery]	865 g
Gross weight (max) [takeoff weight]	1900 g[4]
Flight time (max) – with 4S, 4000mAh battery, hovering, no added payload	20 min[4]
Sustained wind (max)	15 knots[4]
Control distance (max) – with supplied remote control	400 m[4]
Airspeed (max)	15 m/s[4]
Altitude of operation (max) – height above sea level	5000 m[4]
Outside air temperature (min/max)	-0°C/+45°C
ESC and motor – designed and manufactured by Yuneec Input control interface ESC input voltage	Modified for Intel® Aero UART11.1–14.8 V

6.2.1 Drone Monitoring Equipment

It is active or passive; active means to send the signal and after analysis what come back and passive means simply looking and listening and they are perform several function using techniques such as

- ❖ **Detection:** Detection is techniques that use to detect object. It means the technology can detect drones. For example radar can detect drones and it also detects birds.
- ❖ **Classification or Identification:** Classification is most important technique that classification and detection the object like bus, planes, trains, and automobiles. After

classification next step is identification that identifies what type of model or category.

❖ **Locating and Tracking and Alerting:** This technology can find exact location than track the drone and after tracking and give alert, with the help of awareness and ability to understand the drone and the flight controller. Some components or equipment is allowing tracing or tracking real-time location of drone. These classified in four types of drone equipment using monitoring are as follows:

1. Radio Frequency (RF) Analyzers: They examine a radio frequency spectrum with a processor and one or more antennas for capturing radio waves. It is utilized to monitor drone-controller radio communication. It can detect and MAC address of the drone and controller (if drone use Wi-Fi).

Pros: It is low cost. It can detect and identify of multiple drones and controllers.
Passive does not required license.
Cons: Some time cannot locate, track drones and autonomous drones. It is not successful in crowded areas because of limited range.

2. Acoustic Sensors (Microphones): Acoustic Sensors or Acoustic Sensors array detects sound that made by drone and calculates its direction. It is used to find rough triangulation. It is completely passive.

Pros: Microphone detects all drones within the near-field and ground clutter.
Cons: It does not work in loud environments. It's very short-term (max. 400–500 m).

3. Optical Sensors (Cameras): It works as video camera. It working within daylight standard cameras and optical sensors can be used for infrared or thermal imaging.

Pros: It can store and save images for forensic evidence and it use in final hearing.
Cons: It is difficult to use. It has high false alarm rates during the dark or fog so mostly give poor performance.

4. Radar: Radar is a device that uses radio energy to detect an object, it radar sends and receives signal and reflected, that compute the direction and distance (position) of object. Some radar sends their radio signal as a burst and then listens for the 'echo'. Almost all radars are designed to NOT pick up small targets. It is intended for tracking big objects like passenger planes.

Pros: It can handle long range, constant tracking and highly accurate localization targets. It is independent of visual conditions like day, night, fog, etc.

Cons: It is dependent on detection range drone size. It does not differentiate birds from drones. To avoid interference, a transmitter license and frequency check is necessary.

6.2.2 Drone Countermeasure Equipment

Even some technology is important but forbid the use, they have some rule and regulations. So most of the countries prohibit using of following technologies military personnel and law enforcement personnel are given exemption. Countermeasures equipment is classified on base of technologies are follows:

* ❖ Physically destroying the drone
* ❖ Neutralizing the drone
* ❖ Taking control of the drone

There are four types of drone countermeasure equipment are follows:

1. **Radio Frequency Jammers:** Radio frequency jammer is a static device like mobile or handheld, that is transmitted a huge quantity of radio frequency energy in the direction of drone to cover the controller signal.
   Pros:
   It is medium of cost and non-kinetic neutralization.
   Cons:
   It have short range so it can easily affect (and jam) other radio communications. They give unpredictable result on the bases of drone behaviors.
2. **GPS Spoofers:** GPS is hardware device that sends new signal to drone. GPS satellites replace the communication and it used for navigation. In other word the drone has been

tricked into believing it is in another location. In real time GPS coordinates use altering, spoofer controlled drone position for example if once control is gained drone can be going to a safe zone.

Pros: It is medium of cost and non-kinetic neutralization.

Cons: It has short range so it can easily affect (and jam) other radio communications.

3. **High Power Microwave (HPM) Devices:** Electronic devices can be damaged by electromagnetic pulse (EMP)-generating devices. Due to the harmful voltage and currents, the electromagnetic pulse can impair or even totally obliterate the electronic circuitry in drones. It also disrupts with radio communications. An antenna to steer the EMP in a beneficial direction and lessen potential collateral harm may be included in HPM devices.

Pros: It available in different range. Effective drone-stopping is possible. It does not move.

Cons: There is a chance that you'll unintentionally break up conversations or ruin nearby electronics. In actuality, the drone abruptly shuts off and falls to the earth uncontrollably.

4. **Nets and Guns:** When shooting or firing, a net at a drone or somehow getting a net into touch with the drone causes it to halt by preventing the rotor blades from turning. Three main categories exist:

Net Cannon Fired from the Ground: They can be handheld, thrown over the shoulder, or mounted on a turret. Composition changes from 20 to 400 m. For a controlled drop of the captured drone, they can be employed either with or without a parachute.

Net cannon Fired from Another Drone: It surpasses the range's maximum. It can be challenging to record a speeding drone with these. It is typically used in conjunction with a parachute for a regulated drone drop.

Hanging Net Deployed from a Net Drone: Usually, the net drone will carry the trapped drone to a safe location. If the trapped drone is too massive, it can be evacuated either with or without a parachute for a controlled drop.

Pros: Its cannons are highly accurate semi-automatic weapons. There is little chance of civilian casualties. Ground-launched nets and drone-deployed nets both have a maximum range, which helps with actual drone acquisition and is useful for investigations and prosecution.

Cons: Ground-launched nets have a small range and need a lengthy time to reload when used by drones.

5. **High-energy lasers:** It is a high-power optical device to produce beam of light or laser beam. The laser defeats the drone by destroying the structure and the electronics.

   Pros: It physically stops the drone.

   Cons: It is a strong power optical device that can create a laser beam or light beam. The drone is defeated by the laser by obliterating the electronics and the framework. The drone is manually stopped.

6. **Birds of Prey:** In this method eagles trained to capture small unmanned aircraft systems (SUAS) and commercial off-the-shelf (COTS) drones. It is traditional technology but in this technology requires more manpower for training and maintaining the birds.

   Pros: If the birds available on right location, then it catch or pick of drone quickly and accurate without any risk. Military Working Dog (MWD) teams and K9 training teams are similar as Birds of Prey. It's more potentially and its operational solution.

   Cons: Maximum man-power required for training and maintenance of birds.

6.2.3   Collision Avoidance and Obstacle Detection Technology

Collision prevention technologies are currently included on the newest high-tech drones. It is used to clear the area with obstruction sensing devices. The photos are converted into 4D maps using SLAM technology, which enables the drone to sense and avoid objects. To detect and prevent objects, the systems employ one or more of the following sensors: Vision Sensor, Ultrasonic, Infrared, radar, Time of Flight (ToF) and Monocular Vision.

6.2.4   Flight Controllers, Gyroscope Stabilization, and IMU

Gyroscope stabilization technology gives smooth flight capabilities of UAV. The drone is kept flying or hovering extremely efficiently by forces acting against it. The main flight controller receives the crucial directional data from the gyroscope. A part of the inertial measuring unit is the gyroscope

(IMU). It is a crucial part of the flight controller for the drone. The drone's brain, or flight controller, uses one or more accelerometers to reflect the actual rate of acceleration. Using one or more gyroscopes, the IMU monitors changes in rotational characteristics including pitch, roll, and yaw. Magnetometers are incorporated into some IMUs to help with calibrating against position drift.

6.2.5   Drone Propulsion Technology

This technology used for propellers of drone, electronic speed controllers and motors, its help to travel UAV any direction or hover. The configuration of quadcopter has motors and propellers; it has two motors, rotating clockwise propellers (CW Propellers) and two motors rotating Counter Clockwise (CCW Propellers). These are some propulsion systems component and motors of UAV are propellers, windings, bearings, electronic Speed controllers, motor bell (rotor), motor stator, ESC Updater, cooling System.

6.2.6   Real-Time Telemetry Flight Parameters

Ground Station Controller (GSC) or a smartphone app lets users control the drone and see real-time flight data. Remote controller shows telemetry data, including remaining battery power and warnings, UAV range height, speed and GNSS strength. Several drone use FPV (First Person View) of ground controllers, they transmit video with the help of drone to mobile device or ground controller.

6.2.7   No Fly Zone Drone Technology

DJI and other types of drone come in No Fly Zone technology because of flight safety and avoid accidents in No fly Zone or restricted areas. It is synchronized and categorizes by the Federal Aviation Authority (FAA) and manufacturer use UAV firmware to updates and changes No Fly Zone drone technology.

6.2.8   LED Flight Indicators

Led flight indicator placed at the front and rear of the drone. The indicator is red, green and yellow LED of drone. Front indicators is light up to indicate the UAV's nose and rear indicators is light up to indicate the various type of signal like flying, power on and getting a firmware upgrade.

6.2.9   Drones with High Performance Camera

Da-Jiang Innovations (DJI) is stand for Great Frontier Innovations, its latest drones made by Chinese technology company that headquarter in Shenzhen, Guangdong. Frank Wang is the founder of DIJ in 2006 which is manufacture for commercial UAV such as videography, aerial, photography. It can make film up to 4k video still 12 megapixels.

6.2.10   Remote Control System and Receiver of UAV

The 5.8 GHz frequency range is used by the wireless communication equipment known as the DJI Phantom 4. Ground control refers to a remote-control system. UAV Remote Control Receiver location range is 5.8 GHz, approximately all the latest drones can use 2.4 or 5.8 GHz operating frequencies.

6.2.11   Range Extender UAV Technology

Range extender UAV Technology is a type of wireless communication equipment that typically operates at 2.4 GHz. Its function is to increase the drone's ability to communicate with smartphones or tablets in open, unrestricted spaces. The maximum communication range is 700 m. Each range extender has a different system name and MAC address (SSID) [7].

6.2.12   Video Editing Software

All least drone use Adobe DNG raw film, it is excellent video quality software for post processing which means original image hold all information further processing.

6.2.13   Operating Systems in Drone

Most unmanned aerial vehicles (UAV) run Linux, while some do as well. The Drone code project is a 2014-launched initiative of the Linux Foundation. Under a nonprofit agreement with The Linux Foundation, the Drone code project is an open source and collective approach that brings together current and upcoming open source unmanned aerial vehicle (UAV) initiatives.

6.2.14   Drone Security and Hacking

A UAV drone is similar to a flying computer since it has customizable operating systems, aircraft systems, and core signage. Drones have been designed to fly in an area where they can look for additional drones as well as to hack into the wireless network of other drones to detach the operator and seize control of that drone.

6.2.15   Modern Top Technology (Drones with Camera)

DJI is latest technology that takes vast rule of the customer good and expert drone in market. There are some latest emerging technologies of drones are listed:

* **DJI Mavic Air 2:** It is smallest and latest professional drone that used for super HDR video and photos. It is use in Intelligent Flight Modes and APAS obstacle detection and avoidance.
* **AutelEvo 2:** It is newest drone technology with four choices of cameras such as 6k, 8k, and Dual 8k with FLIR thermal.
* **Skydio 2:** It is best tracking drones.
* **DJI Mavic Mini:** It is micro drone so more demand in marketplace. Its weight is 249 grams (8.78 ounces) and video film is 2.7k (2720 × 1540 at /25/40p).
* **DJI Mavic 2 Enterprise** (M2E): In this technology uses zoom and thermal camera models with different accessories like spotlights, beacon and loudspeaker, this type of technology designed special purpose like search, rescue and similar type of work [7].
* **DJI Phantom 4 Pro V2.0:** It is use vision collision avoidance technology that is multipurpose drone such as photography, 4k aerial filming, and photogrammetric.
* **DJI Inspire 2:** This technology is use motors and design patented its multipurpose professional drone such as photogrammetric, multispectral, 5k aerial filming, photography and thermal imaging.
* **Yuneec Typhoon H Pro:** It is patented by Intel Real sense collision avoidance technology for professional filming and aerial photography.
* **Walkera Voyager 5:** It is use latest technology drone of Walkera is incredible; this type of techniques gives camera options such as low light night vision camera, 40x optical zoom and thermal infrared.
* **Walkera Vitus Starlight:** It is the smallest drone in size, with Walkera collision avoidance sensors and a low light night vision camera.

❖ **DJI Matrice 600:** It is profitable multirotor, which use true aerial cinematography platform with the mount seven different Zenmuse cameras.

❖ **DJI Matrice 200 Commercial Quadcopter:** It has duple battery redundancy, IMU and satellite navigation systems; that can mount two cameras in quadcopter (e.g. thermal and zoom camera), it uses Matrice 200 for surveying of bridges and matrice has 200 with six directions collision avoidance techniques using Vision sensors, ToF laser and Ultrasonic.

6.2.16   Intelligent Flight Systems

There are some most recent drones having smart flight controllers and channels or mode likeway points, active tracking, return to home, follow me and others. Phantom 4 Pro is intelligent systems that working with newest intelligent device are Draw waypoint, Beginner mode, S mode (sport), Home Lock, Active Track (Profile, Spotlight, Circle), P mode(for position), Gesture mode, Tap Fly, Obstacle Avoidance, Terrain Follow Mode, Tripod Mode, A-Mode (Attitude), Course Lock.

6.2.17   Drones For Tracking

Tracking technology is most latest exciting technologies in drone because of object tracking capacity like bus, car, person or people, boats, ship etc. also track the location of area a spot or place like road, traffic, bicycle etc. Skydio 2 and Mavic 2 are both most recent drones using tracking vision systems, GNSS systems, specific software, flight controllers, detection sensors, central processing, achieve 100 percent accurate.

6.3   India 2021: The Drone Policy and Rules

6.3.1   India Policy Guideline for Drones

On 15th July 2021 government released new Drone Rules for public conference but the rules take effects on 25th August 2021 and ask for to restore the rigid of government for UAV. This rule is liberal and enhanced to ease increase commerce proceeding for past two iterations on drone's policy and in this guideline highlight operators and manufacturers are for drone. Table 6.6. Discuss the India policy rules as per as all Provision with the help of Provision number and Provision topic.

Table 6.6 Summary of the drone India policy rules 2021 [7, 8].

S. no.	Provision number	Provision topic	Provision
1	2	Applicability of rules	As per as rule the maximum weights up to 500 kg and that registered in India and organism operated over India [7, 8].
2	5	Categories of unmanned aircraft systems (UAS)	There are classifying into autonomously or remotely is based only on weight such as Nano UAS, Micro UAS, Small UAS, Medium UAS, Large UAS
3		Permission Requirements	Type Certificate for each model Unique Identification number is most important for drone manufacturers. Pilot license is necessary for Pilots flying drones
4	14, 15, 16	Registration	A drone operator is necessary to filling register on digital sky platform with fees [7, 8].
5	44	Insurance	Owners required nano drones expected to third party insurance before the use. The Chapter XI rule says motor vehicles act 1988 will apply third party insurance of [7, 8].
6	50	Penalty	Rs 1 Lakh is maximum penalty [7, 8].

(*Continued*)

Table 6.6 Summary of the drone India policy rules 2021 [7, 8]. (*Continued*)

S. no.	Provision number	Provision topic	Provision
7	19, 20, 21, 22, 24, 24	Airspace Maps	On 25th September 2021 government publishes an airspace map on digital sky platform. The map split into red, yellow and green zones. In red zone and yellow zone require permission for flying. Green zone does not required permission or flight path registration.
8	29	Mandatory reporting of an accident	In case of calamity remote pilot informs within 48 hours accident on the Digital Sky platform [7, 8].
9	47, 48, 49, 40, 41	Remote Pilot Training Organization	Application for permission of remote pilot training organization registers on the Digital Sky platform. 10 year valid permission
10	42	Research and Development	In the absence of a remote pilot license, required type certificate, or unique organization identification number. Some organizations seek permission from the federal or state governments. The Department of Industry and Internal Trade has recognized Startup. Any authorized testing organization Any drone manufacturer who has a GST Identification Number [7, 8]

6.3.2   Drone Rules 2021

❖ The UAS Rules have retained the Drone Rules, which are focused on weight divisions for categorizing drones such as nano, micro, small, medium, and large drones (up from the former maximum of 400 kg) and the weight of the drone and payload of 500 kg [7, 8].

❖ The further criteria of optimum flight speed and maximum height that were previously used to classify nano drones were eliminated.

❖ The Digital Sky platform, which will be used to control drone registration and deployment, is an internet registration framework for all drones. There is only one registration process, thus no further departmental approvals are needed. [7, 8].

❖ All drone developers must get type certificates from the Digital Sky Platform for each new model. Drones must be actually delivered to the Quality Council of India or another certified testing agency for testing before the Type Certificate is issued. The conditions for acquiring Type Certificates for different kinds of drones must be governed by guidelines made by QCI [7, 8].

❖ Users will receive a Type Certificate id from the maker that they can use to request a Unique Identification Number for each specific drone via the Digital Sky platform.

❖ Type Certificates are given according to the regulator's requirements, and the Digital Sky Platform has interactive maps. Type certificates will be given out after the process has begun, and after the rules have been announced, a map viewer will be uploaded.

❖ There is no need for a type certificate for a model remotely piloted aircraft system or a nano unmanned aircraft system [7, 8].

❖ Drone imports will be governed solely by the Directorate General of Foreign Trade. Drones can now be owned and operated by foreign companies in India. Future regulations may require drones to have safety features. The previous No Permission, No Takeoff mechanism will not be implemented immediately [7, 8].

❖ The government plans to create an interactive map with four zones: green, yellow, and red. Only the red and yellow zones necessitate prior authorization to fly. The government must release the map by September 24, 2021 [7, 8].

❖ Except for those flying nano or micro drones for noncommercial purposes, all drone operators must obtain a drone pilot license. Drone pilots wishing to obtain a license must first complete training with a remote pilot training organization [7, 8].

❖ To conduct testing in a controlled green zone, recognized research or innovation organizations, academic facilities, start-ups, and other permitted experimental entities are not required to get a UIN, type certificate, remote pilot licensing, or other approvals [7, 8].

❖ The provisions of the Motor Vehicles Act of 1988 concerning third-party insurance will also apply to drones [7, 8].

❖ Despite limitations on these subjects under the old rules (UAS Rules), there is no mention of Beyond Visual Line of Sight activities, drone swarms, pictures taken by drones, drones for delivery, or drone ports in the current rules) [7, 8].

6.4   Unmanned Aerial Vehicle (UAV) or Drone Application

6.4.1   Precision Agriculture

UAVs play is important role in precision agriculture (PA) for crop assessment and tracking [9, 10], weed detection [11], irrigation management [12], disease detection [13], pesticide spraying [9], and data collection from ground sensors (moisture, soil properties, etc.,) [14]. Utilizing UAVs in smart farming is a time and money-saving technique that can assist increase crop yields, agricultural production, and farming systems' viability. UAVs also help to quickly address these difficulties by reducing crop management, weed management, and pest destruction [15]. Figure 6.3 shows the least technology in agriculture uses UAVs task as such including

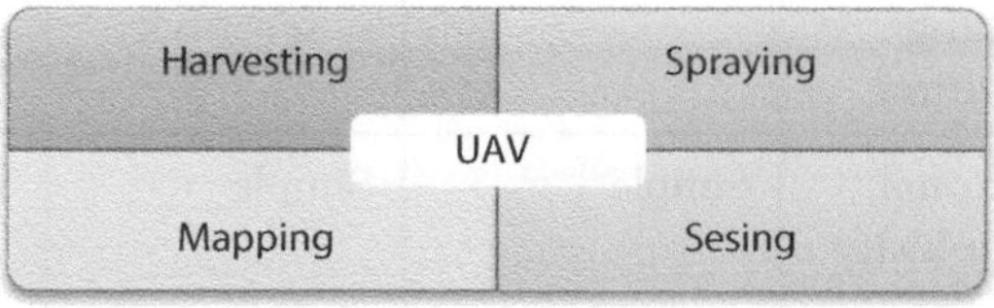

Harvesting	Spraying
UAV	
Mapping	Sesing

Figure 6.3 Agriculture based in UAV.

crop management, irrigation, weed or disease detection and pesticide spraying.

6.4.1.1 Related Work

When compared to traditional drones, drones can be used for crop fields at low altitudes with greater accuracy and at a lower cost. Furthermore, drone suggest and take high resolution crop images, which could be help in crop management like monitoring, detecting the unpredictability crop response to irrigation, weed management, disease detection and bring down herbicides [9, 10, 15, 16]. Table 6.7 shows the comparing both traditional manned aircraft and satellite based system UAV or drone.

Table 6.7 Precision agriculture based similarity among traditional manned aircraft and satellite based system of UAVs.

S. no.	Issues	UAVs/drone	Manned aircraft	Satellite system
1	cost	low	high	very high
2	endurance	short time	long time	all the times
3	availability	when needed	some time	all the times
4	coverage area	small	large	very large
5	deployment time	easy	need runway	complex
6	payload	low	large	large
7	application and usage	thermal cameras and sensors, carry small digital	spraying UAVs for pesticide spraying	high resolution images require specific area
8	whether and working condition	sensitive	low sensitivity	required for clearing sky image
9	Operational Complexity	Simple	Simple	Very Complicated
10	Example	[10, 15]	[16]	[17]

6.4.1.2   Uses of UAV in Precision Agriculture

In Figure 6.4 shows summarized use of UAVs used in precision agriculture applications discussed in below:

- ❖ **Irrigation scheduling:** These classify into four element need to proctor for irrigation are as follows: 1) soil water availability 2) crop water 3) amount of rainfall 4) irrigation system efficiency [18]. These elements measuring by drone such as moisture in soil, plant based temperature and evapotranspiration [19, 20].
- ❖ **Plant disease detection:** The United States concentrated on plant disease-related crop losses in order to recover about $44 billion in annual lost revenue [21]. Drone used for thermal remote sensing to monitoring crop diseases to protect and reduce crop losses [22].
- ❖ **Soil texture mapping:** In soil texture indicate the quality of soil, it revolves influences of crop efficiency and also drone used thermal images to compute soil texture at measuring scale of different field temperature with weather [22, 23].
- ❖ **Residue cover and tillage mapping:** A crop residue necessary to soil management for protects agricultural fields that shield water and wind from soil, so accuracy of crop remains is required for the use of reduced tillage techniques [24].

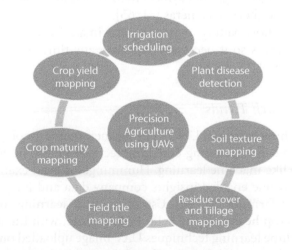

Figure 6.4 Application based UAV in precision agriculture.

❖ **Field tile mapping:** In case of Using tile drainage channels to drain surplus water from fields has both environmental and economic advantages [25], it also measure field temperature and it use thermal drone for images capture for field tile mapping [26].

❖ **Crop maturity mapping:** Drone is used for monitor of crop maturity, harvesting time, that harvesting in all time available [27].

❖ **Crop yield mapping:** Farmers want of harvest and storage requirements, accurate, including crop insurance, Early crop yield assessment is important for a variety of management and cash flow accounting purposes [28, 29].

6.4.1.3   Challenges

❖ Thermal cameras have poor resolution and it is luxurious [30].

❖ Thermal drone images can be precious by various ways like shooting distance, atmosphere moisture, emitted source and thermal radiation reflection [30].

❖ Aerial sensors take temperature readings and cause of crop growth in all stages [30].

❖ Weather is most challenging factor like as storms, extreme wind and rain.

❖ Lightweight UAVs is one of the most challenging tasks, to carry high weight of payload, high-resolution, multiple sensors and thermal cameras [31, 65].

❖ It has short battery life time approximate 1hr and it covers large areas so it necessary to return any times to charging station [9, 13, 15, 65].

6.4.1.4   Research Trends

❖ **Machine learning:** UAVs is next generation technology, will exploit the upcoming emerging technology use in agriculture like machine learning. Hummingbird is an example of that drone enabled insights company data and imagery for smart farming [31, 65]. Use of machine learning to identify crop health on the field. Furthermore, with the help of machine learning techniques UAV image uploaded on cloud and also use web based and mobile app platform to provide information to farmers and crop health. There are following reward of drone using machine learning technology in precision agriculture as follows: detection of crop diseases, Plant growth monitoring, Precision weed mapping, accurate yield forecasting, and Nutrient optimization and planting [31].

❖ **Image processing:** With the help of image processing UAVs to obtain high resolution crops images, farms, which used as a substitute to satellite and UAV imaging system. The Vegetation Indices (VI) is used to prediction with image processing techniques of agricultural like agricultural analysis, weed management and diseases detection and crop yield [16, 32–34]. Vegetation Indices consist of images like NDVI [35], GNDVI [36] and SAVI [37].

6.4.1.5   Future Insights

❖ With stress-free flight rule and regulations policy enhancement methods for georeferencing and categorization in image processing, it provides a vast probable for crop and soil monitoring [30, 38, 65].

❖ UAV sensors is next generation such as 4p sensor [39], they provide inside images and field investigative capabilities and its give farmers on the spot insights information of field, without any internet connection [33, 65].

❖ Furthermore precision agricultural researches are need to design and implementation model of special-type cameras with sensors, that capability to crop remote monitoring, soil detection and another characteristic of agriculture in real life or real time scenarios [40, 65].

6.4.2   Surveillance Applications of UAVs

6.4.2.1   Literature Review

In reference [41, 42] and [65] show the advantages and disadvantages of employ UAVs beside US boundaries for surveillance advantages are follows: (1) UAVs using in border surveillance for improve and reporting to the remote border of the USA. and (2) drone provide a lot reporting and find current position of border surveillance (like stationary surveillance equipment, border agents on patrol, etc.). And disadvantages are as follows

(1) Increases industrial accident rates of drone (for example inclement weather conditions) and (2) Increase major operation costs of a UAV. In [43] authors focused to analysis for UAVs uses within border zones war zones, war area and urban areas in the USA, so benefit of scenarios is safety (means Drones insulates operators) and robustness (means Drones reduce human errors).

6.4.2.2   State-of-the-Art Research

UAVs surveillance classified into five categories based on research work as follows:

- ❖ **Multi-UAV Cooperation:** It is based on increasing fresh drone management designs or systems [65].
- ❖ **Military and Home Security concerns:** It is based on examination and suggestions for the authorities for Homeland/military security [65].
- ❖ **Algorithm Improvement:** It is based on new algorithms of post processing of the data.
- ❖ **New Use Case:** It is integrates UAV system into new application.

6.4.2.3   Product Introduction

It is grown-up techniques implement by an instantaneous UAV monitoring tool.

6.4.2.4   Research Trends and Future Insights

Research Trends: Identify uses of research trends such as machine learning, nanosensors and short range communications technologies etc., in sense of security, efficient and accurate etc.

Future Insights: According to reviews and analysis suggest developing or deploying effective UAV surveillance applications such as Select/design (based on law and budgets), Develop efficient customized algorithms and conduct various experiments on field and to verify newly proposed systems.

6.4.3   Search and Rescue (SAR)

A drone is very useful in search and rescue operation like search, public security, emergency response, and catastrophe control. Vital infrastructure, such as water, floods, terrorist attacks, electricity companies, transportation, and telecommunications systems, can be partially or totally disrupted by man-made or natural disasters like tsunamis.

6.4.3.1   How SAR Operations Utilize UAVs

SAR operation is most primary use case of UAVs, listed as surveying a specific target area with onboard cameras to capture high quality photos and videos (stricken region) [44]. SAR operations using UAVs can be performed autonomously, accurately and without introducing additional risks [45, 46]. UAV can be also used to deliver food, water, and medications can also be delivered to the afflicted using UAV [47]. In the event of a disaster, UAVs can operate as aerial base stations to restore connection to areas that have been damaged by the infrastructure.

6.4.3.2   Challenges

- ❖ **Legislation:** The FAA does not allow the use of swarms of UAVs for commercial purposes in the United States. Its sole purpose is to coordinate the operations of SAR teams [48].
- ❖ **Weather:** Weather is most challenging task of UAVs. In some condition UAVs fail their mission as a result [49].
- ❖ **Energy limitations:** The most significant issue with UAVs is energy conservation, which includes the usage of power batteries for hovering, wireless technology, picture analysis, and data analysis [50].

6.4.3.3   Research Trends

- ❖ **Image Processing:** SAR operation using UAVs is important field of image processing techniques like tracking, identify, detection, monitoring, quickly and accurately targeted objects [51, 52].
- ❖ **Machine Learning:** ML-based approaches, such as convolution Neural Network (CNN) and Support Vector Machine (SVM), can be used to extract images/video frames to detect objects captured by UAVs [53, 54].

6.4.3.4   Future Insights

- ❖ To produce more precise proper technique, forward-looking infrared (FLIR) sensors and heat sensors can be integrated with data fusion and selection fusion algorithms that combine the output of various sensors, such as GPS [51].

❖ Power-efficient decentralized methods for evaluating videos, pictures, and sensor data collected by UAV swarms in real-time [54].

❖ SAR operations rely heavily on an accurate localization and mapping system.

❖ Algorithms for UAV autonomy and swarm coordination are required. Deciding on a flight path, planning it, avoiding collisions, and coordinating swarms are examples of these algorithms [55, 56].

6.4.4   Construction and Infrastructure Inspection

6.4.4.1   Literature Review

UAVs are utilized to examine large electrical power synchronized lines and monitor construction project locations in real time, gas pipelines and GSM towers infrastructure inspection [57–59].

6.4.4.2   Deployment of Drone for Construction and Infrastructure Inspection Applications

❖ Gas/oil and wind turbine inspection
❖ Critical land building inspection
❖ Infrastructure internal inspection
❖ Extreme condition inspection

6.4.4.3   Challenges

❖ In which limited processing, limited energy and short flight time.
❖ Limited payload capacities.
❖ Larger examination scope, greater mistake tolerance, and quicker work completion time present significant challenges.
❖ Most challenging allowing drone indoor environment and access GPS signals.

6.4.4.4   Research Trends

❖ **Machine Learning:** With the increasing importance of artificial intelligence approaches for UAVs to operate, such as improved data processing models, deep learning algorithms assist in obtaining new data finding from existing data.

❖ **Image Processing:** Image processing techniques can be used in construction and infrastructure inspections using UAV components with on-board cameras and sensors. It is capable of monitoring and assessing construction projects such as surveying sites, monitoring work progress, inspecting bridges, monitoring irrigation structures, detecting construction damage, and surface degradation [60, 61].

6.4.4.5   Future Insights

❖ Improve battery life to allow for longer distances and longer flight times for UAVs.

❖ Future research will focus on developing and proposing accurate, autonomous, and real-time power line assessment approaches using UAVs, which will include ultrasonic sensors, TIR or color cameras, image processing, and data analysis tools, as well as developing new techniques for automatically monitoring, detecting, and diagnosing any power line defects

❖ Improving multi-UAV data collection, sharing, and processing algorithms to achieve faster and more efficient results.

❖ To improve UAV autonomy and safety in crowded and indoor environments with no or weak GPS signals.

6.4.5   Delivery of Goods

6.4.5.1   UAVs-Based Goods Delivery System

In UAVs-based goods delivery system is capable for transport food, packages, traveling between pick and drop location with the help of GPS module, Bluetooth and Wi-Fi. UAV communication is used if a package docking device updates its GPS coordinates to the new address, it will transmit a signal containing the new address location. The UAVs, so it easily re-routed to new address and then updates GPS location.

6.4.5.2   Challenges

❖ **Legislation:** The FAA regulation in the United States barred all attempts at commercial use of UAVs, such as the Taco copter Company for food delivery [62]. Package delivery with UAVs is not permitted in the United States as of 2015 [63]. Under these regulations, businesses are allowed to fly commercial UAVs in the US, however there are some restrictions, such as a pilot's line of sight and pilots must obtain licenses, daylight hour restrictions, size of UAV limitation, altitude and speed, and most importantly, common people are not permitted to fly on sight line.

❖ **Liability Insurance:** Some UAVs can weight 25kg and speed approx. 45 m/s. Some problems are reports like eye loss, server laceration and soft tissue injuries caused by UAV accidents.

❖ **Theft:** The main reason of utilizing UAVs for hacking, cyber liability, wireless delivery and data gathering.

❖ **Weather:** It is most important challenging task is flight operation, weather data, direction, duration, weather sensitive, safety, analyzing data.

❖ **Air Traffic Control:** In this challenge air traffic divided into three models are follows low speed restricted traffic, high-speed transit and no fly zone.

6.4.5.3   Research Trends

❖ **Machine Learning:** Unsupervised learning methods is challenging task for unlabeled data collecting and solving real world problems etc.

❖ **Navigation System:** GPS signals is most challenging task for networking.

6.4.5.4   Future Insights

❖ Improving energy density of lithium ion batteries 5%-8% per year and in future expected double 2025 and also implement detect and avoid systems with help of drone avoid collisions and obstacles system [64].

❖ Researcher focused to address localization and navigation and to address UAVs coordination.

6.5 Conclusion

Drones or Unmanned aerial vehicles (UAV) are now a least emerging technology, it continually developing and implementing a new outstanding service provides. In this paper discuss drones, classification of drones or UAV, specification, new policy of UAV, application, challenges and research trends and future insights help of figures and tables. Some several application of UAV is precision agriculture, surveillance, search and rescue (SAR), planetary exploration, real time monitoring, wireless communication, good delivery, networking, civil infrastructure, remote sensor, surface mining and industry mining, etc. AI (Artificial intelligent) is most researching technology in precision agriculture. Furthermore UAVs is an imperative task in precision agriculture using image processing techniques like collecting high-resolution images and classification algorithms. UAV are widely focuses on weed detection, disease detection, crop management and tracking, pesticide spraying, irrigation scheduling and field sensor data collection, and also its research key is very challenging in precision agriculture to lift up the future opportunities and research. In future researcher developed by drone application. In search and rescues operations drone can supply judicious calamity exemplary with help in rescue and healing operations like medical rescue etc. Throughout, also talk about the new emerging techniques use in dronelike image processing, Machine learning, and cloud computing. We classify the issues for UAV or drone in civil or public application like swarming, batteries charging, collision avoidance and networking and security related. So, the moral of conclusion is to introduce an inclusive authorized structure and educational institutions are start learning civil uses of drone and it globally spread services. So we hope to help the key research challenges and opportunities for researcher to develop and implement drone or UAV civil application opportunity in future insights.

References

1. Fahlstrom, P. and Gleason, T., John Wiley & Sons, Hoboken, NJ, USA, 2012.
2. Hassanalian, M. and Abdelkefi, A., Classifications, applications, and design challenges of drones: A review. *Prog. Aerosp. Sci.*, 91, 99–131, 2017.
3. Wikman, D. and Andersson, O., *Propulsion System for a Small Unmanned Aerial Vehicle*, Stockholm, Sweden, 2020 [Online]. http://www.divaportal.org/smash/get/diva2:1438252/FULLTEXT01.pdf (Accessed October 21, 2021).

4. Çoban, S. and Oktay, T., Unmanned aerial vehicles (UAVs) according to engine type. *J. Aviat.*, 2, 177–184, 2018.

5. Torres-Sánchez, J., López-Granados, F., De Castro, A., II, Peña-Barragán, J.M., Configuration and specifications of an unmanned aerial vehicle (UAV) for early site specific weed management, *PLoS ONE*, 8, 3, e58210, 2013.

6. https://www.intel.com/content/www/us/en/support/articles/000024272/drones/development-drones.html

7. https://www.dronezon.com/learn-about-drones-quadcopters/what-is-drone-technology-or-how-does-drone-technology-work/

8. https://www.ikigailaw.com/the-drone-rules-2021-summary-and-key-take-aways/?utm_source=Mondaq&utm_medium=syndication&utm_campaign=LinkedIn-integration#acceptLicense

9. Huang, Y., Thomson, S.J., Hoffmann, W.C., Lan, Y., Fritz, B.K., Development and prospect of unmanned aerial vehicle technologies for agricultural production management. *Int. J. Agric. Biol. Eng.*, 6, 3, 1–10, 2013.

10. Muchiriand, N. and Kimathi, S., A review of applications and potential applications of UAV, in: *Proc. Sustain. Res. Innov. Conf*, pp. 280–283, 2016.

11. Kazmi, W., Bisgaard, M., Garcia-Ruiz, F.J., Hansen, K.D., la Cour-Harbo, A., Adaptive surveying and early treatment of crops with a team of autonomous vehicles, in: *Proc. ECMR*, pp. 253–258, 2011.

12. Gonzalez-Dugo, V., Zarco-Tejada, P., Nicolás, E. *et al.* Using high resolution UAV thermal imagery to assess the variability in the water status of five fruit tree species within a commercial orchard. *Precis. Agric.*, 14, 660–678, 2013.

13. Garcia-Ruiz, F., Sankaran, S., Maja, J.M., Lee, W.S., Rasmussen, J., Ehsani, R., Comparison of two aerial imaging platforms for identification of Huanglongbing-infected citrus trees. *Comput. Electron. Agric.*, 91, 106–115, Feb. 2013.

14. Mathur, P., Nielsen, R.H., Prasad, N.R., Prasad, R., Datacollection using miniature aerial vehicles in wireless sensor networks. *IET Wirel. Sens. Syst.*, 6, 1, 17–25, 2016.

15. Jensen, T., Apan, A., Young, F.R., Zeller, L.C., Cleminson, K., Assessing grain crop attributes using digital imagery acquired from a low-altitude remote controlled aircraft, in: *Proc. Spatial Sci. Inst. Conf., Spatial Knowl. Without Boundaries (SSC)*, Spatial Sciences Institute, Los Angeles, CA, USA, pp. 1–11, 2003.

16. Sullivan, D.G., Fulton, J.P., Shaw, J.N., Bland, G., Evaluating the sensitivity of an unmanned thermal infrared aerial system to detect water stress in a cotton canopy. *Trans. ASABE*, 50, 6, 1963–1969, 2007.

17. Reed, B.C., Brown, J.F., Vander Zee, D., Loveland, T.R., Merchant, J.W., Ohlen, D.O., Measuring phenological variability from satellite imagery. *J. Veg. Sci.*, 5, 5, 703–714, 1994.

18. Rhoadsand, F.M. and Yonts, C.D., Irrigation scheduling for corn: Why and how, in: *The National Corn Handbook (NCH)*, University of Wisconson, Madison,

W.I., USA, 2000 [Online]. http://corn.agronomy.wisc.edu/Management/ NCH.aspx.

19. Gonzalez-Dugo, V., Zarco-Tejada, P., Nicolás, E., Nortes, P., Alarcón, J., Intrigliolo, D., Fereres, E., Using high resolution UAV thermal imagery to assess the variability in the water status of five fruit tree species within a commercial orchard. *Precis. Agric.*, 14, 6, 660–678, 2013.

20. Hassan-Esfahani, L., Torres-Rua, A., Jensen, A., McKee, M., Assessment of surface soil moisture using high-resolution multi-spectral imagery and artificial neural networks. *Remote Sens.*, 7, 3, 2627–2646, 2015.

21. Pimentel, D., Zuniga, R., Morrison, D., Update on the environmental and economic costs associated with alien-invasive species in the United States. *Ecol. Econ.*, 52, 3, 273–288, Feb. 2005.

22. Wang, D.-C., Zhang, G.-L., Pan, X.-Z., Zhao, Y.-G., Zhao, M.-S., Wang, G.-F., Mapping soil texture of a plain area using fuzzy-c-means clustering method based on land surface diurnal temperature difference. *Pedosphere*, 22, 3, 394–403, 2012.

23. Wang, D.-C. *et al.*, Retrieval and mapping of soil texture basedon land surface diurnal temperature range data from MODIS. *PloS One*, 10, 6, e0129977, 2015.

24. Hively, W.D., Lamb, B.T., Daughtry, C.S.T., Shermeyer, J., McCarty, G.W., Quemada, M., Mapping crop residue and tillage intensity using WorldView-3 satellite shortwave infrared residue indices. *Remote Sens.*, 10, 10, 1 22, 2010.

25. Hofstrand, D., Economics of tile drainage. *Ag Decision Maker Newslett.*, 14, 9, 3, 2015.

26. Woo, D., Song, H., Kumar, P., Mapping subsurface tile drainage systems with thermal images. *Agric. Water Manag.*, 218, 94–101, 2019.

27. Huete, A.R., A soil-adjusted vegetation index (SAVI). *Remote Sens. Environ.*, 25, 3, 295–309, 1988.

28. Laliberte, A.S., Herrick, J.E., Rango, A., Acquisition, orthorectification, and classification of unmanned aerial vehicle (UAV) imagery for rangeland monitoring. *Photogramm. Eng. Remote Sens.*, 76, 6, 661–672, 2010.

29. Laliberte, A.S., Winters, C., Rango, A., A procedure for orthorectification of sub-decimeter resolution imagery obtained with an unmanned aerial vehicle (UAV), in: *Proc. ASPRS Annu. Conf*, pp. 08–047, 2008.

30. Khanal, S., Fulton, J., Shearer, S., An overview of current and potential applications of thermal remote sensing in precision agriculture. *Comput. Electron. Agric.*, 139, 22–32, Jun. 2017.

31. Zhang, C. and Kovacs, J.M., The application of small unmanned aerial systems for precision agriculture: A review. *Precis. Agric.*, 13, 6, 693–712, Dec. 2012.

32. Candiago, S., Remondino, F., De Giglio, M., Dubbini, M., Gatelli, M., Evaluating multispectral images and vegetation indices for precision farming applications from UAV images. *Remote Sens.*, 7, 4, 4026–4047, 2015.

33. Burwood-Taylor, L., *The Next Generation of Drone Technologies for Agriculture*, Morges, March 16, 2017 [Online]. https://agfundernews.com/ the-next-generation-of-drone-technologies-for-agriculture (Accessed October 5, 2021).

34. Patil, J.K. and Kumar, R., Advances in image processing for detection of plant diseases. *J. Adv. Bioinform. Appl. Res.*, 2, 2, 135–141, 2011.

35. G.N. Fandetti, Method of drone delivery using aircraft. U.S. Patent 14 817 356, Feb. 9, 2017.

36. Hattingh, P., *Besides Blockchain, What's Missing from the Internet of Things?*, November 4, 2016, [Online]. https://usblogs.pwc.com/emerging-technology/ besides-blockchain-whats-missing-from-the-internet-of-things/ (Accessed October 12, 2021).

37. Huete, A.R., A soil-adjusted vegetation index (SAVI). *Remote Sens. Environ.*, 25, 3, 295–309, 1988.

38. Zhang, C. and Kovacs, J.M., The application of small unmanned aerial systems for precision agriculture: A review. *Precis. Agric.*, 13, 6, 693–712, Dec. 2012.

39. Grassi, M.J., *The SLANTRANGE: 3p Multispectral Drone Sensor*, March 2, 2017, [Online]. https://www.prweb.com/releases/2017/03/prweb14155693. htm (Accessed October 13, 2021).

40. Primicerioetal, J., A flexible unmanned aerial vehicle for precision agriculture. *Precis. Agric.*, 13, 4, 517–523, 2012.

41. Haddal, C.C. and Gertler, J., *Homeland Security: Unmanned Aerial Vehicles and Border Surveillance*, Congressional research service 2010 (Library Congress, Washington, DC, USA, 2010). https://apps.dtic.mil/sti/pdfs/ ADA524297.pdf

42. Maza, I., Caballero, F., Capitán, J., Martínez-de-Dios, J.R., Ollero, A., Experimental results in multi-UAV coordination for disaster management and civil security applications. *J. Intell. Robot. Syst.*, 61, 1, 563–585, 2011. https://doi.org/10.1007/s10846-010-9497-5

43. Wall, T. and Monahan, T., Surveillance and violence from afar: The politics of drones and liminal security-scapes. *Theor. Criminol.*, 15, 3, 239–254, 2011.

44. Estrada, M.A.R., *How Drones can Help in Case of Natural Disasters*, 2017 [Online]. https://www.researchgate.net/publication/315527891_How_Drones_ can_Help_in_Case_of_Natural_Disasters (Accessed October 10, 2021).

45. Alcedo, T., *Student project Report at Swiss Fedual Institute of Technology Swiss*, 2018 [Online]. https://www.alcedo.ethz.ch/# (Accessed October 15, 2021).

46. Joern, J., *Examining the Use of Unmanned Aerial Systems and Thermal Infrared Imaging for Search and Rescue Efforts Beneath Snowpack*, Ph.D. Dissertation. Univ. Denver, Denver, CO, USA, 2015.

47. Jo, D. and Kwon, Y., Development of rescue material transport UAV (unmanned aerial vehicle). *World J. Eng. Technol.*, 5, 4, 720, 2017.

48. Smith, M.L., Regulating law enforcement's use of drones: The need for state legislation. *Harvard J. Legis.*, 52, 423, Sep. 2015.
49. Jordan, B.R., A bird's-eye view of geology: The use of micro drones/UAVs in geologic fieldwork and education. *GSA Today*, 25, 7, 50–52, 2015.
50. Vergouw, B., Nagel, H., Bondt, G., Custers, B., Drone technology: Types, payloads, applications, frequency spectrum issues and future developments, in: *The Future of Drone Use*, pp. 21–45, TMC Asser Press, The Hague, Netherlands, 2016.
51. Rudol, P. and Doherty, P., Human body detection and geolocalization for UAV search and rescue missions using color and thermal imagery, in: *Proc. IEEE Aerosp. Conf*, pp. 1–8, Mar. 2008.
52. Mikolajczyk, K., Schmid, C., Zisserman, A., Human detection based on a probabilistic assembly of robust part detectors, in: *Proc. Int. Comput. Vis. (ECCV)*, pp. 69–82, 2004.
53. Giusti, A. *et al.*, A machine learning approach to visual perception of forest trails for mobile robots. *IEEE Robot. Autom. Lett.*, 1, 2, 661–667, Jul. 2016.
54. Bejiga, M.B., Zeggada, A., Nouffidj, A., Melgani, F., A convolutional neural network approach for assisting avalanche search and rescue operations with UAV imagery. *Remote Sens.*, 9, 2, 100, 2017.
55. Alexopoulos, A., Kandil, A., Orzechowski, P., Badreddin, E., A comparative study of collision avoidance techniques for unmanned aerial vehicles, in: *Proc. Int. Conf. Syst., Man, Cybern. (SMC)*, pp. 1969–1974, Oct. 2013.
56. Pham, H., Smolka, S.A., Stoller, S.D., Phan, D., Yang, J., *A Survey on Unmanned Aerial Vehicle Collision Avoidance Systems*, 2015 [Online]. https://arxiv.org/abs/1508.07723 (Accessed October 14, 2021).
57. Liu, P. *et al.*, A review of rotorcraft unmanned aerial vehicle (UAV) developments and applications in civil engineering. *Smart Struct. Syst.*, 13, 6, 1065–1094, 2014.
58. F. Mohamadi, Vertical takeoff and landing (VTOL) small unmanned aerial system for monitoring oil and gas pipelines. U.S. Patent 8880241, Nov. 4, 2014.
59. Gheisari, M., Irizarry, J., Walker, B.N., UAS4SAFETY: The potential of unmanned aerial systems for construction safety applications, in: *Proc. Construct. Res. Congr., Construct. Global Netw.*, pp. 1801–1810, 2014.
60. Sankarasrinivasan, S., Balasubramanian, E., Karthik, K., Chandrasekar, U., Gupta, R., Health monitoring of civil structures with integrated UAV and image processing system. *Proc. Comput. Sci.*, 54, 508–515, Jan. 2015.
61. Pagnano, A., Höpf, M., Teti, R., Aroadmap for automated power line inspection. Maintenance and repair, in: *Proc. CIRP*, vol. 12, pp. 234–239, 2013.
62. Gilbert, J., *Tacocopter Aims to Deliver Tacos Using Unmanned Drone Helicopters*, New York, NY, USA, updated December 6, 2017 [Online]. https://www.huffpost.com/entry/tacocopter-startup-delivers-tacos-by-unmanned-drone-helicopter_n_1375842 (Accessed November 2, 2021).

63. Franceschi-Bicchierai, L., *FAA Clarifies that Amazon Drones Are Illegal,* Mashable.com, June 24, 2014 [Online]. https://mashable.com/archive/faa-amazon-drones-2 (Accessed October 6, 2021).

64. Cohn, P., Green, A., Langstaff, M., Roller, M., *Commercial Drones Are Here: The Future of Unmanned Aerial Systems,* December 5, 2017 [Online]. https://www.mckinsey.com/industries/travel-logistics-and-infrastructure/our-insights/commercial-drones-are-here-the-future-of-unmanned-aerial-systems (Accessed October 6, 2021).

65. https://www.arxiv-vanity.com/papers/1805.00881/

7

AI Applications of Drones

LNC Prakash K.[1]*, Santosh Kumar Ravva[2], M.V. Rathnamma[3]
and G. Suryanarayana[4]

[1]*Department of Computer Science & Engineering, CVR College of Engineering,
Hyderabad, India*
[2]*Department of Computer Science & Engineering Vasavi College of Engineering,
Hyderabad, India*
[3]*Department of Computer Science & Engineering KLM College of Engineering
for Women, YSR Kadapa, Andhra Pradesh, India*
[4]*Department of Computer Science & Engineering,Vardhaman College
of Engineering, Hyderabad, India*

Abstract

Unmanned aerial vehicles (known as drones) and artificial intelligence are attracting academic and industrial attention. Unmanned aerial vehicles (UAVs) have increased the ability to manage and supervise from remote locations. Unmanned aerial vehicles are employed in a variety of purposes and are becoming extremely popular. Recent advancements in unmanned drones and artificial intelligence have created a new opportunity for independent flying and maintenance. Artificial intelligence and deep learning are already driving the advancement of unmanned aerial vehicles and ensuring their independent future. Computer vision has made significant advances in image feature extraction and classification, as well as object recognition, making this a very appealing research subject when used on unmanned aerial vehicles. Because UAVs learn via techniques and apply them for decision making purposes, artificial intelligence is also vital and beneficial, but it can also be hazardous and severe. Artificial intelligence has decreased the number of obstacles faced by unmanned aerial vehicles while also boosting their capabilities and allowing them to enter new markets. The combination of Unmanned Drones and Artificial intelligence has resulted in quick and accurate results. The integration of UAVs and artificial intelligence has aided in real-time surveillance, data collection and analysis, and forecasting in computer/wireless network

Corresponding author: klnc.prakash@gmail.com

Sachi Nandan Mohanty, J.V.R. Ravindra, G. Surya Narayana, Chinmaya Ranjan Pattnaik
and Y. Mohamed Sirajudeen (eds.) Drone Technology: Future Trends and Practical Applications,
(153–182) © 2023 Scrivener Publishing LLC

infrastructure, smart cities, the armed services, agricultural production, and mining. This study obtained and consolidated studies on the usage of UAVs, along
with artificial intelligence and related algorithms, in various domains and countries. The goal was to illustrate overall scope and significance of artificial intelligence models in improving UAV performance, problem solving, and a variety of
application domains.

Keywords: Unmanned aerial vehicles (UAV), artificial intelligence, computer
vision, applications

7.1   Introduction

Drones are a form of unmanned aerial vehicles (UAV) that may be operated remotely by people or flown independently by machines. UAV stands
for unmanned aerial vehicle, which functions similarly to an aircraft in
that it employs aerodynamic principles to raise itself and may transport
cargo like as camera systems, firearms, as well as other items (Permenhub
PM 180, 2015). Drones were first developed and utilized by the defense to
strike adversaries. In general, this drone's initial application is in the armed
services. However, although technology evolves, it may be applied to many
fields. Drones are flights that don't have an operator or passengers. This
vehicle is remotely operated by operators on land or in other locations with
service regions, or automatically by computer programs that have been
built. Originally, a UAV was a remote-controlled plane, but automated
machines are gradually becoming more common. Multi-rotor, fixed-wing,
single-rotor helicopter, and fixed-wing hybrid VTOL are the four primary
structural categories of corporate drones. The amount of mass that each
form of drone can transport, as well as the effectiveness and period of the
flight, are all affected by the varied body types. Now a days drown are made
strengthen by embedding with artificial intelligence. Artificial Intelligence
(AI) is a type of computer intelligence that mimics human behavior or
thoughts and may be educated to address the issues. AI is a mix of Machine
Learning and Deep Learning approaches. Artificial intelligence is used in a
variety of ways in this era. It will become increasingly important in today's
world as it can efficiently handle complicated issues in a variety of areas,
including medicine, entertainment, banking, and academia, our living
conditions are becoming more pleasant and efficient because of artificial
intelligence. Figure 7.1 represents the AI-enabled drone flying to capture a
photograph using a camera attached with it.

Artificial intelligence is quickly evolving today, with new technological
inventions being released on a daily basis. Small activities, such as image

Figure 7.1 AI-enabled flying drone with a camera attached to it.

recognition, automobile operating, and other modest responsibilities, are incorporated into today's computer systems. The major objective of artificial intelligence, on the other hand, is to create sophisticated and more complicated systems that can exceed people in any way. This involves more difficult skills such as chess and completing mathematics. The creation of a super AI will be the most technique that allows in human history. As a result, the development of more modern technology has aided in the abolition of wars, the correct combating of diseases, and the development of suitable prevention strategies. In addition, technological advancement would be extremely beneficial in the battle against poverty. AI algorithms that have been trained with large amounts of data may make smart judgments. Technologically speaking, artificial intelligence and drones are a great match because the amount and type of data produced by drones can be easily analyzed by artificial intelligence algorithms. In general drones are remotely operated vehicles that may be utilized for a wide range of tasks. These gadgets were first operated manually and wirelessly. Drones now frequently include artificial intelligence, which automates part or all functions. Drone providers may gather and apply sensory and physical data using information from the sensors linked to the drone, thanks to AI.

Artificial intelligence (AI) in drones is a software application which can imitate human intellect in machine operations such as decision-making, problem-solving, and forecasting (Russel & Norvig, 2016). The drone's purpose becomes increasingly complicated when artificial intelligence is added, allowing it to assist people with challenging tasks. This information enables autonomous or assisted flying, which simplifies operations and improves usability. Therefore, drones seem to be a component of the intelligent transport services that are now offered economically to companies and consumers. Drones powered by AI rely heavily on computer vision.

Drones can identify things while traveling, as well as analyze and collect data on the surface, according to this innovation. A neural network is used to do high-performance integrated image analysis in computer vision. A neural network is a common framework for implementing machine learning technique. Drones can recognize, classify, and manipulate objects with the help of neural networks. Drones can prevent accidents and detect and track objects by combining this data in real time. Investigators would first educate machine learning techniques to predict and accurately classify items in a range of scenarios before using neural networks in UAVs. This is accomplished by providing the classifier with specifically labeled photos. These visualizations educate the neural network what characteristics distinct categories of items possess and how to distinguish one sort of thing from the other. While functioning, more complex neural networks keep learning without monitoring.

Drones may be utilized for a variety of activities, including infrastructure surveillance, transporting commodities, forest burning, mining resources extraction, farming area mapping, manufacturing area mapping, entertaining toys, and respiration checks. Drones, as watching equipment, may fly extremely low and near to things, allowing them to acquire high-resolution photographs, particularly in tough regions without interrupting design process. Drones can fly and then transfer items very swiftly and effectively as a delivery service. As a firefighter forestry, a drone can assist in extinguishing burnt woods, particularly in hard-to-reach forest locations, while also providing protection for firemen. As a mining material exploration tool, the drone collects photographs that track the growth of the mining region to maximize surveillance operations on mining operations. Drones capture photos of farmlands to be utilized as prediction data for crop yields, and they aid in the procedure of spraying liquid fertilizer and administering pesticides to crops. Drone mapping of industrial regions is like mining material investigation, including observation of industrial locations for monitoring. The drone is being used as an enjoyable toy and may become a pastime for its owners. Finally, as a body temperature monitoring device, the creation of a drone with an infrared sensor function to measure a human's body temperature. This drone is required so that doctors may measure body temperature in the event of a pandemic, such as the present COVID-19 outbreak. These are some potential drone uses that will emerge as technology advances.

The fast advancement of unmanned drones necessitates the upcoming deployment of drones employing artificial intelligence. Drones are mostly operated remotely by people now, but they will be operated by Artificial Intelligence in the future (Rajit, 2018). Artificial intelligence in

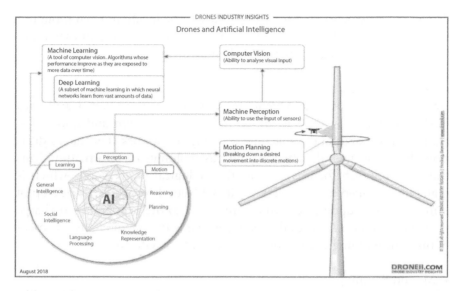

Figure 7.2 Drones and artificial intelligence (Source Droneii.com).

drones enables drones and other robots to determine and function independently in flights. Artificial intelligence in UAVs enables drones and other technologies to make judgments and function independently while in flying. Artificial Intelligence (AI) might assist people in overcoming challenging hurdles and solving issues. It must be considered while developing AI-powered drones so that they do not do more damage than good. Artificial intelligence, in its most basic form, refers to a computer's capacity to execute complicated activities those are similar to those performed by humans, which including thinking, issue handling, strategizing, learning, and comprehending and interpreting natural language. Machine Learning, Deep Learning, and Motion Planning are examples of artificial intelligence applications in drones, as illustrated in Figure 7.2 (Schroth, 2018).

• Machine Learning
The characteristics that can be differentiated are optimized using Machine Learning. Whenever connecting to current data, the Machine Learning model is supposed to learn and continue to get better. The design is not the same as technology that has been professionally created and executes operations based on precise instructions (like Computer Vision software). Drones will be able to obtain current information according to their own orders, as well as examine data and execute particular jobs, like when a

drone might be instructed to return to a specific location. Even though it has travelled to the farthest location, land with an auto sometimes without supervision to the beginning point when first traveling.

- Deep Learning

Deep Learning allows us to blend and enhance what we have learnt with fresh information. Deep Learning employs information and knowledge analysis techniques and a machine learning framework that makes decisions based on neural networks and vast volumes of information. This learning approach focuses on how the biological brain works, which is made up of linked neurons. The Artificial Neural Network is made up of numerous layer upon layer, which are linked to the next and is in charge of different duties. As a result, deep learning is the ongoing advancement of machine learning used by drones to execute specific jobs.

- Motion Planning

Motion Planning is a technique for detecting and recognizing things like people, bikes, and automobiles, and then determining the best flight paths. This allows the drone to calculate the distance to its target without having to determine what is in the surroundings. As a result, the drone can distinguish things in its field of view, reducing the likelihood of accidents with many other drones or items. It also has the ability to search for things based on certain activities.

In contrast, a development methodology is required for the implementation of artificial intelligence in drones. System models are usually done by a team of professionals with experience in hardware, programming, and machine design (Singh, Muthukrishnan, & Sanpini, 2019). Equipment, programming, and mechanical parts must all work together to ensure that data processing is as fast and quick as possible. Drones that use artificial intelligence can digest data much faster, which will aid in decision-making. Drones will frequently create vast volumes of data in order to gather data, while artificial intelligence will assist in processing data quicker, more accurately, and with acceptable and readily analyzed outcomes. This technology's use is to assist the industry in increasing effectiveness and make choices, as well as generating additional insight so that this innovation may bring even more advantages. Because practically every company is dealing with data collection and analysis, drones can assist in data acquisition according to market demands.

Drones incorporating artificial intelligence have a more sophisticated purpose. Drones not only capture video, but they also interpret data based on objects, behavior, feelings, colors, and noises, much like intelligent behavior. Drones will be able to undertake autonomous monitoring by monitoring the specified area, estimating the location about someone recognized, triggering an alert if someone enters a prohibited region, and so on. Artificial intelligence in drones enables drone operation more convenient for consumers. The drone can maintain a stable trajectory, preventing it from collapsing. If you hit the corresponding buttons, the drone will back to its former place. Drones may also transmit information whenever the battery is depleted and afterwards resume to the initial check location.

7.2   Review of Literature

Indonesia, being an industrialized country [1], necessitates the use of drones in a variety of businesses. So far, the usage of drones in Indonesia has been limited to a few businesses, even though there have been prospects for other sectors. Commercial drones may assist in decision-making, enhance performance of the companies, decrease operating costs, boost analyze efficiency, enhance security, and bring new information. In addition to this this article denotes the advances in artificial intelligence in UAVs which may overcome complicated hurdles and find solutions that people find hard to address. The usage of artificial intelligence-powered drones in industry can assist people in doing challenging tasks by making it more effective and productive [2]. This survey provided and summarized information on the usage of UAVs, as well as advanced analytics and associated techniques, in various fields and countries. The goal was to synthesis the extent and significance of machine learning algorithms in improving Surveillance Aircraft capacities, issue resolutions, and variety of application domains. This study [3], focuses on wetland mitigation as portion of a road project using AggieAir Unmanned Aircraft Drones (UAVs) and a cutting-edge wetland categorization approach. The research site is a wetland on the Southern Parkway building site in Washington County, Utah, USA. The navigation UAV footage of the region was classified using the multiclass relevance vector machine method. Article [4] in this research, researchers provide an outline of AI Simulations and Application Programs for Drones. Initially, they examine shortly the usage of AI simulations, notably on UAVs, often referred as quadcopters or unmanned aerial vehicles. Next, they contrast the simulated environment and software platforms employed in the construction of their software system. In this paper [5], researchers provide divergent approaches for observing smart cities and vast oceans using many heterogeneous devices UAVs. Because the needs and essential surroundings of a modern city and a vast ocean are so unlike, they investigate two distinct techniques. For smart cities, they propose a tight plane-based structure that makes advantage of current public transit, such as buses and urban trains, as well as their itineraries, to enable time-sensitive monitoring.

In this study [6], an intruder recognition system based on Bayesian game theory is developed to address the challenge of UAV information security. A Bayesian game is one in which game players have not clear knowledge on the player's profit function. In this study, there is partial availability of resource that the IDS agent is unsure of the sort of intruder and that the attack is unsure whether its neighbor node is an Intrusion unit. The computational and empirical analysis demonstrates that the suggested system produces a high intruder detection accuracy. Furthermore, the network's communication overhead is the minimum. The research [7] examines UAV machine intelligence at three stages: single aircraft smart flying, multi planes smart collaboration, and mission independent intelligence. The study [8] discusses the major stages of AI technologies on UAVs and recommends particular artificial intelligence uses in military and commercial UAVs. The bibliography [9] examines the effects of smart unmanned aircraft on military capabilities through three perspectives: altering the actual combat structure, entire fighting capabilities, and battle winning methodology. Article [10] is research in which it describes work, problems, and risks associated with deploying artificial intelligence onto UAVs. It also tells that unmanned aerial vehicles (UAVs) are increasingly used in a variety of applications. Advancements in unmanned aircraft and artificial intelligence have created the new opportunity for independent flying and administration. Artificial intelligence and machine learning are already powering the advancement of unmanned aerial vehicles (UAVs) and ensuring their independent future. The CNN approach in [11], was used to come up with a system for recognizing sufferers of environmental hazards, which was built on a Raspberry Pi and can recognize sufferers of environmental hazards using streamed cameras mounted on UAVs. The Mobile-net SSD model is used in this study to implement the Convolutional Neural Network (CNN) technique with complete accuracy for ranged objects of 1-4 m.

The research [12], provide a complete drone approach to detect drones using machine learning-based research. This technology is made to work with drones that have cameras. Depending on machine categorization, the algorithm determines the position of the camera photos and the drone supplier type. An unmanned aircraft (UAV) with accurate sensors may fly right from over fire area to capture video of the fire line and thermoelectric pictures, which are then geo-tagged and sent in actual time to mobile control centers via the surveillance and organizing systems [13]. In [14], Drones, from the additional line-up, offer additional benefits they could save life, reduce time & expense on constructions, enable secure infrastructures administration and repairs, and other sectors also.[15] investigates visual examination in many areas such as constructions, infrastructures, and utility companies, advantages of using UAVs for thorough observation, time-sensitive assessment activities, lower costs of aircraft service and maintenance, the worth of Intelligence visual inspection in 2020. Article [16] explains how to use DT to control catastrophe and risk linked with the COVID-19 epidemic in the physical environment. Even during COVID-19 epidemic, [17] concentrated on future developments of drone navigation investigation, and results reveal many uses of drones utilizing multiple models and kinds. Drones are being utilized this year to carry COVID vaccination to underserved areas in Telangana and Northeast India. If the scenario remains (current phase of the third wave of COVID-19/or continues) in 2021, it raises additional questions for future study on the usage of AI enabled drones/UAVs in building.

Drone technology has been viewed as a tool to bring outcome and methods into the twenty-first century in a variety of businesses. Corporate and public-sector needs are accelerating progress in aerial drones [18]. The author [19] released a statement saying, "drones are indeed a newer concept, but investigation is extending its wings." Not just that, but he said that architecture and engineering organizations were first to use Drone technology. We anticipate that they can always lead the way the drive for novel application scenarios that utilize quicker cellular services, edge computing, computer vision, artificial intelligence, and pattern recognition. Many authors [20–22] have emphasized the importance of developments such as technology solutions, Intelligent systems, BIM and electronic twins, AR/VR, wearables, adaptable, remote, and prefabricated constructions, robots and drone videography. The relevance of each trend is fully discussed, and 12 building trends, incorporating drones, are projected to prevail [23]. Intelligence is a new developing technology has the potential to transform the CI. Advanced CT allows major safety enhancements. Article [24] comprises image processing techniques for flower detection, a mechanical MPR technology, and an unmanned aircraft pollination system. For independent artificial blooming, a six-configuration MPr framework has been described. QWinOut DIY F450 OpenMV Webcam Unmanned aerial pollination experimental quality assurance, set a framework for smart unmanned drone pollination using CIAD. New AI methods, like the heredity method, are being created (HA).

The research [25], represents, improving irrigation effectiveness by determining agricultural water needs and monitoring crop water condition utilizing proximal and distant sensors. Investigating creative ways to improved irrigation control and more precise crop water demand estimation. It also mentions the future enhancements such as Optimizing forecasting accuracy at bigger levels attempts to close gaps in knowledge and acquire reliable dynamic location model parameters for AI systems. Unmanned aircraft enable academics and watershed management to acquire vast volumes of data in a safer, better expense, and more informative manner than prior methods. Novel technological techniques and procedures will be created as unmanned aircraft and sensing technologies progress. The article [26], describes the advantages of drone methods of farming and their wide applications, showing how UAVs operate on farms with instances and highlighting the many characteristics of drone for agricultural production by using accurate sensors to spot insects and water shortages. For farming purposes, different smart agriculture approaches, such as the use of unmanned aircraft vehicles, UAVs are a hybrid of developments in information technology, as well as robotics, artificial intelligence, big data, and the Internet-of-Things. The essay finishes by urging more farmers to use drone technology to enhance their farming performance. In this section, we look at recent existing research in agriculture UAVs, control system, equipment, and advancement. The research [27], describes a unique method for distinguishing between different field farming processes using an RGB-D sensor. The given approach is simple to incorporate into available on the market Unmanned Aircraft (UAVs). Two separate measuring methods were created to properly classify the ploughing strategies.

In the paper [28], researchers show how UAVs with artificially intelligent (AI) abilities may be used for automated cargo transfer. On the surface, a methodology for target recognition with numerous stages is devised, and once the objective is recognized, the payloads attached to the drone are released. A 3D printed mechanism depending on pinion gears is used to release the cargo. Furthermore, an auto flying program is being created to allow the drone to travel autonomously. A tiny drone termed as "Phantom DJI" is utilized as concrete evidence to deliver a.6 kg independent cargo along a specified route to a destination address. [29] suggest an innovative and effective drone delivery architecture In particular, designers present the multi-layer AI (MLAI) architecture, that enables the easy integration of ad hoc Machine learning and artificial intelligence. We used

MLAI-based deep neural networks to monitor and recognize objects to show the advantages of the new system. The article [30] represents that drones, also called unmanned aerial vehicles (UAV), are being employed in a variety of purposes. Drones are already being taught with AI and ML for a variety of reasons. Drones are much more efficient and robust than other technologies since they fly via air and execute tasks in a fraction of the time. Also, this research says how UAVs have previously been used in various nations, including Iran, Turkey, and Russia. India is presently developing AI drones. Drones are now utilized in military defense and healthcare systems, but they will play a significant role in every industry in the future. With rising prices for supplies in China, the flaws of the present short-term logistics industry are unproductive, and worker occupancy has become more visible. The fraction of people who were involved in shipping and manufacturing workers who assure the movement of supplies keeps rising, but the overall social labor command's industrial production falls. In this article [31], a special unmanned aerial vehicles architecture framework is developed rather than relatively brief logistic support manufacturers, with the goal of adjusting relatively brief logistic support connections, trying to improve operational efficiencies, and optimizing integrated logistics methods to eliminate the number of employees required to maintain constant logistics transport services. In [32], it describes the recent advancement of artificial intelligence (AI) in several domains and an influence on unmanned drones as well. However, most of the methods presented either totally rely on enterprise applications or provide a poor interconnection mechanism that prevents the creation of further approaches. As a result, this research suggests an advanced and effective architecture for unmanned drones. Designers present the multi-layer AI (MLAI) architecture, which enables the integration capabilities of commercial Ai technologies. They used MLAI-based deep neural networks to monitor and recognize obstacles to show the advantages of the new system.

Article [33] describes a method for predicting thermal currents using neural networks to aid outcome in the flying of unmanned drones using this type of energy to complete their operations. It is based on the empirical measurements of climatic conditions, geographical position, period, and movement taken while flying. Brain-computer interaction (BCI) interprets brain impulses to determine intent of the user. As the need for drone command grows, recent progress in the BCI-based drone control scheme has just been made. Drone swarming management associated with brain impulses might supply numerous businesses such as serving in the military or disaster response. Using only a various visual framework, this research [34], shows a model of a central nervous system interaction process for a variety of circumstances. Designers created an experimental setting that could gather signals from the brain while running a drone swarming management simulator. Researchers acquired electroencephalogram (EEG) readings from four distinct scenarios using the device. In the investigation, seven disciplines assessed performance of the classifier that use the basic data mining techniques.

Based on the idea that now the signal intensity emitted by various drones has varied periodic qualities, this work [35], investigates the use of rhythm-based classifiers as initial intent to ADI. Designers develop an ADI technique based on rhythm-based characteristics to investigate and evaluate their method. Researchers compare their findings to a benchmark investigation using a freely available online drone sound database. Therefore, their recognition system enhances the binary classification tasks by 3.47 percent and the multiclass classification by 2.97 percent, yielding average accuracy of 0.9985 and 0.9591, respectively. Even though the gains are limited, they emphasize that auditory characteristics have a high ability to improve ADI, particularly in dense drone-based scenarios wherein drone recognition is critical. The Internet of Drones (IoD) is a multi-layered security management framework developed primarily for organizing unmanned aircraft accesses to regulated airspace and delivering navigational activities among nodes. The IoD offers general services to multiple of drone applications, including delivering packages, vehicle tracking, rescue operations, and others. In this [36] work, researchers develop a conceptual framework of how such an infrastructure may be arranged, as well as the qualities that an IoD system built on such design should include. To do this, designers collect essential ideas from three additional large-scale networks, such as the air traffic control networks, the wireless connection, and the World Wide Web, and investigate their linkages to our unique drone traffic enforcement framework. In this research [37], researchers present a quadrotor modulation technique that use reinforcement learning to command a quadrotor aircraft based on previous enhanced partial state inspections. The reinforcement learning approach might allow the model to establish a strategy that maps perceptions to regulate instructions, which in current study is the quadrotor actuation instruction. Furthermore, we apply our technique inside the results in making through modeling, with outstanding results.

The challenge of lively picture identification is addressed in the article [38], the major method for solving this challenge is the employment of deep learning artificial neural networks or convolution networks. A summary of the properties of these networks, as well as its fundamental framework, is provided. The surface and characteristics of local area networks at their usage is demonstrated in difficulties of autonomous operation of drones based on technical vision technologies. Techniques for learning such GPS technology as the challenge of forming training examples is not less complex than the establishment of networks of deep learning and platforms of automated driving on their basis are provided. Based on unmanned drones, this research [39], proposes a method for dynamically charging hydrodynamic identification of composite insulation. A drone, a sensor gadget, a humidification device, and an incorporated artificial intelligence system module make up the detecting technique. The water spray gun's fineness and length may be adjusted using the water spray gadget. The sprinkler equipment spray liquid on the composite's insulator, and the photography device captures a snapshot after sprinkling. The smart algorithm method includes the picture to evaluate the form of water droplets on the insulation, and then produces the ageing percentage result. This approach increases the effectiveness and awareness of chargeable hydrodynamic examination of compound insulator, which is critical for assuring the safe functioning of composite insulators. It also has a lot of potential for utilization.

7.3   AI in Drone Navigation

Drones that use AI as well as automatic steering are still being researched. [39–43] are several contemporary publications that concentrate on the idea. In [39] and [40], the researchers utilized imitation behavior to manage the drones and obstacle avoidance. Researchers in [43], [42], and [41] have also employed feed-forward deep neural networks to operate drones and avoid collisions. Article [43] discusses developing a database from 11,500 accidents that would be used to build a typical deep learning model. The network generates binary categorization and gives instructions on how to avoid a collapse. It's a straightforward self-supervised system that works well in a populated area. The designers of [42] describe an approach for calculating altitude that used a simple forward-facing sensor that is input into a developed CNN network. The training model was successfully evaluated in both a physical and virtual context. Likewise, the researchers of [41] used a neural network model to learn the drone to recognize woodland pathways with an only one monocular camera. Additionally, the database used to learn the drone is publicly accessible for access, as is the database utilized for experiments in [41]. The developers of [40] are attempting to teach the UAV to navigate across a modeled room in congested surroundings using mimic training and recurrent neural networks. The authors of

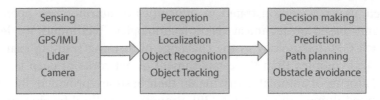

Figure 7.3 Autonomous vehicle architecture [44].

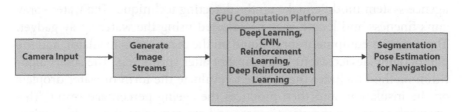

Figure 7.4 Demonstration of various parts of AI components for a smart approach [45].

[39] utilized a similar scheme to travel in a forest and restricted indoor environment using a forward-looking sensor and imitation modeling. Figure 7.3 describes the architecture of autonomous vehicle that uses AI for navigation.

This article analyses how AI may be utilized to aid obstacle detection and navigation, that necessitates knowledge of the various elements of AI and whether they could be employed in UAVs. The broad notion of AI, searching issues, neural networks, artificial neural network, reinforcement learning, and deep reinforcement having to learn are all used in this part. It also uses the programming called caffe and numerals, tools for creating the convolutional network lastly, it highlights existing AI development for drones. Figure 7.4 summarizes the AI framework employed in many of the implementations.

7.4   Companies that Use the AI Drone to Solve Big Problems

UAVs and artificial intelligence are a perfect match in terms of technology. Ground-level controllers can get a sense of sight by combining AI's real-time machine learning and artificial intelligence with the exploring capabilities of drone aircraft. Drones, on the other hand, can only really show whatever their cameras collected until recent. Drones nowadays can detect its environment due to AI techniques that allows them to scan places, track things, and deliver analyzing timely feedback. Some businesses that integrate Artificial intelligence in drones are listed below.

- DroneSense

DroneSense is a public security drone technology platform located at Austin, Texas that transforms raw drone information into relevant information for policing, firefighting, and other emergency response teams. The DroneSense OpsCenter allows several UAV users to work together, monitor what every UAV observes, and even track a drone's flight path in actual environments. Dozens of groups have been using the DroneSense public security framework to address a variety of attacks to public security. SWAT teams could use Intelligence technology to collect image information, analyze damages after storms and tornadoes, and even use imaging technology to identify missing people.

- Neurala

Neurala is a deep-learning platform to help UAVs that detect and recognize people of interest in groups located at Boston. It could also check and inform on massive industrial infrastructure, such as telecommunication antennas, in actual environments. The business says that instead of taking hours or even days to search masses for an individual, their AI-powered technology just takes 20 minutes to interpret the image of a person. The Lindbergh Foundation fights elephant illegal hunting in Africa with drones operated by Neurala. The industry's image analysis software is used by the artificial intelligence-based drones to watch elephant populations and discover potential poachers kilometers ahead they approach the animals.

- Scale

Scale is located at San Francisco and employs artificial intelligence and machine learning to assist in the training of drones on aerial photos. Drones can use machine learning techniques to recognize, classify, and map it all from small things like vehicles to entire neighborhoods. Insurance businesses such Liberty Mutual utilize the Scale machine learning management system for drone training. The UAVs are often used to detect and evaluate insurance payments.

- Skycatch

Skycatch is located at San Francisco and develops software that takes, evaluates, and analyses drone input from aerial photographs automatically. To acquire a comprehensive perspective of the terrain getting inspected, the company's software converts such aerial footage into requirement needed, 3D models, or thermal photographs. Komatsu, a Japanese home builder, employs Skycatch-enabled drones on even more than 5,500 labor locations.

The business claims to be able to produce 3D images that really is accurate to within 5 cm. The integrated software typically takes approximately 30 minutes to analyze aerial photographs, compared to days for people that do the same job.

- Alive

The Alive Framework located at Doral and Fla, employs artificial intelligence to assist drones in inspecting machinery. Drone equipment from the business includes 2D/3D modeling, topography mapping, picture recognition technologies, and even soil test detectors, allowing drones to monitor everything including agricultural output to wind turbines. Millions of drones across a range of industries, from green power to transportation, are using the Alive technology platform. People on construction projects use the technique to analyze aerial photographs and as a monitoring device during the construction stage.

- Applied Aeronautics

In their fully independent "Albatross" drones, Applied Aeronautics employs AI. Aerial surveys, infrastructure monitoring, disaster mitigation, and rescue operations are among the diverse applications for the company's drone. The unmanned aerial vehicle (UAV) has been utilized on each and every country so far, assisting with anything from aquatic life protection to humanitarian assistance. The unmanned aerial vehicle can fly beyond 100 miles and achieve faster speeds to 90 miles per hour, according to the firm. Advanced Aeronautics' drone is used by NASA and other internationally known organizations for scientific activities.

- Skydio

Skydio's unmanned drones placed at Redwood City in California, records surveillance video using AI powerful computers and 13 camera systems. The self-flying drone can record video in a variety of ways (like stationary position and track style), and none of them needs human intervention. Drones like the Skydio are utilized in a variety of areas, especially athletics. Marathon, cyclists, and climbers may select a video recording mode, and Skydio will record each step they take. The drone can indeed recognize and track the object moment on its own.

- Airlitix

Airlitix develops AI and machine learning supported drones located at San Jose in California, for greenhouses maintenance which can fly over greenhouse and gather information about the temperature, moisture, and levels of carbon dioxide to guarantee that plants thrive in the best possible environment. The Airlitix AI drones can assess environmental and crop monitoring in additional to environmental measures, ensuring that crops are disease-free and able to flourish unimpeded.

- Orby

Orby is a company that creates unmanned drones and robotic systems for enterprises. Unmanned aerial vehicles using artificial intelligence aid in the mapping and planning of huge warehousing and industries. Drones can also check inventories utilizing AI that analyses existing supply in detail via aerial footage. Orby employs drone technology and AI algorithms to check cabinets and dynamically monitor and acquire fresh supplies in its trial program. Unmanned aerial vehicles flights may even be scheduled to complete work during off-peak hours and then return.

- Above

Companies and politicians utilize Above's unmanned UAV to gather data from the environment it is there at San Francisco. The zeppelin-like UAVs can fly placed above a white forest, hillsides, factories, and farmland to evaluate losses from natural catastrophes or environmental degradation, then come back and comment by using technology platform. Apart than collecting data from the environment, Above's UAV can monitor railways, electricity lines, and oil and gas pipelines for good maintenance. Above's UAVs may also be routed precisely by controllers. The aircraft will next fly the route, video what it observes, and compile a study based on its observations.

- Shield AI

The Shield AI UAV aids surface soldiers and first attackers in data collection and investigation. The company's UAVs can interact with each other using exclusive "Hivemind" technology to instantly detect environment and recognize persons in an urgent situation. In aerial vehicles, the Shield AI "Hivemind Nova" drone assists law enforcement and armed forces. To obtain ground-level information, the robots may get entry to GPS-restricted places such as indoor spaces and subterranean infrastructure.

- AeroVironment

AeroVironment employs artificial intelligence to drive its unmanned drone technology. The Switchblade®, which is outfitted with a highly precise attack payload for military activities, is a three-foot-long surveillance aircraft that is untraceable. AeroVironment produces a variety of Intelligence UAVs for various military applications. Agriculture uses the industry's UAVs to monitor agricultural area, detect crops health risks, and assess watering difficulties.

7.5 Drone Applications Using AI

Drones are a fast-growing business with a near unlimited variety of uses, due to the large part to artificial intelligence (AI). Drones could gather vital information, react about what they're being shown in timely manner, and relay that information to grounded outcome, and therefore be trained to make such a choice for themselves because of AI. Drones are a low-cost, high-efficiency option for companies to enhance efficiency, save expenses, and make their workplaces safer for existing employees. The following are some innovative AI drone technologies.

• Surveillance

Drones can be outfitted with a variety of monitoring devices to capture video stream and still photographs at all hours during the day and night. Unmanned aerial vehicles can be outfitted with equipment that allows them to listen in on phone conversations, track GPS movements, and collect vehicle license data. The high payload flexibility enables something like a variety of surveying instruments, including laser scanners, multidimensional and multispectral sensors, and more — 24 hours a day, with minimal labor and expenses. Drone monitoring is the capturing of still photographs and videos by unmanned aerial vehicles (UAVs) in order to acquire data regarding particular goals, which can be persons, groupings, or locations. Drone monitoring allows for the collection of information about such a location from a range or position while remaining undetected. Unmanned aerial vehicles monitoring allows for the covert gathering of data about a destination from a range or position. Of course, drone utilization in this broad business (or area) goes much further than such basic and uncomplicated parameters. Drone technology is used by public authorities, policemen, as well as other security professionals. As firms and academics discover new ways to utilize machine learning to evaluate current clips, automated monitoring will become more ubiquitous. A recent experiment led by scientists in the United Kingdom and India demonstrates one prospective application for this innovation: using webcam drones to detect violent behavior in gatherings. It employs a cheap Parrot AR quadcopter to broadcast video evidence for actual research through a mobile internet access. The opinions of persons in the film are estimated by a deep learning algorithm, which compares them to "violent" positions identified by the scientists.

• Weather Forecast

The environment is changing. Natural disasters are the same way. Drones, obviously, fall short of satellite imagery's accuracy in anticipating natural calamities. They are, nonetheless, able to offer crucial aid in the event of a tragedy. Government agencies and insurers are becoming more aware of the possibility for employing them to estimate post-disaster losses, particularly at places that have not been designated as safe for people to enter. Drones collecting atmospheric samples are a significant improvement over conventional methods of collecting data, but it has the potential to significantly enhance the accuracy of climate estimation methods. The significance of this is increasingly precise simulations have an impact on both every day and every big picture. It enables weather forecasters to provide improved 10-day weather predictions, but it also means being able to provide greater advanced notice for events such like tornadoes, as well as determining when and where hurricanes could impact. The boundary layer, which is the lowest layer of the atmosphere, is where the majority of our climate occurs. The climate data that affect each of us are influenced by a bewildering array of variables and influences. Trying to anticipate what the climate does next depends on extremely intricate climate estimation methods; however the forecasting model' result is as good as the information supplied into it. Obtaining quality data is also more difficult than expected. Weather drones can help with this. Environmental drones can acquire critical temperature and humidity, precipitation, air density, and wind velocity by flying across the whole vertical layer of the border layer of the atmosphere. Temperature, moisture, and air density detectors connected directly to the drone is one method. Sending sensors known dropsondes from a high altitude with something like a parachute may be another method to acquire data. The dropsondes gather information as they move down through the vertical components of the boundary layer. Visual imagery, such as images and video, is another significant way in which weather drones acquire data.

Many key efforts are currently underway to make better use of and enhance drones for climate data gathering. Something like this is inextricably linked to researchers at the NOAA. They're doing brief field research in specific parts of the United States to learn more about how geography and land surface elements influence climate changes. The objective was to find out what and how the earth's surface influences climate, including how to include that variable in environmental and climatic simulations.

• Putting Out Fires

Drones are better at finding, assessing, and even putting out flames. Drones give firemen with "big picture" insight on the extent and progress of a wildfire, as well as topography and meteorological factors, which is important information for deploying personnel and evacuating civilians. Drones can explore and detect harmful areas despite smoke and flames utilizing their camera systems, making firefighters activities easier altogether (Figure 7.5).

Figure 7.5 UAV putting out the fire [21].

- Emergency Services and Policing

Police, fire services, as well as other legislation organizations or rescue groups can examine data collected by drones. Drones may be designed to cooperate and exchange data through with a service center, allowing for an organized and real-time reaction. Capabilities involve aiding rescue workers in examining criminal investigations, environmental disaster, and even searching for misplaced individuals utilizing infrared sensors.

- Facial Detection for People Recognition

Drones could float across gatherings seeking for people of interest utilizing face recognition techniques. Drone sensors could also capture human expressions and emotions, which might be beneficial while hunting for frightened or unpleasant people. The drone may be configured to recognize predictable and unpredictable suspects, capture video, and recognize still images.

- Locations of Constructions

AI drones are used by building contractors to monitor and analyze their construction projects. Actual views at the job site, available 24 hours a day, seven days a week, make it so much easier to monitor work progress and resource consumption, resulting in more precise planning and development work. Drones are also utilized to detect dangerous places and workplace behaviors, as well as to offer protection.

- Situational Awareness

In the event of a tragedy, UAVs could be the first "eyes mostly on earth." They may interact and exchange data in order to immediately evaluate an urgent scenario and deliver essential data to first authorities. Drones may also infiltrate fallen constructions and provide resources to disaster relief organizations and sufferers.

- Water Quality Monitoring

Drones are being utilized all over the globe to assess water sources. South Korea has devised a framework that tracks the incidence of algal blooms using drones and multispectral photography. Drones also can identify leaky water pipelines by detecting temperature and moisture changes in the earth. SmartTerra, an Indian company, is utilizing artificial intelligence to detect water loss caused by leaky equipment or unlawful usage.

- Evaluation of Infrastructure

Physical inspection of infrastructure may be risky as well as time consuming. UAVs remove this danger by checking infrastructure in seconds for fractures or other physical problems. Technology for climate data, such as soil water content, comprises 3D modeling, picture identification, and analysis. Reservoirs, bridges, electric lines, transit, and utilities are all good candidates for drone inspection.

- Agriculture with Intelligence

Drones could be used in almost every farming sector, ranging watching from climate patterns and relative humidity conditions to determining the ideal moment to sow. The amount of fertilizer required may be determined by analyzing the ground conditions. Drones could identify sick plants and sometimes even establish trees by delivering seedpods into prepared soil at specific intervals.

- Action Taking

Drones may visit locations that are too risky or unreachable for humans (Figure 7.6), giving photographs and information that protect searchers secure and aid in lifesaving. Drone technology is evolving rapidly and will facilitate lots of new advantages throughout a multitude of areas, together with commercial, procurement, shipping, agriculture, and ecologic, as

Figure 7.6 UAV that evaluates the infrastructure [21].

well as self-use, due to the amazing customization options offered by the Internet of Things (Figure 7.7).

- Military Exercises

Drones deliver surveillance data and can transport provisions, armaments, and other equipment to armed forces. They may also be set up to kill opposing soldiers. According to published accounts, Israel used the first-ever AI drone swarms to detect, recognize, and target Hamas terrorists without any human involvement. To detect and identify their destinations, the drones connected and exchanged their AI findings.

- Smart Cities and Urban Development using Drones

Smart cities are built using cutting-edge technologies such as well-connected house and infrastructure facilities with operation in the palm of the user's hand. Most of the infrastructure is completely or largely mechanized, with improved safety monitoring using AI-based camera systems for rapid face recognition software and tracing undesired things. Drones also assist in urban management by providing overhead view modeling of urban houses and urban site planning, enabling civil architects to create the most practicable planned arrangement. Furthermore, it is important for monitoring system and defining the city path for a comfortable and confusion transportation.

- Maintenance of Farm Animals

Through an Intelligence drone technology, Folio3 has assisted several big names in the animal industry with the optimization of numerous tasks associated in livestock management. Measuring livestock, for instance, may take a long time. Implementing this method mechanized with artificial intelligence-powered drones has enabled such firms significantly

Figure 7.7 UAV in agriculture using AI [21].

reduce the time and effort necessary for the operation. Likewise, physically recognizing ill animals and implementing the required measure to avoid additional injury might take a long time.

• Mapping the Terrain

Civil engineering has also benefited from the deployment of AI-assisted drones. UAVs for mechanized terrain mapping could enable this technique faultless, which is a critical necessity in the engineering profession. Folio3 has created Ai technologies specifically for this need. Drones use 3d cameras to capture live video of the region under examination. This input should be used to obtain information for 3D model building using various Computer programs. The drone is released well over field or region within examination, outfitted with a 3D sensor and a LIDAR sensor.

• Supervision of Railway Corridors, Highways, and Electrical Wires

Drones are employed to examine governmental infrastructure such as roads, railways, electrical wires, and irrigation, among others. The quality of the track may be monitored and maintained by keeping a constant eye on it. Drone surveys are highly useful for railway route mapping since they give ultra-high-resolution pictures all along corridor and subsequently aid in project execution using artificial intelligence.

• Coastal and Border Protection Organization

Unmanned aerial vehicles (UAVs) have been well for playing a vital part in several nations' security strategies. Previously, these robotic flying aircraft were only utilized by developed, richer countries; however, technological improvements using AI and a competition in the market have allowed numerous countries to include drones into its defensive strategy, becoming them a regular appearance for sea, ground, and immigration enforcement. Even a vast country like India must secure its boundaries and crucial sea routes, particularly since it borders some countries that are not the more friendly of neighbors.

• Healthcare and Swachh Bharat Implementations

In various places of India, local municipal officials and police are using drones for monitoring as part of the government's Swachh Bharat Abhiyan, a clean India initiative. Drones are used in both urban and rural regions to avoid contamination of water resources. Drones have shown to be quite useful in bringing essential medications to isolated areas and therefore saving people. Drones might also assist address the need for blood components in the pre-hospital situation promptly and affordably by delivering pricey and seldom used medications for snake bites.

• Emergency Communication Systems in Outlying Regions

AI enabled Autonomous drone projects to have been piloted by humanitarian organizations around the world for several purposes, namely bringing materials to difficult locations such as conflict zones and disaster areas. Modern UAV implementations and subsequent intelligent transportation breakthroughs provide a look towards what would become the accepted norm in emergency management and rehabilitation. Civilian intelligent drones are commonly used in relief efforts to map and scan disaster zones to identify whichever area needs the most infrastructural restoration as well as where recovery efforts are most critical. The Florida Air National Guard recycled combat drones to conduct survey work of the disaster region affected by Hurricane Irma, ability to identify what regions needed the greatest assistance.

• Filming

Drone pictures are ubiquitous — there isn't a TV show or news story that doesn't incorporate these gorgeous photographs from the skies, and AI-drones are increasingly affordable for professional videographers as well. Drones, also known as unmanned aerial systems (UASs), have quickly become a must-have camera addition for many filmmakers. They present the prospect of outstanding flying cinematography to a far lower cost than employing a helicopter. Drones also enable shooting in areas that helicopters cannot reach. Drones may be flown to mountain peaks, through the jungle, and over lakes. The audience can be taken to places they would never go on their own.

• Transmission Wires for Submarines

A bank in an Asian country could not transmit money to Saudi Arabia paying for oil if there were no underwater cables. Military authorities in the United States would have difficulty communicating with forces combating militants in Afghanistan and the Middle East. A undergraduate in Europe would have been unable to Skype his American parents. Each day, large quantities of revenue are sent over the wires, as well as the internet also transports most conversations between the United States and Europe, as well as other parts of the world. These internet lines are monitored and protected by AI enabled aquatic drones.

7.6   Issues in the Integration of AI with Drones

Despite the fact that AI is widely used in drone modeling and other aspects of life, there are certain drawbacks with it. Among these concerns are:

- Extremely expensive: Integrating AI in automated devices, particularly drones, can be prohibitively costly. And that doesn't include normal maintenance procedures, repairing, and replacement [46].

- Human error: While AI-powered devices may identify errors, they are not immune to producing them themselves. They make judgments based on learned data and may be unable of making smart decisions that people would make in certain key situations. This suggests that they might be flawed and lead to errors, particularly when taught with incorrect data.

- Security and privacy: AI in particular disrespects personal confidentiality and protection. These technologies have enabled cyber thieves in invading other people's privacy [47].

- Adaptability: It is conceivable that drones using AI would be unable to adjust to changes in their surroundings. Drones using AI, for instance, may struggle to navigate in unfamiliar areas [48] because they are taught with inputs.

- Necessary experience: As AI grows, there should be basic software and many other domains of AI competence. Such professionals are in short supply since they have worked hard to learn the skills [49] necessary for development of the system.

- Accuracy: Furthermore, some studies should be regarded with a grain of salt, especially when it comes to its guarantees of reliability. Some claim claimed their method is 94% effective in detecting violent attitudes, but they point out that the higher the number of individuals in the photograph, the lesser the accuracy.

- Surveillance: Using AI to recognize body positions is a prevalent issue, with major internet companies such as Facebook producing extensive research on the subject. With the advent of low-cost drones and high-speed mobile internet, capturing and transmitting continuous video footage has never been simpler. It's not difficult to put these elements together to construct efficient surveillance.

7.7 Conclusion

This research provided a detailed structured literature assessment of unmanned aerial operations in design, engineering, and manufacturing, with an emphasis on implementing technology. In comparison to prior evaluations, this research thoroughly analyzed existing studies and classified UAV implementations using AI into groups that effectively represented its various applications, while also analyzing the technological utilization within each division. The findings of this study are likely to help the professionals and academics uncover cutting-edge innovation on Application domains in the above said sectors. Infrastructure, facilities and structural examination, mobility, cultural and historical tracking, urbanization and urban planning, assessment processes, post-disaster, and constructions protection were the themes from which the study was organized. This study also denotes that the experts frequently supported UAV incorporation with Artificial intelligence in the different areas since it provides equivalent, if not better, results than traditional approaches in relation to time, quality, protection, and prices. Another benefit was the UAV's ability to reach lofty or unreachable regions. Climatic (season and illumination circumstances, wind patterns, daylight reflectance) and technological defects in the equipment and system elements were the most reported problems (GPS-deficiency, magnetic interferences, UAV batteries, and image quality). Drone flight methods were nearly equally balanced among human and automated navigation, but the latter is the most widely employed kind of air traffic control.

UAVs with AI have proven to be a secure, yet effective, option in a field that is constantly growing and incorporating new emerging technology. UAV technology, together with future equipment and software advances, is expected to be increasingly frequently used in all the sectors. More research is required to improve methodology and best practices for present and future UAV integration into existing AI processes. Recognizing the benefits, as well as the operational hurdles, and comparing them to standard UAV-less procedures should be researched further. The current study includes the use of drones in development of various fields, as well as a summary of previous evaluations and numerous kinds of literature mentioned during several years. Different varieties of drones equipped with AI are described, along with the responsibilities of each necessary element in brief. This study evaluates various papers from recent years, and many applications, in addition to the primary uses of UAVs in the construction sector; the usefulness of drones in development businesses is emphasized.

This article also presents different applications and advantages and limitations such as, Drones' ever-improving features and cost enable associated with projects efficiency by decreasing waits, redesigns, and safety problems. Drone costs, flying laws and regulations, operator expertise, flight duration, and weather conditions are all limitations. Drone adoption would reduce project costs, enhance productivity, generate new employment, and provide value to many of the industry domains. Drones can quickly collect data in remote and dangerous regions. Drones give real-time data, significantly improving inspection reliability and increasing total manufacturing and successful innovative. In context of digital documenting, the implementation of various drone technologies plays a major role in the pre-, during, and post phases of all domains.

References

1. BPPT, Di puspiptek, jokowi nonton drone 'alap-alap' terbang, October 02, 2015. https://news.detik.com/berita/d-2885550/di-puspiptek-jokowi-nonton-drone-alap-alap-terbang
2. Wagner, I., *Projected Commercial Drone Revenue Worldwide 2016-2025*, Statista GmbH, J.B. Platz 1 20355 Hamburg, Germany, 2019.
3. Zaman, B., Mckee, M., Jensen, A., *UAV, Machine Learning, and GIS for Wetland Mitigation in Southwestern Utah*, USA, 2017.
4. Olawale, O.P., Dimililer, K., Al-Turjmanc, F., Chapter six-AI simulations and programming environments for drones: An overview, in: *Drones in Smart-Cities, Security and Performance*, pp. 93–106, 2020.
5. Kim, H., Mokdad, L., Ben-Othman, J., Designing UAV surveillance frameworks for smart city and extensive ocean with differential perspectives. *IEEE Commun. Mag.*, 56, 98–104, 2018.
6. Sun, J., Wang, W., Da, Q., Kou, L., Zhao, G., An intrusion detection based on Bayesian game theory for UAV network. *Proceedings of the 11th EAI International Conference on Mobile Multimedia Communications*, pp. 56–67, September 2018.
7. Fan, B.K. and Zhang, R.Y., Unmanned aircraft system and artificial intelligence. *Geomat. Inf. Sci. Wuhan Univ.*, 42, 11, 1–7, Nov. 2017.
8. Wang, Z.R., Application analysis of artificial intelligence in UAV field. *Sci. Technol. Innov.*, 15, 56–57, 2020.
9. Zhao, Z.P., Lu, R.M., Wang, J.C., Development and prediction on intelligent unmanned aerial vehicle technology. *Tactical Missile Technol.*, 3, 1–7, 2017.
10. Taberkit, A.M., The importance of applying artificial intelligence on unmanned aerial vehicle. *4th International Conference on Electrical Engineering and Control Applications, ICEECA 2019*, Constantine, Algeria, vol. 682.
11. Hartawan, D.R., Purboyo, T.W., Setianingsih, C., Disaster victims detection system using convolutional neural network (CNN) method. *2019 IEEE International Conference on Industry 4.0, Artificial Intelligence, and*

Communications Technology (IAICT), pp. 105–111, 2019, doi: 10.1109/ICIAICT.2019.8784782.

12. Lee, D., Gyu La, W., Kim, H., Drone detection and identification system using artificial intelligence. *2018 International Conference on Information and Communication Technology Convergence (ICTC)*, pp. 1131–1133, 2018.

13. Roberts, M.R., 5 drone technologies for firefighting. *Fire Chief*, March 20th, 2014. https://www.firechief.com/fire-products/communications/articles/5-drone-technologies-for-firefighting-8wLtpDrDLmgDEReO/

14. Joshi, D., Drone technology uses and applications for commercial, industrial and military drones in 2020 and the future, 2019. https://www.businessinsider.in/tech/news/drone-technology-uses-and-applications-for-commercial-industrial-and-military-drones-in-2020-and-the-future/articleshow/72874958.cms

15. Tatum, M.C. and Liu, J., Unmanned aerial vehicles in the construction industry. *3rd Annual International Conference Proceedings*, http://www.ascoro.ascweb.org, 2017.

16. Wahab, A., Applications of drone technology in the management of disaster and risk associated with COVID-19 pandemic in the built environment: The Nigerian experience. *INTREST*, 14, SI, 73–81, 2020. www.utm.my/interest.

17. Husien, I.A., Borisovich, Z., Naji, A.A., Covid-19: Key global impacts on the construction Industry and proposed coping strategies. *E35 Web Conf.*, 264, o5056, 2021. Available: https://do.org/10.1051/e3sconf/2021.

18. Ewoldsen, B., Drones are still a new technology, but the research is spreading its wings, 2021. https://www.nationalacademies.org/trb/blog/drones-are-still-a-new-technology-but-the-research-is-spreading-its-wings

19. Patil, G., Applications of artificial intelligence in construction management. *International Journal of Research in Engineering, IT & Social Sciences (IJREISS)*, 2018, vol. 9, 2019.

20. Israr, A., Ghulam E. Abro, M., Ali Khan, S., Farhan, M., Zulkifli, S.A. B M, Internet of Things (IoT)-enabled unmanned aerial vehicles for the inspection of construction sites: A vision and future directions, *Math. Probl. Eng.*, vol. 2021, Article ID 9931112, 15 pages, 2021. https://doi.org/10.1155/2021/9931112

21. Jones, K., Top 7 construction technology trends for 2021, 2021. https://www.constructconnect.com/blog/7-construction-technology-trends-to-watch-in-2021

22. Patil, G., Applications of artificial intelligence in construction management. *International Journal of Research in Engineering, IT & Social Sciences (IJREISS)*, 2018, vol. 9, 2019.

23. Ansari, T., Construction trends that are likely to dominate 2021, 2021. https://housing.com/news/construction-trends-that-are-likely-to-dominate-new-year/

24. Chen, Y. and Li, Y., Intelligent autonomous pollination for future farming: A micro air vehicle conceptual framework with artificial intelligence

and human-in-the-loop. *IEEE Access*, 7, 119706–119717, 2019. https://doi.org/10.1109/ACCESS.2019.2937171.

25. Cancela, J.J., González, X.P., Vilanova, M., Mirás-Avalos, J.M., Water management using drones and satellites in agriculture. *Water*, 11, 5, 874, 2019. https://doi.org/10.3390/w11050874.

26. Ranal, V. and Mahima, Impact of drone technology in agriculture. *Int. J. Curr. Microbiol. Appl. Sci.*, 9, 1, 1613–1619, 2022. doi: https://doi.org/10.20546/ijcmas.2020.901.177.

27. Tripicchio, P., Satler, M., Dabisias, G., Ruffaldi, E., Avizzano, C.A., Towards smart farming and sustainable agriculture with drones. *2015 International Conference on Intelligent Environments*, pp. 140–143, 2015.

28. Alshanbari, R., Khan, S., El-Atab, N., Hussain, M.M., AI powered unmanned aerial vehicle for payload transport application. *2019 IEEE National Aerospace and Electronics Conference (NAECON)*, pp. 420–424, 2019.

29. Togootogtokh, E. *et al.*, An efficient artificial intelligence framework for UAV systems. *2019 Twelfth International Conference on Ubi-Media Computing (Ubi-Media)*, pp. 47–53, 2019.

30. Kartik, M.V., Ranjan, S., Kaur, S., Introduction of 5G to artificial intelligence drones. *2021 5th International Conference on Information Systems and Computer Networks (ISCON)*, pp. 1–4, 2021.

31. Hua, H. and Zhang, Z., Application of artificial intelligence technology in short-range logistics drones. *2019 8th International Symposium on Next Generation Electronics (ISNE)*, pp. 1 4, 2019.

32. Togootogtokh, E. *et al.*, An efficient artificial intelligence framework for UAV systems. *2019 Twelfth International Conference on Ubi-Media Computing (Ubi-Media)*, pp. 47–53, 2019.

33. Catuogno, C., Catuogno, G., De Yong, D., Magnago, F., García, G., Bosso, J., Prediction of thermal currents for autonomous drone flight with artificial intelligence. *2021 XIX Workshop on Information Processing and Control (RPIC)*, pp. 1–7, 2021.

34. Jeong, J.-H., Lee, D.-H., Ahn, H.-J., Lee, S.-W., Towards brain-computer interfaces for drone swarm control. *2020 8th International Winter Conference on Brain-Computer Interface (BCI)*, pp. 1–4, 2020.

35. Svaigen, A.R., Bine, L.M.S., Pappa, G.L., Ruiz, L.B., Loureiro, A.A.F., Automatic drone identification through rhythm-based features for the internet of drones. *2021 IEEE 33rd International Conference on Tools with Artificial Intelligence (ICTAI)*, pp. 1417–1421, 2021.

36. Gharibi, M., Boutaba, R., Waslander, S.L., Internet of drones. *IEEE Access*, 4, 1148–1162, 2016.

37. Shan, G., Zhang, Y., Gao, Y., Wang, T., Chen, J., Control of quadrotor drone with partial state observation via reinforcement learning. *2019 Chinese Automation Congress (CAC)*, pp. 1965–1968, 2019.

38. Zhilenkov, A.A. and Epifantsev, I.R., The use of convolution artificial neural networks for drones autonomous trajectory planning. *2018 IEEE Conference*

of Russian Young Researchers in Electrical and Electronic Engineering (EIConRus), pp. 1044–1047, 2018.

39. Ross, S., Melik-Barkhudarov, N., Shankar, K.S., Wendel, A., Dey, D., Bagnell, J.A., Hebert, M., Learning monocular reactive uav control in cluttered natural environments, in: *2013 IEEE International Conference on Robotics and Automation (ICRA)*, IEEE, pp. 1765–1772, 2013.

40. Kelchtermans, K. and Tuytelaars, T., How Hard is it to Cross the Room?– Training (Recurrent) Neural Networks to Steer a UAV, SAGE Publishing, 1-18, 2017, arXiv preprint arXiv:1702.07600.

41. Giusti, A., Guzzi, J., Ciresan, D.C., He, F.-L., Rodrıguez, J.P., Fontana, F., Faessler, M., Forster, C., Schmidhuber, J., Di Caro, G. *et al.*, A machine learning approach to visual perception of forest trails for mobile robots. *IEEE Robot. Autom. Lett.*, 1, 2, 661–667, 2016.

42. Chakravarty, P., Kelchtermans, K., Roussel, T., Wellens, S., Tuytelaars, T., Van Eycken, L., CNN-based single image obstacle avoidance on a quadrotor, in: *2017 IEEE International Conference on Robotics and Automation (ICRA)*, IEEE, pp. 6369–6374, 2017.

43. D. Gandhi, L. Pinto and A. Gupta, Learning to fly by crashing, *2017 IEEE/RSJ International Conference on Intelligent Robots and Systems (IROS)*, 2017, pp. 3948-3955, doi: 10.1109/IROS.2017.8206247.

44. Liu, S., Tang, J., Zhang, Z., Gaudiot, J.-L., Computer architectures for autonomous driving. *Computer*, 50, 8, 18–25, 2017.

45. Pokhrel, N. and Jung, A., *Drone Obstacle Avoidance and Navigation Using Artificial Intelligence*, Degree Programme of Computer Science and Engineering, Aalto University, School of Science, April 20, 2018.

46. DataFlair Team, Pros and cons of artificial intelligence–a threat or a blessing? https://data-flair.training/blogs/artificial-intelligence-advantages-disadvantages/

47. Dilek, S., Çakır, H., Aydın, M., Applications of artificial intelligence techniques to combating cybercrimes: A review. *Int. J. Artif. Intell. Appl.*, 6, 1, 21–39, 2015.

48. Huang, L., Qu, H., Fu, M., Deng, W., Reinforcement learning for mobile robot obstacle avoidance under dynamic environments, in: *Pacific Rim International Conference on Artificial Intelligence*, Springer, Cham, pp. 441–453, 2018.

49. Tuomi, I., The impact of artificial intelligence on learning, teaching, and education, Cabrera Giraldez, M., Vuorikari, R. and Punie, Y. (Eds.), EUR 29442 EN, Publications Office of the European Union, Luxembourg, 2018, ISBN 978-92-79-97257-7.

8

Applications of Drones—A Review

Swathi Gowroju* and Santhosh Ramchander N.

Department of Computer Science & Engineering, Sreyas Institute of Engineering and Technology, Hyderabad, India

Abstract

Reducing costs and expenditures while improving the efficiency of technological solutions is essential for large-scale industries. Industries such as mining, ports, gas, and large plants involve processes and applications where human power may be at risk or impossible to use. Areas of inspection may include infrastructure maintenance, leak detection, equipment monitoring, security surveillance, and quick hazard monitoring. In such areas, one can rely upon drone technology to assist in better visibility, investigation, and quick alerting of failures in various systems. Hence, large-scale industries are quite interested in investing in drones to maintain systems and infrastructure for efficiency and productivity. In this chapter, we identified various types of drones and drone technologies to help readers better understand the technology. Furthermore, note that challenges to drones or unmanned aerial vehicles (UAVs), especially obstacle detection, battery life, and attackers, need to be overcome for them to fly or operate effectively. Here, we discussed UAV categorization, automation systems, and future areas of interest in business and research. Various control techniques for their construction process, manufacture, and analysis are also explored in depth. Moreover, obstacles to UAV function or operations are highlighted and examined, including fast charging, emergency braking, and security.

Keywords: Unmanned aerial vehicle, large scale industries, drone technologies, attackers

Corresponding author: swathigowroju@sreyas.ac.in

Sachi Nandan Mohanty, J.V.R. Ravindra, G. Surya Narayana, Chinmaya Ranjan Pattnaik and Y. Mohamed Sirajudeen (eds.) Drone Technology: Future Trends and Practical Applications, (183–206) © 2023 Scrivener Publishing LLC

8.1   Introduction

Drones are very popular nowadays. Having unmanned aerial vehicles (UAVs) is a dream of many industries as their use can help overcome man-power challenges and increase system efficiencies. An UAV or a drone is an aircraft without a pilot or a flying device or robot operated remotely or autonomously through software-controlled commands or plans installed in its system. It possesses sensors, trackers, and many other features according to the usage of application or industry. Drones were initially designed in 1930 as a radio control device. Later, their applications were initiated for information retrieval in the military by attaching cameras and sensors to them. Currently, drones are attracting big industries such as oil, gas, and plants, and companies such as Google, Amazon, and Facebook with drone applications geared towards carrying and delivering parcels and other goods, respectively, and with their systems connected to the Internet.

Drone works with a controller for launching, navigating, and landing, among others. It establishes the communication channel using radio waves with a frequency of 2.4 GHz. There are drones that work using Wi-Fi net-works for their movements. Most of the features of smart phones are also carried by drones. They work intelligently in the air and can cover very long distances and can perform spying tasks very easily. A number of drones are popular for their camera features. A comparison of various drones is given in Table 8.1. DJI Mavic 3 is one of the advanced drones in imaging and flight technology. DJI Air 2S is designed for adventurous visual creations. DJI FPV is famous for recording with goggles in aerial travel. DJI Mini 2 is a unique drone as its features include the ability to automatically return to its place of take off. DJI Mavic Air 2 is an affordable drone and is capable of long-range flights. DJI Mavic 2 Pro is a photographic and portable drone just like a smart camera. DJI Mavic 2 Zoom and DJI Inspire 2 are usually used in making films and for outdoor shooting. DJI Phantom 4 Pro V2.0 has obstacle sensors.

Because of the high demand for UAVs in a variety of applications such as crop monitoring, coast guard surveillance activities, mobile connec-tivity, and domestic spying. Several different kinds of UAVs have already been established, each with its own set of specifications for volume, gravity, range and endurance, engine power, and configuration. Because of this wide assortment, work flow and interpretation depend on the nature of UAV, with several control-technique opportunities to facilitate

Table 8.1 Drones available in the market and their configuration (collection of decade data) (source: Amazon.com).

Drone	Configuration	Comments	Image
DJI Mavic 3	4/3 CMOS Hasselblad and 1/2-inch-sensor tele camera	46 min of flight time with 15 km transmission range	
DJI Air 2S	1 inch CMOS sensor, 5.4K video recording	With the feature of Master Shot, it offers 12 km of video transmission using 1080p	
DJI FPV	4K/60 FPS video recording	Emergency operations such as, break, hover features	
DJI Mini 2	4k Camera and can fly for 31 minutes	Functions such as automatic takeoff and returning to the home are available	
DJI Mavic Air 2	48 MP photos, 4K/60 FPS at 120 Mbps and 8K Hyperlapse using OcuSync 2.0	Provides HD feed long range shooting	
DJI Mavic 2 Pro	10-bit Dlog-M with camera of Hasselblad	Vivid colors and camera options are available	

(*Continued*)

Table 8.1 Drones available in the market and their configuration (collection of decade data) (source: Amazon.com). (*Continued*)

Drone	Configuration	Comments	Image
DJI Mavic 2 Zoom	48 MP Super Resolution	It's a Creative Filmmaking Drone	
DJI Phantom 4 Pro V2.0	1-inch 20 MP camera, 4K/60 FPS video, 14 FPS Burst Mode	It's a Professional 4K Drone	
Typhoon H Plus	1080 Mbps 60 FPS video capture, 20 megapixel still images, 5 rotor fail safe mode and 1 sensor	With touch screen facility	
Yuneec H520E/520	5 rotor fail-safe, Up to 28 minutes of flight time, Integrated 7" touch screen, global rotation up to 330 degrees	Radio communication up to 3 km range, swappable cameras	
ANAFI Ai	4G robotic UAV, 48 MP cam, 32x zoom thermal camera	Integrates the sensor, visual behavior technology of animal's	

(Continued)

Table 8.1 Drones available in the market and their configuration (collection of decade data) (source: Amazon.com). (*Continued*)

Drone	Configuration	Comments	Image
Aircraft — 3DR H520-G	H520-G hexacopter, E90 camera, controller-ST10C, battery capacity of 5250 mAh 4S/15.2 V (79.8 Wh)	Better durability, and compliance with flight time of 28 min, transmission of 2.4 GHz RC	
H111 NANO Q4	720P High-definition camera	2.4 GHz control, flight time of 5 min. rolls of 360 degrees. Can travel 50 meter distance	
EVO II dual 640T	Resolution of 640x512@30 FPS with thermal imaging sensor, 13mm lens	Covers the objects from 100 m distance.	

efficient AUV flights, without obstacles and potential accidents, along with finding the shortest distance to save battery life through optimization techniques. However, to travel smoothly, UAVs must overcome various obstacles, including emergency braking, battery life, and trespassing. This chapter discusses the taxonomy of unmanned aerial vehicles (UAVs).

There are a number of drones available in market with various configurations. Industries can choose drones that may be best suited to them based on their needs. Three types of drones are available in the market today. The first type has drones with fixed wings and are used to cover larger distances. These drones are used to photograph images while running or in operations. Note that they cannot take images while fixated in the air. Single-rotor drones are the second type. These are similar to helicopters, and the people who operate them need to be trained. These drones are powered by gas and they could do long-term activities. These drones can be fixated and can stay in the air for longer periods. They can also lift loads of up to 40 kg and can travel quite fast. Their image-capturing capacity is quite high when compared to the first type.

The third type falls under multi-rotor drones. These drones are built using the theory of vertical take-off and landing (VTOL) that requires less free or available area for take-off. They consume high energy to fixate themselves in the air. They are also not good load carriers. These drones are generally used for building construction, power lines inspection, etc.

8.2   Drone Hardware

A UAV (Unmanned Aerial Vehicle) can be operated with or without assistance of humans. For few types of UAVs the professional assistant is must which depends upon the UAV technology, its size and resources it is using. Generally UAV have camera, radar, sensors, and GPS tracking capacity inbuilt with it. They were categorized into different types of drones based upon their flying mechanism as discussed in the introduction section, or with its flying altitude. Figure 8.1 shows the categorization of UAVs.

Based upon landing property UAV's are divided into two types. HTOL requires runway to take-off and they generally use fixed wing property. VTOL types of drones can take off vertically with fixed wings and

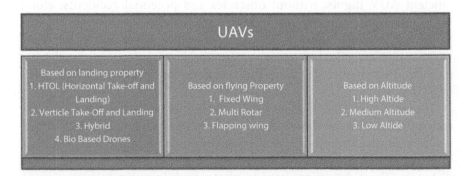

Figure 8.1 Types of UAVs.

multi-rotor wing combination. Like a normal helicopter they can take-off vertically which enables them to fly from smaller place without the need of runway. Few drones such as WingtraOne can carry loads also in it. Hybrid drones carry the mixed characteristics of HTOL and VTOL drones which uses fixed as well as multi rotor type wings for taking-off. Bio based drones contains flapping wings like birds.

Based on altitude (elevation of drone above the ground) is generally held at 2000 ft from the ground till 60,000 ft to the maximum. Certain navy drones can fly up to 65,000 ft and they can stay at that height for around 1 and half day. Medium altitude UAV can fly up to 30,000 ft and can stay there for a day. Low altitude drones will fly up to 400 m.

8.3   Components of UAV

Frame: Frame is the main component of UAV, which describes the material with which a UAV is made of. Usually the material will be either one of polystyrene or plastic or aluminum or fiber (made of carbon). Fixed wings are made of plastic, multi rotors are made of carbon fiber.

GPS Navigation: GPS navigation is another main component which allows the drones to switch to autopilot mode using GPS controls. Dual frequency constellation is used to share raw data to the other systems.

Flight Control: Flight control is core component which works along with GPS navigation to verify trajectory of UAV with respect to main UAV. It connects different sensors of UAV. Inertial measurement unit is installed on board along with micro electro mechanical systems, which is the component that is connected to the GPS to estimate the angle of drone navigation.

Power system is another element of UAV to provide fuel to the system. electric system is commonly used energy source for the drone. Along with it, Wakel rotary engine and fuel cells are also used to supply energy to the drone. Sometimes lithium polymer batteries also used to supply the power.

Sensors are another component which is composed of payload of camera and other sensors. The sensors may be fixed, stabilized or can be controlled remotely. Positioning sensors consists of GPS/GNSS receivers which are operative in the range of 1 km.

Range is regarded as one of the most essential factors in the evaluation of UAV performance. It is simple to compute using various parameters. Other UAV factors, particularly cargo weight, influence range. It determines the maximum distance a UAV may fly from its ground stations. Another key parameter is the UAV's stamina, which specifies how long a UAV typically fly before needing to be recharged or sometimes refueled. In other words, it specifies the amount of time a UAV may fly. Depending on the operation, UAV endurance might range from 1 to more than 36 hours. The volume of the fuel as well as battery load has an impact on the UAV's endurance and efficiency. The fuel's mass might range between 10 and around 50% of the UAV, which has a direct impact on the UAV's durability. As a result, the quantity of fuel should be proportional to the UAV's longevity [21]. The range and endurance characteristics are connected since the further a UAV would fly, the greater its coverage. As a result, many classes depending also on range and duration of UAVs were offered. The US military created a categorization system for unmanned aerial vehicles (UAVs) predicated on range and lifespan.

8.4   Literature Survey

There are plenty of applications on which drones can be used. This chapter gives the overview of few applications based upon state-of-the-art methods found on drones.

8.4.1   Applications of Drones in Aerial Systems

Many researchers are still inventing drones with various properties, many researchers also proposing various techniques to build drones to improve efficiency and performance. In this review, the author gathered the recent proposals of drones and their usages. Li *et al.* [1] proposed animal herding system using barking drones. These drones are similar to DJI Mavic 2 Zoom drones which uses loud speaker to record audio and also generates a dog barking sound audible from around hundreds of meters. Proposed system consists of two different drones, one is to detect and track the animals, second is to track them using sound. The Artificial Intelligence algorithm is used to detect the animals. According to the proposed system, the drones create a convex hull of polygon that surrounds the herd, where each edge of the polygon is in one-to one correspondence. Consider H_i as the position of hull vertex and H_iH_{i-1} and H_iH_{i+1} are two neighboring edges, I is the intersection of the parallel lines drawn against each edge at a distance of 'd' from the drone, then the angle of intersection (α) can be calculated as,

$$\alpha = \frac{\left(H_i\vec{H}_{i-1}\right)\left(H_i\vec{H}_{i+1}\right)}{\left|H_i\vec{H}_{i-1}\right|\left|H_i\vec{H}_{i+1}\right|}$$

With the angle of intersection the hull polygon is constructed to surround the circumstance of drone coverage. With the help of two laws, fly to edge and fly on edge of polygon the motion of drone is controlled. Fly to edge is the moving point inside the polygon which may change over time depending upon the sliding-mode control laws specified by Utkin *et al.* [2] proposed fly on edge guidance law and edge sliding guidance law, drone motion and brake system is controlled. Table 8.2 gives the survey of literature of types of drones and their working principles.

The other usage of drone is photography. Ng *et al.* [4] surveyed on multi-scale aerial and terrestrial photogrammetric image interpretation for archaeological site located at Argentinian Cordillera (Tacuil, Argentina). Two approaches were studied in this case. They are ground and UAV photogrammetry. Phantom 4 Pro UAV drone is used to study the area with the focal length of camera of 8.8mm and the image resolution of 4864x3648 pixels. 3 drones were used to cover the area about $0.6km^2$ from above 80 m. Using the drones 11 ground control points were identified successfully in the span of half day. Two *pukara* surfaces were detected with exact dimensional area in the surveying. About 70 unique structures were identified on the geometry. Sixteen mortars were detected. Few of them were found on stone and few are isolated. The measure of each structure was in the range of 4 to $6m^2$ as smallest and $20\text{-}25m^2$ as largest. The findings helped the archaeologists to explore the habitats of populations of pre-Hispanic era and their enemies, resources used physically and symbolically.

Alexey *et al.* [5] team worked on non-holonomic swarm UAV to travel in 3D space with the directions and altitude given. The system considered the swarm UAV proposed in Dubins vehicle model. Every drone moves with regular speed in 3D with limited magnitude. The moments are wheeled by yawing and pitching rates. The three units i.e., roll, yaw, and pitch are considered to build a vector. The roll constructs the movement in X-direction, pitch of UAV in Y-direction and yaw in Z-direction were employed to create a model. The roll motion is created using $r_i = v_i u_i$, where $u_i = B_i \in R^3$ with the upper bound R and magnitude $v_i > 0$. For each i of *essential neighbors*. Proposed inclinometer accesses the frames of UAV. To control the motion of targeted iso-surface when the plane is unsteady, the spatial gradient of the field is associated with location with an angle of α_h = arcsin $(N; h)$ where N is the unit vector of location r at time t.

Table 8.2 Applications of drones in aerial industry.

Author	Drone type	Working principle	Findings of drone
Li et al. [1]	Barking drone	Animal sounds were imitated to gather the herd, flock and group of animals	In helping the foresting without human intervention
Ng et al. [4]	Survey drone	Archaeological survey of valleys	Findings helped in checking the traces of pre-Hispanic era
Alexey et al. [5]	Swarm UAV	3D space survey	Drone travelling in 3D space when the directions are given.
Seongjoon Park et al.	Multi-robot SLAM	Feature maps	Generates the 3D feature map of constructed area.
Ding et al. [24]	Positioning Drones	Network positioning using Global Navigation Satellite Systems (GNSS)	Derives network based locations from Wi-Fi to investigate android network location engines.
Georgios et al. [8]	Flying Drones	Drones serving as network base stations	Drones help in creating base station to provide Wi-Fi facility to increase the availability of network services to the devices around the base station.
Yang et al. [25]	Fixed wing drones	Creating trajectory for the other drones	The drone's trajectory positions were trained with set of drones to create a better trajectory position for other drones in aerial system.

The technology of SLAM (Simultaneous Localization and Mapping) is used in military applications [6] to accurately localize the location from the map. The web camera is used to take the images and locate each feature of the image. The real time mapping using this will generate the 3D images slowly from each key frame. The criteria to process map data is time taking since the sessions of SLAM are slow. To overcome this limitation, Park et al. [7] used simultaneous localization and mapping in UAVs for sharing map data. Proposed system used distributed ledger technology for enabling map feature in drones. It collects global 3D data for navigation. It generates the DAGmap to continuously share 3D map data using distributed ledger technology (DLT) which manages the single point of failure problems, invalid transactions, and malicious transactions using sequential validation broadcasts. Hence the applications of such drones can be named as field reconnaissance, large scale mapping, and obstacle recognition. Due to DLT system and inherent features of DLT results in identifying the target location of the map. According to local ledger, the incoming transaction is validated and allowed to broadcast to perform ledger synchronization later. In local SLAM, DAG map is created using 3D feature validation to broadcast and share in 3D map. In multi-robot SLAM, the map based localization is performed followed by 3D feature extraction repeatedly until full 3D map reconstructed. Then map merging process is followed by constructing global map.

Amponis et al. [8] used drones to locate Android network using NLP processing. The drone generated altitude measurements with which the correlation of various visible access points was calculated to analyze the latency and updating rate. From the network provider the dimensions of longitude and latitude were collected for positional accuracy. From the access points different set of numbers were taken to process each scenario with Root Mean Square to get better accuracy. The horizontal and vertical accuracies were calculated in different regions such as rural, suburban, and urban to get the accuracy. With the help of Wi-Fi, urban scenario got the best horizontal accuracy and vertical accuracy was not promising. In rural scenario, where ever the cell locations are available the accuracy of system was good. In suburban region, the vertical accuracy was good compared to horizontal accuracy using NLP. The accuracies of 1637 m, 38 m, and 32 m were recorded in the three scenarios respectively. It comes around 68% of probability error.

Shahmoradi et al. [9] investigated on cellular enabled drones in the cellular networks using FANET (Flying Ad-hoc Networks). In the proposed system, drones serves in two purposes. One as consumers that takes the service of 5G Core technology to communicate with end-user and two as they serve as flying base stations to offer services to other drones. Though the smart technologies such as 5G infrastructure is the most mature orchestration technologies, there were certain drawbacks in this. It cannot perform slicing services and virtualization functions. Hence, drones can be used as base stations which are cost effective and can extend their connectivity to the next generation. It enables the connectivity of existing territories with cellular networks which can adjust dynamically with the altitude to establish LOS links and gNB instances. The author proposed the functioning of base stations in terms of extending the coverage and capacity of terrestrial network, support MANETs, and beam forming tasks. In high end technology build on will be aligned for 3GPP standards. This will reduce the complexity to increase more bandwidth of network signals.

Ulin *et al.* investigated on group of UAVs for hierarchical formation to act as trajectory planners to reference actual UAV. To improve the converging speed of error the sliding mode technique is integrated with an ANN (Artificial Neural Network). The author studied two-layer Neural Network with the trajectory reference $T_{i,p}^d$ where 'i' is ith UAV and d is reference of position, and p is the desired angle. The weight matrix w_i is calculated using, $\hat{W}_i = \hat{w}_{1\psi}^T (y_1, \vartheta_1)$.

8.4.2   Applications of Drones in Oil and Gas Industries

There are specific drones in the market (Source: www.globenewswire. com) in VTO1 UAV which can be used for surveying, surveillance, security and disaster recovery drones [11–18] etc. the minimum endurance required for such drones is 20–40 min which can cover 2–4 km of range. They start from 3.5 kg in weight up to 6 kg. Less weight drones can resist the wind up to 36 km/h and can launch with an altitude of 3000 m. Few Q-series type drones can carry the payload of 1280 x 720 pixels in the daylight and 15+ MP of HD payloads. Few have the thermal payload of 320 x 420. As discussed in the types of drones, these drones can return home on low-battery status and upon high wind status and if there is any redundancy in GPS tracking.

Other types of drones which can be used in oil industry are, Netra V series drones are portable drones and they are very small and fully autonomous. They will respond intelligently in worse situations. They endure for 40–60 min and covers 4–5 km range. They weigh up to 6 kg with size 1x1 meter in length and breadth. Most of their wings can be fixed wings. These types of drones are water resistant and can launch maximum in 3000 m. The image payload of such drones can be 720p with 10x zoom. HD payload is same as Q-series payload. The thermal payload varies as 640 x

480 pixels. Fall safe proof will be available for such drones. Like Q-series drones, these drones also return home upon high winds and on low battery status [19–22].

8.4.3   Applications of Drones in Military

Threat of terrorism is always unpredictable in the developing countries like India. It's always needed to counter terrorism and save nation from the attacks. Anti-terrorism tasks such as developing awareness, keeping surveillance and investigation missions are always needed for preserving safety. How efficient these strategies may be, it is always insufficient to relay upon. In such cases, drones are used as life-saving devices under anti-terrorism operations as they are more feasible and can generate accurate information gathering as expected. They provide advanced surveillance using AI (Artificial Intelligence) techniques and can make decisions using HCI (Human Computer Interaction). They are a game-changer in anti-terror operations because of their ability to execute flawlessly in difficult weather and terrain, as well as their swappable payload feature. During challenging hours, these drones assisted security personnel by delivering real-time intelligence.

One among such type of drones is Netra V Series drones. The NETRA V Series UAV is a small unmanned aerial vehicle that is fully autonomous and highly portable. The drone can be launched from small spaces and cover a large area that is out of human visual and auditory range. The NETRA V Series is IP53 certified for ingress protection and features target tracking with zoom-in capability. NETRA V Series can intelligently respond and adapt to unexpected situations thanks to multiple fail-safe features. There are other types of applications suggested in [22–25] through segmentation which can aid in the numerous applications of military and security fields. These techniques can aid drone technology in identifying the persons and capturing and identifying age of unknown persons.

8.4.4   Applications of Drones in Mines

Drone use in deep mines has been limited despite advances in drone technology. This is due to the difficulty of using drones in subsurface mines. Drones flying in harsh subsurface conditions have numerous challenges. Flying a drone in underground working locations is problematic due to confined space, poor sight, air velocity, dust concentration, and the lack of a wireless connection system. Furthermore, a drone operator's access to inaccessible and dangerous regions in deep mines is almost impossible.

Drones in underground mines could be used for a variety of health and safety purposes. Surface roughness mapping, rock mass stability analysis, ventilation modeling, hazardous gas detection, and leakage monitoring are some of these uses.

8.4.4.1   Underground Mine Geotechnical Characterization

In most underground openings, rock mass data gathering requires the inspector(s) to physically survey the rock mass. Personnel safety is jeopardized when they are present in unsupported regions such as open stopes and freshly blasted working faces. Drones are instruments that are better suited for monitoring unreachable locations in underground mines. Drones' small size and mobility allow them to reach hard-to-reach regions in underground mines without endangering miners' lives. Photogrammetry and FLIR (forward-looking infrared) imaging techniques can be used to characterize rock masses. Photogrammetry can offer data for geological models as well as structural data for kinematic and numerical analysis. FLIR images can also be utilized too.

8.4.4.2   Underground Mine Rock Size Distribution Analysis

Drilling and blasting are the most common methods of rock extraction in underground hard rock mines. The size distribution of the rock after blasting is a crucial measurement for future production. Phases (i.e., loading and hauling). There are a few methods for determining the size distribution of rocks. Including an expert's ocular observation, sieve analysis, and picture processing Analyzing images for rock fragmentation measurements, procedures are quick and relatively precise.

8.4.4.3   Underground Coal Mine Gas Detection

A system of sensors will be installed to continuously monitor air factors and gas concentrations. Enable the use of a drone to detect harmful gases in subterranean mines Lucila and Masami are a couple. In underground coal mines, an unmanned aerial vehicle was utilized to detect gas.

8.5   Analysis and Discussion

Ulin *et al.* referred UAVs as a technology that has recently been integrated into the distribution of products, which are packages in this study. It has the potential to improve distribution. System in areas where there is heavy traffic congestion Furthermore, UAVs can assist in Deliver small packages between warehouses by using them as an alternative distribution method. The invasion is of a new technology, in this case, the use of drones for a new delivery system. In order to improve a university distribution system, given that in recent years, Companies have concentrated on the use of logistics operations to boost productivity as well as delivery times. The investigation presents a mathematical model based on the Problem of the Traveling Salesperson (TSP).

Drones (UAVs) are a new technology that has recently been integrated into the distribution of goods, implementations (Johnsen *et al.*). It can improve the distribution mechanism in situations where there is a high volume of site visits. Further, using UAVs as a distribution possibility, UAVs could help to deliver tiny applications across warehouses (Johnsen *et al.* [13]). The encroachment is of a growing era, in this situation the use of unmanned aerial vehicles, for a newly designed travel machine that shall allow you to improve a college transfer machine, in view of the fact that in recent years, corporations have focused on the use of logistics services for the progress of performance employee and travel times.

The examiner offers a mathematical version, primarily based totally at the Problem of the Travelling Salesman (TSP) for the making plans of a direction to growth the performance within side the distribution method at Ciudad Universitaria, an Universidad Autónoma de Nuevo León (UANL) campus that contemplates the usage of the rising era of UAVs, and the conventional technique of the usage of trucks. Table 8.3 explains the various perspectives of drone's applications in recent years. The version considers regulations on the usage of drones, along with the difficulty of tour instances and most distance.

Khelifi *et al.* [14] worked on Archaeological sites are constantly being discovered and monitored. For many nations, these aims have long been essential national goals. Early detection of changes is critical for preventative conservation. Archaeologists have long pondered utilizing service drones to gather data on and just below the ground level of ancient sites, with economic and technological limitations standing in the way of widespread deployment. Thermography, depth image processing, drones, and machine intelligence advancements have reduced costs while improving the quality and scale of data gathered and analyzed. proposed work is final platform for detecting and monitoring archaeological sites utilizing independent service drones. It is accomplished by installing RGB, depth, and thermal cameras aboard an aerial vehicle for low-altitude data collection. The author presents two-stage multimodal intensity and thermal-to-RGB

Table 8.3 Analysis of drone applications.

Sl. no.	Author	Year of publication	Application	Comment on analysis
1	Ulin et al. [10]	2020	Distribution network using emerging technologies	Businesses have focused on increasing productivity through the utilization of supply chain.
2	Mourtzis et al. [11]	2021	UAV for industrial applications	Introducing an intelligent framework for real-time machine appropriate steps to prevent on the "Industrial Internet of Things" (IIoT).
3	Elghaish et al. [12]	2020	Digitalization in the construction industry	It helps builders, construction planners, designers, scholars, architects, and architects optimize construction operations for higher efficiency and performance.
4	Johnsen et al. [13]	2020	Safety and secure implementation of drones in oil and gas industries	UAVs for distribution can be employed for certain delivery tasks which need a high level of severity, regional inaccessibility, or a variety of hazards that people may experience when delivering products.

(Continued)

Table 8.3 Analysis of drone applications. (*Continued*)

Sl. no.	Author	Year of publication	Application	Comment on analysis
5	Khelifi *et al.* [14]	2021	Monitoring of archaeological sites	RGB, depth, and infrared cameras are mounted to an aerial vehicle for low-altitude data collecting.
6	Agudo *et al.* [15]	2018	Multispectral and thermal aerial images to enhance detection of archaeological cropmarks	Six drone detectors were studied and evaluated for their capacity to visually recognize excavated artifacts in archaeological communities.
7	Shahmoradi *et al.* [16]	2020	Applications in mining industry	3D modeling of the underground environment such as mining, ore management, mapping of rock discontinuities, measures of postblast rock fragmentation, and incurring losses stability monitoring.
8	Chung *et al.* [17]	2018	Purpose – Drone delivery	Heuristic method is implemented to solve the problem. Research design, data, and methodology.

(*Continued*)

Table 8.3 Analysis of drone applications. (*Continued*)

Sl. no.	Author	Year of publication	Application	Comment on analysis
9	Urbanová *et al.* [18]	2017	Lowered the cost associated with aerial imagery	Forensic practitioners now have quick, minimal amount exposure to aerial pictures taken in isolated areas.
10	Goodchild *et al.* [19]	2018	Carbon dioxide (CO2) emissions	Various ArcGIS technologies and pollution criteria were used within a context of distribution and logistics assumptions to discover that emission outcomes vary widely and are highly reliant on the drone's energy consumption.
11	Puttock *et al.* [20]	2015	Beaver-impacted ecosystems monitoring	At a site where otters have now been reintroduced, a small hexacopter equipped with a basic digital camera may be utilized to provide aerial photograph and digital model of the object (DSM) data products.

(*Continued*)

Table 8.3 Analysis of drone applications. (*Continued*)

Sl. no.	Author	Year of publication	Application	Comment on analysis
12	Merkert *et al.* [21]	2020	Deployment of airborne	Unrestricted drone use might cause issues for other airspace users such as airports and rescue services.
13	Stolaroff [22]	2018	Deliver commercial packages	More versatile when it comes to transporting packages because these vehicles do not require the use of roads, such as highways Drones can travel in a straight line with no obstacles in their route.

mosaicking strategies to align and combine gathered pictures. The detection method is used to the stitched pictures in order to recognize change regions and provide a user interface to track specific regions over time. The results were obtained by superimposing matching thermo and depth data on RGB mosaics of ancient sites. The mean square error (RMSE) of the detection method is 0.04.

Agudo *et al.* [15] highlights research conducted in the Roman Democratic city of La Caridad (Teruel, Spain), where several technologies were used to produce multispectral and thermal aerial pictures to aid in the detection of archaeological cropmarks. Two separate drone systems were used: a Tecnitop SA (Zaragoza, Spain) Mikrokopter and an eBee manufactured by SenseFly Corporation (Cheseaux-sur-Lausanne, Switzerland). As a result, the author has integrated in-house manufacture with commercial items in this study. Six drone sensors were tested and evaluated for their capacity to visually recognize excavated artifacts in archaeological communities. The sensors' spectral bands and spatial resolutions differ. This study explores photos recorded with several spectral range sensors to assess the technology's potential for archaeological benefits. The approach for comparing the tools was referred to as direct visual inspection, just like in traditional airborne archaeology. Our goal was to identify which drones and sensors produced the greatest results in the depiction of archaeological cropmarks by interpreting the resultant data. As a result, the experiment at La Caridad highlights the effectiveness of deploying drones with various sensors are used to monitor historic cropmarks for a much more cost-effective assessment, highest spatial resolution, and multitrack recording of buried archaeological sites.

Shahmoradi *et al.* [16] intends to give a thorough examination of the present status of drone innovations and advancements in the coal industry The mining industry is becoming more interested in using drones for normal operations. To mention a few, these technologies include 3D modeling of the mine site, ore handling, stone discontinuities tracking, post blast boulder fragmentation measures, and tailing stability tracking. The page provides an overview of drone types, specs, and uses for widely viable mining drones. Finally, the research requirements for the design and execution of subsurface mining drones are reviewed.

Beavers are sometimes referred to be ecological engineers because of their capacity to change the shape and flow of watersheds and build complex wetland habitats with dams, pools, and canals. As a result, beaver activity has ramifications for a wide variety of natural ecosystem services, including species, flood risk reduction, water quality, and long-term drinking water supply. With the present discussion about the readmission of badgers into the UK, it is vital to be able to measure the environmental effect of beavers. This work offers the first proof-of-concept findings demonstrating how a lightweight hexacopter equipped with a modest digital camera may be utilized to generate photogrammetric and digital surface model (DSM) data items at a reintroduced beaver site. Early results show that fine-scale (0.01 m) aerial photograph and DSM analysis may be utilized to determine influences on ecosystem structure, such as the area of dams and related ponds, and changes in plant structure caused by beaver tree-felling activities (Puttock *et al.*). UAV data collecting provides a powerful toolbox for regular repeat surveillance at fine granularity, which is essential for monitoring quickly changing and difficult-to-access beaver-impacted habitats.

According to Merkert *et al.* [21], commercial and private aerial drone deployment is revolutionizing numerous ecosystems. Our comprehensive literature review findings imply that historic problems such as privacy, acceptability, and security are rapidly being replaced by operational factors such as contact with and effects on other airspace users in order to identify significant challenges and research needs. Recent occurrences demonstrate that unfettered drone use might cause issues for other airspace users such as airlines and rescue services. A summary of current regulatory approaches reveals the need for additional policy and management responses to both maintain rapid and efficient drone usage growth and facilitate innovation (e.g., intraurban package delivery), with low altitude airspace management (LAAM) mechanisms for all drone use cases being one promising strategic response.

According to Stolaroff *et al.* [22], the use of autonomous, unmanned drones (drones) to transport commercial items is on the verge of becoming a new industry, dramatically changing energy usage in the transportation sector. The primary parameters the present feasible radius of multi-copters to still be roughly 4 km with current renewable technology, necessitating the establishment of a new infrastructure of urban stores or waystations as backup. It demonstrates that, while drones use less energy per package-kilometer than delivery vehicles, the additional storehouse necessary energy and the greater distances flown by drones each package significantly increase the life-cycle consequences. Nonetheless, in the majority of scenarios studied by Swathi *et al.* [23–27] in various military and border applications, the implications of delivery services by tiny drone are smaller than those of ground-based delivery. Drone-based delivery might cut greenhouse gases and energy usage in the freight industry, according to the findings. To reap the potential impacts of drone delivery, authorities and businesses should focus on reducing additional storage and regulating drone size. It was determined that there was a large disparity between a drones' route and a vehicle route; the prediction cost was much lower than the usual cost. The paths offered by each scenario indicate that a drone may arrive sooner than that of a truck and that it is more adaptable when it comes to carrying items as these vehicles must not demand the use of roadways such as highways. Unlike trucks, drones can follow the same path with no obstructions in their path.

Conclusion

Different types of drones are described, along with the functions of each essential component, in brief. Design and planning are the most common, followed by procurement, well before construction, until completion. Drones or unmanned aerial vehicles (UAVs) provide construction stakeholders with extensive, reliable, and exact geographical analysis. Land surveying, examination, progress tracking, worker deployment, waste recycling, map annotation of study areas and pictures, calculation of tangible types and volume replenishment of various warehouse items, and increased safety are all part of UAV capabilities.

Moreover, UAVs enable rapid, consistent, and cost-effective monitoring. In construction, this is especially important when it comes to examining the impact of beavers, as research and field observations have revealed the rapid rate of ecosystem change caused by dam and canal construction works. In particular, UAV surveying alleviated many of the challenges that come with ground-based observation in these surroundings, reducing habitat disturbance and personal safety risks inherent in physically accessing wetlands. The low-altitude overflights of drones also allow for the collection of high-resolution imagery.

References

1. Li, X., Huang, H., Savkin, A.V., Zhang, J., Robotic herding of farm animals using a network of barking aerial drones. *Drones*, 6, 2, 29, 2022.
2. Utkin, V., II, *Sliding Modes in Control and Optimization*, Springer Science & Business Media, Berlin, Germany, 2013.
3. Park, S. and Kim, H., DAGmap: Multi-drone SLAM via a DAG-based distributed ledger. *Drone*, 6, 2, 34, 2022.
4. Ng, Z.Y. and Phang, S.K., Development of simultaneous localization and mapping algorithm using optical sensor for multi-rotor UAV, in: *AIP Conference Proceedings*, vol. 2233, AIP Publishing LLC, p. 030007, 2020.
5. Matveev, A.S. and Semakova, A.A., Distributed 3D navigation of swarms of non-holonomic UAVs for coverage of unsteady environmental boundaries. *Drones*, 6, 2, 33, 2022.
6. Lee, D.-K., Nedelkov, F., Akos, D.M., Assessment of android network positioning as an alternative source of navigation for drone operations. *Drones*, 6, 2, 35, 2022.
7. Park, S., Kim, H.T., Lee, S., Joo, H., Kim, H., Survey on anti-drone systems: Components, designs, and challenges. *IEEE Access*, 9, 42635–42659, 2021.
8. Amponis, G., Lagkas, T., Zevgara, M., Katsikas, G., Xirofotos, T., Moscholios, I., Sarigiannidis, P., Drones in B5G/6G networks as flying base stations. *Drones*, 6, 2, 39, 2022.
9. Shahmoradi, J., Talebi, E., Roghanchi, P., Hassanalian, M., A comprehensive review of applications of drone technology in the mining industry. *Drones*, 4, 3, 34, 2020.
10. Ulin Hernandez, E.J., Martinez, J.A.S., Saucedo, J.A.M., Optimization of the distribution network using an emerging technology. *Appl. Sci.*, 10, 3, 857, 2020.
11. Mourtzis, D., Angelopoulos, J., Panopoulos, N., UAVs for industrial applications: Identifying challenges and opportunities from the implementation point of view. *Proc. Manuf.*, 55, 183–190, 2021.

12. Elghaish, F., Matarneh, S., Talebi, S., Kagioglou, M., Hosseini, M.R., Abrishami, S., Toward digitalization in the construction industry with immersive and drones technologies: A critical literature review. *Smart Sustain. Built Environ.*, 10, 3, 2020.

13. Johnsen, S.O., Bakken, T., Transeth, A.A., Holmstrøm, S., Merz, M., Grøtli, E., II, Jacobsen, S.R., Storvold, R., Safety and security of drones in the oil and gas industry, in: *Proceedings of the 30th European Safety and Reliability Conference and the 15th Probabilistic Safety Assessment and Management Conference*, ESREL2020-PSAM15 Organizers, Singapore, 2020.

14. Khelifi, A., Ciccone, G., Altaweel, M., Basmaji, T., Ghazal, M., Autonomous service drones for multimodal detection and monitoring of archaeological sites. *Appl. Sci.*, 11, 21, 10424, 2021.

15. Agudo, P.U., Pajas, J.A., Pérez-Cabello, F., Redón, J.V., Lebrón, B.E., The potential of drones and sensors to enhance detection of archaeological crop-marks: A comparative study between multi-spectral and thermal imagery. *Drones*, 2, 3, 29, 2018.

16. Shahmoradi, J., Talebi, E., Roghanchi, P., Hassanalian, M., A comprehensive review of applications of drone technology in the mining industry. *Drones*, 4, 3, 34, 2020.

17. Chung, J., Heuristic method for collaborative parcel delivery with drone. *J. Distrib. Sci.*, 16, 2, 19–24, 2018.

18. Urbanová, P., Jurda, M., Vojtíšek, T., Krajsa, J., Using drone-mounted cameras for on-site body documentation: 3D mapping and active survey. *Forensic Sci. Int.*, 281, 52–62, 2017.

19. Goodchild, A. and Toy, J., Delivery by drone: An evaluation of unmanned aerial vehicle technology in reducing CO2 emissions in the delivery service industry. *Transp. Res. D Transp. Environ.*, 61, 58–67, 2018.

20. Puttock, A.K., Cunliffe, A.M., Anderson, K., Brazier, R.E., Aerial photography collected with a multirotor drone reveals impact of Eurasian beaver reintroduction on ecosystem structure. *J. Unmanned Veh. Syst.*, 3, 3, 123–130, 2015.

21. Merkert, R. and Bushell, J., Managing the drone revolution: A systematic literature review into the current use of airborne drones and future strategic directions for their effective control. *J. Air Transp. Manage.*, 89, 101929, 2020.

22. Stolaroff, J.K., Samaras, C., O'Neill, E.R., Lubers, A., Mitchell, A.S., Ceperley, D., Energy use and life cycle greenhouse gas emissions of drones for commercial package delivery. *Nat. Commun.*, 9, 1, 1–13, 2018.

23. Gowroju, S. and Kumar, S., Robust deep learning technique: U-net architecture for pupil segmentation, in: *2020 11th IEEE Annual Information Technology, Electronics and Mobile Communication Conference (IEMCON)*, IEEE, pp. 0609–0613, 2020.

24. Ruijin, D., Gao, F., and Shen, X. S., 3D UAV trajectory design and frequency band allocation for energy-efficient and fair communication:

A deep reinforcement learning approach. *IEEE Transactions on Wireless Communications*, 19, 12, 7796–7809, 2020.

25. Kang, Y., Yang, G. Y., and Huang Fu, S. I., Research of control system for plant protection UAV based on Pixhawk. *Procedia Computer Science*, 166, 371–375, 2020.

26. Swathi, A., and Sandeep Kumar. A smart application to detect pupil for small dataset with low illumination, *Innov. Syst. Softw. Eng.*, 17, 1, 29–43, 2021.

27. Swathi, A. and Kumar, S., Review on pupil segmentation using CNN-region of interest, in: *Intelligent Communication and Automation Systems*, pp. 157–168, CRC Press, Florida, USA, 2021.

28. Gowroju, S. and Kumar, S., Robust pupil segmentation using UNET and morphological image processing, in: *2021 International Mobile, Intelligent, and Ubiquitous Computing Conference (MIUCC)*, IEEE, pp. 105–109, 2021.

29. Rani, S., Gowroju, S., Kumar, S., IRIS based recognition and spoofing attacks: A review. *2021 10th International Conference on System Modeling & Advancement in Research Trends (SMART)*, pp. 2–6, 2021.

9

Drones Enable IoT Applications for Smart Cities

R. Santosh Kumar[1]*, LNC Prakash K.[2] and G. Suryanarayana[3]

[1]Department of Computer Science & Engineering, Vasavi College of Engineering, Hyderabad, India
[2]Department of Computer Science and Engineering, CVR College of Engineering, Hyderabad, India
[3]Department of Computer Science and Engineering, Vardhaman College of Engineering, Hyderabad, India

Abstract

Around the world, the concept of a smart city is becoming an increasingly important research area. The amount of data collected and the number of mounted sensors, the surveillance equipment, and other gadgets that must be installed in a smart city are so large that using a mobile platform to monitor and keep replacing them can save effort and materials. Currently, people are experiencing rapid evolution in every field, owing to continuing developments in communication technology and other advances such as fully independent autonomous drones. These developments have already led to an improvement in people's lives, such as in wellbeing or quality of life, in the energy sector, in the shipping industry, in the supervision and surveillance guidelines for large households and commercial projects. As such, this chapter discusses the assimilation of the most widely used communication technology trend, the Internet of things (IoT), with drones or unmanned aerial vehicles (UAVs). The uses of IoT in UAVs to examine numerous buildings in smart cities are evaluated and the use of such IoT-enabled self-governing aerial vehicles is reinforced for guaranteeing smart city applications and safety measures. The chapter also describes the important drawbacks and limitations of conventional technologies for the same reason, such as optimization methods in trajectory tracking, portable artificial intelligence (AI) and machine vision methodologies, collaboration in IoT correspondence, and network infrastructure

*Corresponding author: santosh@staff.vce.ac.in

Sachi Nandan Mohanty, J.V.R. Ravindra, G. Surya Narayana, Chinmaya Ranjan Pattnaik and Y. Mohamed Sirajudeen (eds.) Drone Technology: Future Trends and Practical Applications, (207–242) © 2023 Scrivener Publishing LLC

scalability. Here, we suggest an open-source smart IoT technology that would enable the deployment of a variety of sensible or smart city services, including smart homes, smart unmanned drones, smart energy grids, smart transportation networks, and other prospective services that could be developed in response to market demands. We hope that readers will find the discussions on multiple open research issues relevant and helpful.

Keywords: Unmanned drones, IoT-enabled, artificial intelligence, network infrastructure

9.1 Introduction to Smart Cities

From the last several years the progress of hardware and software design has resulted in accelerated expansion of information and communication technologies (ICTs). The increased effectiveness of municipal operations due to ICTs; the cities have been named with several titles like, "wired city", "digital city", "electronic city", "information security". Smart city is the broadest term used among all the labels. In general, a city is considered as smart city where the traditional networks and services can become more adaptable, efficient, and long lasting without effecting much the environment. The important goal of the smart city is to make smart use of telecommunication technologies for the advantage of the people who live there. In a smart city, digital technologies translate into advanced public operations for residents and businesses in a more efficient use of resources with less environment impact. Some definitions of the smart city are as follows: "A smart sustainable city is an innovative city that uses ICTs to improve life quality, urban operations while also meeting the economic, social, and environmental needs of current and future generation". Another definition is the following: "To harness the city's collective knowledge, connect the physical infrastructure, information technology infrastructure, social infrastructure, and commercial infrastructure. All or any mixture of components can be capable of designing the city as smart. Based on the city's available cost and technology some components are chosen to design the smart city.

The smartness of the city is measured in terms the quality of life, safety of the public, pollution control, traffic monitoring, disaster management and the services [1]. The primary goal of the smart city design is to compliance with the quality infrastructure and costs of the services at reduced cost. Importantly, any smart city design should consider the future aspects of the life and trends in the information technology trends [2].

After the global financial crisis, the interest of the smart city design is crucial for the quality life of at the reduced cost. By 2050, the population is expected to be doubled and the design of the smart city faces the even more difficulty. So the interest is mainly resided as the well integration of communication technologies and related objects. The communication technologies like artificial intelligence, Wireless sensor networks, and robots solved the some of the problems in the smart city design. Artificial Intelligence helps the some of the smart city design problems like waste management, water management, traffic congestion, reduced pollution etc. Wireless sensors can be used in natural calamities, traffic flow monitoring, and measuring and detecting floods. The robots can help the smart design and reduced the operational costs. The need of smart components to design smart city and challenges are addressed in [3, 7, 9, 10, 15, 17, 18, 20, 21, 23, 28, 29, 35, 37, 40–42, 53, 56, 67, 76, 80, 84–87, 90–93, 108].

Standards address the climate change, traffic monitoring, security concerns, and quality of water services. Standards also take the concern of environmental impact by monitoring resource management and smart city's performance. IEEE set the smart city standards for many components such as smart grids, IoT, e-Health, and intelligent transportation systems (ITS). ISO 37120, for example, is a standard that defines 100 city quality measures. Among the 100 indicators, 46 are core measures and 54 are supporting measures. This new standard provides basic information for smart city designers to plan given the advanced technologies.

The main contribution of this paper is to provide the Drone enables IoT techniques for different components in the smart city design. In section 9.2, we introduce the characteristics of the smart cities. In section 9.3, we broadly highlight the role of IoT in smart cities. In section 9.4, we discuss the general approach for IoT enabled process in smart cities. In section 9.5, we present the weaknesses of the IoTs in smart city design. In section 9.6, we introduce the unmanned aerial devices and its types. In section 9.7, we highlight the opportunities and challenges of drones in smart city design. In section 9.8, we outline the drone enabled IoT techniques for improving the quality in the smart city design. In section 9.9, we present the conclusion and future scope.

9.2 Components and Characteristics of Smart Cities

Cities are concentrating on becoming smart cities in order to deal with rising population, hyper-urbanization, and globalization, as well as to assure economic and environmental stability. By utilizing technologies and connected data sensors, the idea of improving and becoming more powerful in terms of infrastructure and city operations is known as a "smart city". Smart city can be viewed in several ways [11–14, 19, 22] and in general the components considered for smart city are the following [4–6, 47, 52].

9.2.1   Smart Healthcare

The healthcare industry is the most often discussed smart city application. The need of smart healthcare and frame work is discussed in [33, 34, 36]. Wearable smart devices are the way of the future. Patients can now communicate with doctors or physicians via visits, teleconferences, and text messages. Doctors had no way of regularly monitoring their patients' health and making accurate suggestions. Remote monitoring will be possible thanks to smart wearable devices. Patients' data will be constantly monitored, which will aid in keeping them safe and healthy while also empowering doctors to provide superior treatment. Hospital visits and stays are reduced with remote monitoring. The cost of healthcare will also be greatly decreased. The implementation will benefit patients, clinicians, hospitals, and insurance companies [120]. The components in smart healthcare depicted in Figure 9.1.

9.2.2   Smart Transportation

Transportation is a major issue in larger cities. Every day, everyone requires some form of transportation. Everyone wants to be on time at their destination. The solution would be smart transportation. Smart transportation includes cameras, sensors, smart traffic signals, smart toll booths, smart automobiles, and GPS devices. For data processing, all of these devices communicate with each other and with the gateway. Communication between vehicles is also feasible. It will assist in diverting traffic away from crowded highways while also assisting in avoiding roads where accidents have happened. Vehicles that run red signals or do not obey road safety

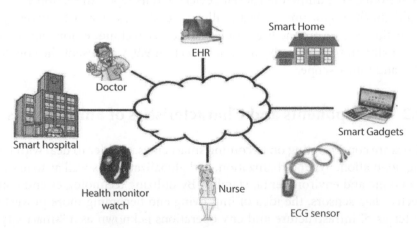

Figure 9.1 Components in smart healthcare.

Figure 9.2 Need of smart transportation at various locations.

standards will be captured by smart cameras. Smart sensors will also assess the health and fitness of the car and issue a report before it breaks down. Smart sensors in public transportation will provide accurate position, anticipated arrival time, and the amount of unoccupied seats. All of these devices will contribute to increased efficiency, reduced pollution, wasted time, and road mishaps. Transportation is a major issue in larger cities. Everyone requires some form of transportation. As the population of cities grows, so does the amount of traffic and automobiles on the road. As a result of this growth, a parking system is required, one that can save time, fuel, labor, and, most crucially, space. Sensor modules and low-power lights are strategically placed throughout the environment. If open space is available, these devices send data about it. When a vehicle enters a parking lot, the software looks for an empty spot and the lights switch on, pointing the driver to it. The motorist can then park the car and go. A device would be installed to gather vehicle information, including the time of entry, which would be necessary for invoicing purposes. The locations where the smart transportation is needed is depicted in Figure 9.2. The model would undoubtedly save the cyclist time and fuel [121].

9.2.3 Smart Pollution Monitoring System

Pollution in the air, water, and soil is increasing as a result of industrialization and automobiles. Industrial chimneys, car exhausts, highways, canals, and other smart sensors sense and communicate data to the gateway for further processing. The sensors in the chimneys would detect the type of vapors and, if any, would alert dispensaries. If the trash does not meet industrial standards, the appropriate actions can be performed. If the vehicles emit more exhaust waste, the owners should be warned, and if they do not rectify the problem, they should be penalized. Sensors installed along highways would continuously check air quality, and required actions, such as the planting of more green areas near polluting areas, would be done. Sensors would offer sufficient data. Sensors would offer sufficient information to aid in the reduction of soil and water pollution [122].

9.2.4   Smart Infrastructure and Building

A city's physical components, such as its buildings, roads, pathways, parks, temples, and bridges, that make it possible for the city and its citizens to function are referred to as its infrastructure. However, any physical, electrical, or digital infrastructure that forms the foundation of a smart city can be referred to as infrastructure in this context. The backend of the smart infrastructure is the ICT infrastructure that "smartens" up the physical infrastructure. ICT infrastructure is critical to the development of smart cities and is dependent on aspects such as availability and performance. Physical infrastructure, sensors, firmware, software, and middleware are all possible components of smart infrastructure. The "middleware," which is a sort of software, is usually critical in the automation and quick reaction of smart infrastructure.

Middleware collects data and brings it together in one place for critical findings. The smart infrastructure information is made available rapidly and can be obtained by operation personnel and management to have a quick influence on smart city operations, thanks to middleware and ICT. A smart power grid, or smart grid as it is more commonly known, is an example of smart infrastructure.

9.2.5   Smart Building

Different hardware, software, and sensors can be found in a smart building for various automated processes such as data network, power optimization operations etc. Green buildings are not the same as smart buildings. Green buildings are energy-efficient, water-efficient, and interior environmental control structures with the aim of minimizing carbon emissions and providing optimal energy performance. Smart buildings can communicate with other smart buildings, people and technology, the global environment, and smart power grids with ease. Smart buildings make good use of the information that exists beyond their accessible area. The Internet of Things (IoT) delivers integrated systems that can collect and analyze enormous volumes of data, allowing smart buildings to operate more efficiently and save energy. The technologies needed for smart city design is depicted in Figure 9.3.

9.3   The Role of IoT in Smart Cities

The role of IoT in smart design provides wide variety of advantages in smart city design ranging from simple to complex tasks. The impact of IoT and its methodologies addressed in [112, 113]. Applications of IoT in smart city design are depicted in Figures 9.3 and 9.4.

9.3.1   Road Traffic

Smart cities ensure that residents go from one place to another place by in more secure manner. Municipalities use IoT development and smart traffic systems to do this. To assess the path of vehicles, smart traffic solutions use several types of sensors as well as GPS data. To control traffic flow, central traffic management platform receives real-time traffic flow information from road-surface sensors and closed-circuit video cameras. The platform analyses the data and warns platform users via desktop user apps about traffic congestion and traffic light problems. In addition, the city is building a network of smart controllers that will automatically modify traffic signals second by second, according to changing traffic circumstances in real time.

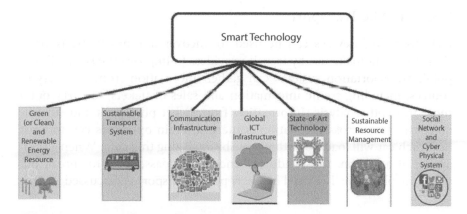

Figure 9.3 The technologies needed for smart city design.

Figure 9.4 Applications of IoT in smart city design.

9.3.2 Smart Parking

Smart parking systems assess whether parking places are occupied or available using GPS data and build a real-time parking map. Drivers receive a message when the nearest available parking spot is obtained, and instead of driving around aimlessly, they take help from the map to find the available parking spot.

9.3.3 Public Transport

Data from IoT devices can be used to uncover findings in transportation. This data make use by transit providers to improve the safety. Smart public transportation systems can gather information from a variety of sources, including traffic information and sales of tickets, to carry out a more in-depth analysis. Information from ticket purchases and surveillance cameras across the station is merged. Train operators can forecast how each car will load up with people by analyzing this data. When a train arrives at a station, the conductor encourages passengers to spread out along the platform. The need of smart pubic transport is discussed [16].

9.3.4 Utilities

By having more control over their domestic utilities, residents of IoT-enabled smart cities can reduce their utility costs. Different methods to smart utilities are enabled by IoT.

9.3.4.1 Billing and Smart Meters

Using a system of smart meters, authorities help consumers with cost-effective connectivity to IT businesses systems. Smart linked meters can now transfer data can send to a consumer location over a telecommunications network, delivering accurate meter readings. Utility companies can bill correctly for the quantity of energy, gas, and water used by each residence thanks to smart metering.

9.3.4.2 Disclosing Consumption Habits

Utilities firms can acquire a better understanding of customer usage of electricity and water by using a network of smart meters. Utilities businesses can analyze demand in real time via smart meter connectivity.

9.3.4.3 Remote Surveillance

Utility management services can also be provided via IoT smart city technologies. Citizens can use their smart meters to monitor and control their usage from afar with these services. A homeowner, for example, can use their phone to switch off their home's central cooling. Also, utilities firms can notify homeowners and dispatch personnel to resolve the issue.

9.3.5 Waste Management

Most garbage collection companies have predetermined times for emptying containers. This is inefficient because it results in inefficient trash container utilization and wasteful fuel use by waste collection trucks. Smart city systems powered by the Internet of Things can assist optimize waste collection schedules by measuring waste levels. Each garbage container is equipped with a sensor that collects information about the waste level in the container. The waste management system analyzes the records and notifies to truck driver's mobile app when it gets close to a specified threshold [111].

9.3.6 Environment

Smart city solutions powered by the Internet of Things can track parameters that are important for a good environment. A city can construct a network of sensors and connect them to a cloud system to monitor water quality, pH, dissolved oxygen, and dissolved ions are all measured using sensors. If there is a leak and the chemical mixture of the water changes, the cloud system generates an output that the users specify. Another application is air quality monitoring. A network of sensors is installed along major roadways and surrounding factories to accomplish this. Sensors collect information on CO, nitrogen, and sulphur oxide levels, while a central system analyses and use the data to identify places where air quality is a problem and make recommendations to citizens.

9.3.7 Public Safety

IoT-based smart city solutions provide analytics, decision-making capabilities, and real-time monitoring for improving public safety. By merging data from sound sensors and CCTV cameras located at different places in the city and with social media feeds and analyzing it, public safety systems can identify possible crimes. In turn it is enable the police to apprehend or track down suspected criminals. A gunshot detection solution, for example, is used in over 90 communities across the United States. A citywide network of connected microphones is used in the solution. The information from the microphones is sent to a central platform, which analyses the statistics of gunshot. By measuring the time, the platform may infer where the gun is. When a gunshot is detected and its location is determined, cloud software sends a mobile app notice to the police.

9.3.8 Security and Privacy for Smart Cities

Despite the fact that the aforementioned innovations in smart cities have contributed significantly to the overall benefit of society, practically every smart application is subject to hacking via current techniques, such as background knowledge attacks eavesdropping attacks, spam attacks, likability attacks, inside curious attacks, outside forgery attacks, and identity attacks, knowledge attacks etc....

9.4 General Approach to Implement IoT Solutions in Smart City Design

Smart city applications include a wide spectrum of topics. The strategy to implementation is what they share in common. If a municipality wants to increase the breadth of smart city usage for the upcoming generation, new tools and technologies can be added to the existing architecture without having to rebuild it. The need of Green IoT is discussed in [25, 26, 46, 48]. In general, the majority of approaches follow the approach which consist the following steps [45, 66, 75, 79]:

Stage 1: A smart city platform based on IoT
The design of a basic model for a smart city should be the first step, as it will act as a springboard for future expansions and allow the addition of more services without compromising functional performance. The basic IoT solution for smart cities consists of four modules:

The Smart Things Network: A smart city uses sensors and actuators. Sensors' primary objective is to gather the data and transmit it to a central system. Actuators enable equipment to do tasks such as changing the lights etc.

Gateways: Every IoT system consists of a "tangible" component made up of IoT devices and network nodes as well as a cloud component. Field gateways streamline data collection and compression by sorting and cleaning data before transmitting it to the cloud. The cloud gateway of a smart city solution ensures secure data transfer. Every IoT system consists of a "tangible" component made up of IoT devices and network nodes as well as a cloud component. The cloud gateway of a smart city solution ensures secure and safe data transmission.

Data Lake: A data lake's role is to keep data. Data lakes store data in its basic form. The data is extracted and sent to the warehouse when it is needed for relevant insights.

Big Data Warehouse: A single data storage system is a huge data warehouse. It contains exclusively structured data, unlike data lakes. Data is extracted, converted, and loaded into the big data database once its worth has been determined. It also saves information about items, such as when sensors were started and the commands that control programs send to device actuators.

Stage 2: Basic analytics and monitoring
Using data analytics, it is able to measure the surroundings of devices and define rules for control programs to perform out a specified operation. By looking at data from moisture in the soil sensors set all over a smart park, cities can create rules for electronic valves to close or open based on the observed moisture content. Sensor data might be shown on a uniform platform dashboard, allowing users to view how another park zone is performing at any given time.

Stage 3: In-depth analysis
City administrations can use IoT-generated data for basic analytics to find patterns and hidden relationships in sensor data. The techniques such as machine learning (ML) and statistical techniques are used in data analytics. In order to find trends and construct models based on them, machine learning algorithms evaluate the available data stored in the warehouse. Control apps that transmit commands to IoT device actuators employ the models. Here's how it works in real life. A smart traffic light, has to display a specific signal for a set length of time, can modify signal timings. Machine Learning algorithms are applied to past sensor data to identify traffic trends and change signal timings, allowing for faster average vehicle speeds and fewer traffic jams.

Stage 4: Intelligent control
By transmitting commands to the actuators of smart city devices, control software ensures enhanced automation. Essentially, they "inform" actuators what to do in order to complete a task. Control applications might be rule-based or machine learning-based. Manually specified rules are used in rule-based control platforms, whereas models incorporated by machine learning algorithms are used in ML-based applications. These models are analyzed by data analysis, and they are evaluated, authorized, and updated on a regular basis.

Stage 5: Interacting with residents in real time via user applications
Along with an option of automatic control, consumers have the ability to affect the behavior of smart city apps. User apps are in charge of this task. Citizens can connect to the central smart city administration platform through user applications to monitor and operate IoT devices. A smart traffic management solution, for example, can detect a traffic jam using GPS data from users' smartphones. To avoid even more congestion, the system sends an automatic notification to nearby drivers, asking them to choose a different route. Employees that utilize a desktop user app at a

traffic control center receive a 'congestion alert' at the same time. They send a notice to the traffic light actuators to modify the signals in order to relieve congestion and reroute some traffic.

Stage 6: Combining multiple solutions
Developing "smartness" is an ongoing process, not a one-time event. Municipalities should consider what services they might want to implement in the future as they develop IoT-based smart city solutions now. Let's use the example of a smart city traffic monitoring technology to demonstrate this functional scalability. A city implements a traffic management tool to identify traffic jams and adjust traffic signals to alleviate congestion in congested regions. After some time, the city decides to link with traffic management technology with a smart air quality monitoring solution to ensure that city traffic does not hurt the environment. Controlling both is possible with cross-solution integration. Cross-solution integration enables for dynamic traffic and air quality control in the city.

9.5   Challenges in IoT Solutions to Use in Smart City Design

Even though some of the problems are solved by IoT approach but still it has some problems to implement in smart city design [38, 39, 68, 82, 83, 109–111].

i)   Compatibility issues [94, 95]: The different vendors follow their own standards to design IoT devices. Communicating and exchanging the data between these IoT devices is a concern in the smart city design.

ii) Poorly designed IoT devices [96, 97]: Poorly planned and implemented smart city systems may have a significant impact on network resource utilization and overall smart city operations.

iii) Interoperability [32, 98, 99]: Limited communication between different transport protocols such as Ethernet, Wi-Fi, and Zigbee between devices made by different manufacturers. There are no clear norms or standards at the application level, making it impossible to combine and supplement data from various sensors and devices.

iv) Lack of Standards [100]: Improper regulatory requirements make it difficult to structure and manage enormous
amounts of unstructured data.

v) Data security and privacy concern [101–103]: IoT devices
handle data containing confidential data about citizens' behavior. People can be profiled as a result of a less
strengthen security protocol or a data leak.

vi) Difficult networking plan implementation [104, 105]: The
large number of devices connected to networks places a
substantial load on it, and network implementation is the
main difficulty in this area. The energy efficient IoT architectures discussed in [49–51, 54–57, 61, 65].

vii) Lack of Internet skill [106, 107]: Internet skills are important for IoT use IoT organizations confront a talent shortage
when it comes to planning, implementing, and maintaining IoT systems on the market.

viii) Sharing more than just open data [117]: The first wave of
smart city services was built mostly on the basis of open
data. If the data isn't regarded sensitive, it's usually shared,
and the success of services built on top is frequently small.
Private datasets, such as IoT-generated data or personal
data of residents, have the ability to provide more value for
smart city operations, but they are not publicly available.

ix) Encouraging more agile policy making [118]: Because
licensing models aren't well understood or developed,
there aren't enough incentives, market confidence, or trust
for organizations and individuals to share fresh datasets.
A crucial ecosystem underpinning for such a market place
has yet to be established. Private datasets, such as IoT-
generated data, closed company data, or citizen personal
data, have the potential to provide more value for richer
smart city services, but they are not yet available. Future-
proof market mechanisms that would extend open data's
commercial viability beyond existing licensing models are
also lacking.

x) Acquiring data from low level access areas [119]: In disaster management, acquiring the data through IoT devices is
difficult because the strength of the signal and low battery
life.

xi) Security and privacy issues with IoT [58, 114–116]: The
vulnerabilities that exist in each layer of a smart system,

the design of these smart applications may bring significant security and privacy issues. Unauthorized access, Sybil, and denial of service (DoS) attacks can all lower the quality of intelligent services.

9.6 Introduction to Unmanned Aerial Vehicles

Aerial vehicles (UAVs) or drone can be visualized as an aircraft with no pilot on board. Initially their invention is used in military applications, but later on they have used in the significant areas to improve the design the smartness of the city. They can be integrated as a part of smart city design and increase the quality of life by incorporating them to find solutions of the problems like package delivery, supervising construction works, monitoring the pollution, monitoring the traffic control, and in disaster management. Drones have become more affordable because to advancements in sensor, data processing, and rechargeable battery technology. Drones can be serving as aerial base stations (BS) for subscribers on the ground, delivering communication operations. Drones' agility capabilities have made them indispensable in the Internet of Things (IoT) framework. Drones can also serve as aerial base stations (BS) for subscribers on the ground, delivering communication services (both uplink and downlink). Drones' agility and line-of-sight (LoS) capabilities have made them indispensable in the Internet of Things (IoT) framework. Smart houses, smart streets, smart parking, smart power grids, and other smart city applications have all benefited from the Internet of Things. The main premise behind IoT is that everything, including smart phones, buildings, home appliances, automobiles, and even natural objects, may connect and communicate through the Internet, resulting in a smart world. The creation of those applications is vital to our way of life, economics, and ecology.

Green technologies, in addition to economic advancement, support the entire development of geographical objects by protecting the environment, conserving natural resources, mitigating the effects of climate change, and reducing pollution and electricity usage. For modern smart city infrastructure, a large number of IoT devices that can cooperate, communicate, and share information is required. Because a high number of networked devices are necessary, operating cost minimization and power conservation solutions are becoming increasingly important. Load balancing is also important for extending the life of the network by minimizing the amount of energy consumed by smart devices.

Drones (Aerial Unmanned vehicles) embedded with IoTs solves some the problems listed above. Unmanned flying aircraft that can be controlled by a human pilot are known as drones. This is no longer the case, thanks to advancements in drone technology. Self-flying drones, as well as autonomous and AI-powered underwater UAVs, are now referred to as self-flying drones. Drones are referred to in a variety of ways. Remotely piloted aircraft, remotely operated aircraft (ROA), remote-operated aircraft system (UAS), unmanned aeroplane (UA), and unmanned aviation (UA) are some of the various terms used to describe aircraft that do not have a flight crew. There are several types of drones available: i) Single-rotor drone, ii) Multi-rotor drone, iii) Fixed-with drone, iv) Fixed-wing drone, v) Drones for small areas, vi) Tactical drones, vii) Micro drone, viii) Reconnaissance drone, ix) Large Combat drone, x) Non-Combat drone, xi) Target and decoy, xii) Photography, xiii) GPS Drones, and xiv) Racing drone. The comprehensive survey of UAV in communications can be found in [154], and UAV with automatic battery replacement is discussed in [153].

9.7   Opportunities and Challenges of UAV's in Smart Cities

UAV's can be used to improve the design of smart city and adapting UAV's with other technologies will definitely benefit to the smart city design. Here we discuss the areas where UAV's can be incorporated to better service. The challenges are addressed in [69–73, 81, 88, 89].

    i)   Geo-Spatial and surveying activities: Survey technology is the main improvement with the UAV's over the traditional surveying methods. With the GPS technology and high resolution camera the UAV's referred as unmanned aerial systems (UAS). With the advanced components of software and hardware drones add significant improvement for their ability to provide 3D site modeling, aerial photogrammetry, aerial video, and photo comparisons.

    ii)   Civil security control: Drones can be used as a vehicle which can improve the collaboration between construction workers, improve the safeties of site, employ the site measurements of broad areas, and produce structures, using principles of aerial photogrammetry. Establishing images from building components, overviews the challenges with the obtained images [30].

    iii)   Traffic management: Drones transmits the real-time video to a central station to assist the employees for the road observance, traffic guidance, statistical data of traffic activity, monitoring license number plates. Drones can give on-the-ground situational awareness and collect evidence in the event of crises such as road accidents or oil leaks. Drone data can be evaluated to improve traffic flow and safety on the road.

iv) Crowd management: Drones provide a wider-field perspective of large crowds and provide real-time data to crowd control teams in crowd monitoring. Drones can zoom in on a specific region of interest, giving the team valuable information about what's going on the ground. Such data can assist ground people in making vital judgments. In the case of an emergency, the team's in-depth situational awareness allows them to respond quickly and efficiently. This increases the teams' effectiveness in crowd management.

v) Disaster management: Technological reinforcements in disaster management and control go beyond conventionalism, implying growth and advancement toward safety. While reiterating the possible protective measures, it is critical to comprehend the event and its causes. Public safety is one of the government's top priorities, and advances in disaster relief, assistance, and help the people. Critically, the significant improvement in catastrophe control brought about by technology improves the afflicted situations even more. Drones, quadcopters, and unmanned aerial vehicles (UAVs) are increasingly being used to help manage and maintain the public during such trying times. The importance of using drones to efficiently provide assistance to the poor and affected has dramatically improved disaster management.

vi) Environmental management: Environmental management options have vastly increased in recent years. Sensors for air and water contaminants, as well as subsets of the electromagnetic spectrum, have shrunk in size, cost, and are increasingly combined into comprehensive devices. Low-altitude unmanned aerial vehicles (micro-drones) have also increased aerial sensor platforms, although their usage in populated areas is becoming increasingly restricted for safety and privacy reasons. UAVs monitored for air pollution and environmental management are addressed in [62–64].

vii) Security: Drone surveillance arose from military engagement before shifting to the civilian sector with lightweight cameras and better-than-average images. In numerous situations within their territories, municipalities around the world realized the benefits of swift reaction and cost savings [31].

viii) Big data processing: Drone use will continue to accelerate due to rising demand for big data for commercial purposes, technological breakthroughs, and greater venture capital funding. Agriculture, real estate, construction, and traffic safety are just a few of the businesses that are looking into this information. Drones are valuable for the photographs and video footage they acquire, but they are also increasingly capable of storing other types of data, such as radio signals, soil moisture, manufacturing emissions, and geodetic data etc....

ix) Coordination between heterogeneous systems: Data gathered at different locations by different drones should communicate themselves and obtain the relevant data.

9.8   Drone-Enabled IoT

In this section, we outline the research work contributed towards the drone-enabled IoT techniques. Collaboration of UAV with IoT is discussed in [8].

9.8.1   Drone-Enabled IoT for Disaster Management

Monitoring the disasters and notifying the people in affected areas requires the efficient and reliable system. IoT can address the key problems which arise in disaster management. Unmanned aerial vehicles with the IoT can further improve the efficiency of system. In general any drone enabled IoT Should deal with efficiency of the IoT network, techniques for data collection and analytics, communications technologies. IoT devices have the capability of process, sense the data among the physical objects. IoT offers utmost connectivity that can be used for controlling, monitoring the users in disaster. Further the capability of IoTs in communications improved by using unmanned aerial vehicles. UAVs are recently enabled with IoTs to improve the connectivity to ground IoT network which will help the people where existing infrastructure no longer works. It will help by reducing the response time. During disasters, unmanned aerial vehicles

(UAVs) are used in search and rescue operations to discover survivors and provide assistance. In the absence of traditional network infrastructure, human worn IoT devices establish connectivity with UAV mounted base stations to send data. disaster-recovery messages A multi-UAV system is being considered for searching and rescuing persons who have been affected. UAV-based aerial catastrophe assessment delivers accurate and real-time situational evaluation. In disaster circumstances, a lack of situational awareness is investigated, and it is suggested that the use of UAVs in conjunction with a sensor network in the impacted area can offer accurate and fast information. The techniques which work on the disaster management can be found in [123–128]. Activity scheduling protocols for UAV are discussed in [59].

For effective disaster management, data regarding the amount and degree of damage caused by a disaster is essential. UAVs acquire video, photos, and sensor data from the impacted region, which is then processed and analyzed. UAVs are employed in to estimate the damage to a power network caused by a severe weather event. After an earthquake, UAVs are employed to acquire high-resolution films and photographs from inaccessible places to estimate damage. One can also structure-from-motion technique to create three-dimensional (3D) models based on the acquired data to assess the damage condition.

9.8.2   Drone-Enabled IoT for Public Safety

For search and rescue operations, a public safety network between first responders and victims is critical. A successful public safety communication network requires capable and quick emergency communication. Effective communication technology will save lives and promote connectivity during public safety communications. If a calamity strikes, the technologies like Wi-Fi, LTE will be ineffective. As a result, when terrestrial communication networks are destroyed, space technology is the finest alternative for disaster recovery.

Wearable internet of public key things (IoPST) can be used as first responders to share the information among the parties. Parties can be law enforcement agencies in the case of crimes, rescue teams in the case of disasters. Parties make correct and appropriate decisions to help the people. The authors of [129] looked examined how a drone deployed as a BS could supply communication services to a specific area. It was also considered how heterogeneous devices and drones may cohabit. Basic operations and approaches for improving were introduced. The authors of [127] looked at disaster management approaches and IoT availability. Reina et al. [128] discussed catastrophe preparedness and the value of IoT and big data.

9.8.3   Drone-Enabled IoT for Data Collection

The central aspect of the smart city design depends on the quality of the acquired data and the analytic techniques on that data. Data should collect, store and process efficiently to take the required decisions to improve the design of the smart city. The IoT devices can be placed in several areas to gather the data but these are having the limited capabilities. The low battery power and weak signal strength are main energy constraints to use in the smart cities [130–132].

Recently, drones show the other alternative in the data collection. They can be placed in different locations to acquire the data. Drones and IoTs together shows main advantages like better connectivity, cost efficiency, and high QoS. A drone can have the onboard material such as sensors, high end cameras, and communication technologies LTE, Wi-Fi, 5G. These services can be used to capture and deliver services effectively and efficiently.

Wireless communication technologies, such as 4G/5G networks improve the quality of data by incorporating sensors, GPS, and cameras to drones. In this case the energy consumption is minimized and also drones with this equipment can be used as base ground station for a longer duration. Drones with this equipment can be guides to rescue teams for safe routes.

Drones can also play an important role in IoT device communication by collecting information from devices like health monitoring equipment, environmental sensors, and etc. [135–137]. The authors of [138] looked at how drones and IoT devices could be used together in a critical scenario and included features like incident findings and network improvement. Drones with IoT devices may collect, store, and interpret data, allowing them to perform complicated tasks more quickly and efficiently. Since the energy is necessary to analyze IoT data and conduct IoT tasks in a timely manner, Koulali *et al.* [165] proposed that data processing be done locally first, and then the processed data be transferred to the cloud.

Drones with IoTs can be used in different ways in the smart cities: 1) Drone can be used to collect the data from different IoTs in the smart cities [133] 2) Drones with IoT equipment can be used to gather data from different IoTs [134] 3) Drones with IoT equipment can gather data.

The data which gathered from IoT enabled drones can be used for different applications such as pollution control, water quality, estimation of yield, crop monitoring, highway patrol. The different data analytics tools can be used to take necessary actions to improve the quality.

9.8.4   Drones and IoT for Improving Life Quality

Two major concerns in the smart city with respect to life quality are air quality and water quality. Sensors have been used to collect the data about air quality, but they are unable to send the data in real time. Drones equipped with IoT sensors can gather data and send the data to larger distances [139, 140]. Villa *et al.* [141] provided the optimal points for four different sensors. They implemented the drone which has NO_2, NO, CO, and CO_2 sensors for measuring the dangerous emissions. Hamilton [142] also designed the drone which is solar powered with a CO_2 sensing system and WSN.

The authors of [143] examined existing drone systems for environmental monitoring. Furthermore, the authors [144] offered drones for monitoring duties that were fitted with off-the-shelf sensors, but they did not address the guidance system. The same approach is used to solve this problem. The user might monitor a specific area and operate the system using a meta-heuristic and PSO approaches by concentrating on the most contaminated areas [145]. Another relevant example the authors of [146] proposed employing drone equipment.

9.8.5   Drone-Enabled IoT for Energy Efficiency

Standalone IoT devices consume more energy if data is sending to long places. If drones equipped with IoT device then drones can go physically in to the covering area of IoT devices. Mozaffari *et al.* [147] computed the altitude values for small drones which can provide the minimal transmit power and maximum coverage area. The WSN enabled drone was introduced in [148], which has fixed leaders, drones sink and sensor nodes. They argued that this method reduced the complexity of the leader selection and energy consumption. The efficient data collection with WSN enabled drones are discussed in [149]. These methods reduce the latency, data collection, and flying time. Drones and mobile agents with WSN have been proposed in [150] to reduce the energy of sensor nodes and time for gathering data. The mathematical model for the energy efficient management is discussed in [151]. Tracking algorithms for UAVs discussed in [43, 44]. Sharma *et al.* [152] introduced the WSN enabled drone which provides the following: 1) drone routing with less delay 2) large coverage area 3) better battery life. The above models do not take into account the speed of the drone and load on the drone. Load balancing WSN is discussed extensively in [74]. An automatic battery replacement mechanism is discussed in [155]. An advantage of automatic battery replacement can avoid the human intervention. Drones with small cells in the clouds is discussed in [60].

9.8.6　Privacy and Security Issues in Drone-Enabled IoT

From the standpoint of a cyber-physical system (CPS), Wang et al. [114] explored the security and privacy challenges of UAV networks. The authors looked at the major modules of UAVs which have the weakest points to apply the cyber-attacks from both the cyber and physical domains. A similar paper [154] discussed the security issues faced by UAV communication networks and identified key security needs. Shakhatreh et al. [155] looked at the civil uses of UAVs and the major obstacles they face. UAV cyber security issues were investigated by Krishna et al. [156]. The authors suggested a taxonomy to describe various forms of cyber-attacks involving unmanned aerial vehicles. Shafique et al. [157] have published a study on UAV security protocols and vulnerabilities. Syed et al. [158] reviewed the literature for new technologies to address the security and privacy issues in UAVs. The authors of [159] conducted a survey of commercial drone security, privacy, and safety. They discovered severe weaknesses, cyber and physical dangers, as well as potential attacks that could cause the drone to crash during a flight mission. Similarly, the authors of [160] looked into emerging cyber assaults and issues that commercial drones face. The writers of [24] looked into current dangers and harmful uses of drones in civilian applications. Nassi et al. [116] has published a systematic literature assessment of commercial drone security and privacy issues. Nassi et al. [116] has published a systematic literature assessment of commercial drone security and privacy issues. The researchers looked into the dangers of wireless communications in commercial UAVs, including as Wi-Fi-based UAV communications, in [161]. They also brought up the issue of privacy invasion caused by UAVs via aerial photography. The Security challenges of UAV in Defense addressed in [27].

The key security challenges of UAV-assisted cellular communications were examined by Fotouhi et al. [162]. According to Mishra et al. [163], the integration of UAVs with cellular networks such as 5G creates security challenges that the research community must thoroughly study. From a communication standpoint, Hayat et al. [164] examined the safety, security, and privacy aspects of UAV networks. The authors of [77, 78, 165] offered a thorough examination of the most recent UAV communication technology, as well as the need to safeguard the data captured and transferred to the Ground Control Station (GCS). Hassija et al. [166] conducted a survey that covered all of the important topics. A survey on the Internet of Drones was conducted by Boccadoro *et al.* [77, 78, 167]. They talked about the security and privacy concerns surrounding drone-drone communications, as well as existing solutions. In another paper, Noor *et al.* [168] taken into account the concerns of security and privacy the creation of drone networks One of the most significant challenges is ad hoc communication between numerous UAVs. Flying is the term for this form of communication.

9.9   Conclusion and Future Scope

The techniques for drone enables IoT recently improving by using the new technologies of IoT and drones with the advanced techniques of Machine learning and Artificial Intelligence. These advanced techniques mainly contributed towards the data analytics and security. Machine learning (ML) techniques aim to provide the prediction and inference of data in less amount of time which is gathered through the drones. Artificial Intelligence (AI) techniques provide the improvement of smartness in IoT and drone devices which can work independently and intelligently. Machine learning (ML) techniques also helps in providing the security by using adversarial set ups. Much research has to be done to provide the security and privacy for drone enables IoT with the help of advanced techniques in cryptography and ML. Also there are some issues with communication technology in IoTs and drones such as signal strength, time of data gathering, and time for data sending to the base station. Battery life time and independent battery replacement should be addresses in the future. Before deployed Drone enables IoT, the above issues should address in some extent.

References

1. Barrionuevo, J.M., Berrone, P., Ricart, J.E., Smart cities, sustainable progress. *IESE Insight*, 14, 14, 50–57, 2012.
2. Cocchia, A., Smart and digital city: A systematic literature review, in: *Smart City*, R. Dameri, (Ed.), pp. 13–43, Springer, Cham, 2014.
3. Javidroozi, V., Shah, H., Amini, A. *et al.*, Smart city as an integrated enterprise: A business process centric framework addressing challenges in systems integration, in: *Proceedings of 3rd International Conference on Smart Systems, Devices and Technologies*, Paris, IARIA, Wilmington, DE, pp. 55–59, July 20–24, 2014.
4. Lytras, M.D. and Visvizi, A., Who uses smart city services and what to make of it: Toward interdisciplinary smart cities research. *Sustainability*, 10, 6, 1998, 2018.

5. Ijaz, S., Shah, M.A., Khan, A., Ahmed, M., Smart cities: A survey on security concerns. *Int. J. Adv. Comput. Sci. Appl.*, 7, 2, 612–625, 2016.

6. Yin, C., Xiong, Z., Chen, H. *et al.*, A literature survey on smart cities. *Sci. China Inform. Sci.*, 58, 10, 1–18, 2015.

7. Mohanty, S.P., Choppali, U., Kougianos, E., Everything you wanted to know about smart cities. *IEEE Consum. Electron. Mag.*, 5, 3, 60–70, July 2016.

8. Alsamhi, S.H., Ma, O., Ansari, M.S., Survey on collaborative smart drones and internet of things for improving smartness of smart cities. *IEEE Access*, 7, 128125–128152, August 2019.

9. Coe, A., Paquet, G., Roy, J., E-governance and smart communities: A social learning challenge. *Soc. Sci. Comput. Rev.*, 19, 1, 80–93, 2001.

10. Martínez-Ballesté, A., Pérez-Martínez, P.A., Solanas, A., The pursuit of citizens' privacy: A privacy-aware smart city is possible. *IEEE Commun. Mag.*, 51, 6, 136–141, Jun. 2013.

11. Chourabi, H., Nam, T., Walker, S., Gil-Garcia, J.R., Mellouli, S., Nahon, K., Pardo, T.A., Scholl, H.J., Understanding smart cities: An integrative framework, in: *Proc. 45th Hawaii Int. Conf. Syst. Sci. (HICSS)*, pp. 2289–2297, 2012.

12. Lytras, M.D. and Visvizi, A., Who uses smart city services and what to make of it: Toward interdisciplinary smart cities research. *Sustainability*, 10, 6, 1998, 2018.

13. Bibri, S.E. and Krogstie, J., Smart sustainable cities of the future: An extensive interdisciplinary literature review. *Sustain. Cities Soc.*, 31, 183–212, May 2017.

14. Hollands, R.G., Will the real smart city please stand up? Intelligent, progressive or entrepreneurial? *City*, 12, 3, 303–320, 2008.

15. Jara, A.J., Genoud, D., Bocchi, Y., Big data in smart cities: From poisson to human dynamics, in: *Proceedings of the 28th International Conference on Advanced Information Networking and Applications Workshops (WAINA)*, pp. 785–790, 2014.

16. John, R.M., Francis, F., Neelankavil, J., Antony, A., Devassy, A., Jinesh, K.J., Smart public transport system, in: *2014 International Conference on Embedded Systems (ICES)*, pp. 166–170, 2014.

17. Guedes, A.L.A., Alvarenga, J.C., Goulart, M.D.S.S., Rodriguez, M.V.R.Y., Soares, C.A.P., Smart cities: The main drivers for increasing the intelligence of cities. *Sustainability*, 10, 9, 3121, 2018.

18. Cooley, R., Wolf, S., Borowczak, M., Secure and decentralized swarm behavior with autonomous agents for smart cities, *IEEE International Smart Cities Conference (ISC2)*, pp. 1–8, 2018. arXiv:1806.02496. [Online]. Available: https://arxiv.org/abs/1806.02496.

19. Maeda, A., Technology innovations for smart cities, in: *Proc. of Symposium on VLSI Circuits (VLSIC)*, pp. 6–9, 2012.

20. Zanella, A., Bui, N., Castellani, A., Vangelista, L., Zorzi, M., Internet of things for smart cities. *IEEE Internet Things J.*, 1, 1, 22–32, Feb 2014.

21. Harrison, C., Eckman, B., Hamilton, R., Hartswick, P., Kalagnanam, J., Paraszczak, J., Williams, P., Foundations for smarter cities. *IBM J. Res. Dev.*, 54, 4, 1–16, 2010.
22. Guedes, A.L.A., Alvarenga, J.C., Goulart, M.D.S.S., Rodriguez, M.R.Y.R., Soares, C.A.P., Smart cities: The main drivers for increasing the intelligence of cities. *Sustainability*, 10, 9, 3121, 2018.
23. Mohammed, F., Idries, A., Mohamed, N., Al-Jaroodi, J., Jawhar, I., UAVs for smart cities: Opportunities and challenges, in: *Proc. Int. Conf. Unmanned Aircr. Syst. (ICUAS)*, pp. 267–273, 014.
24. Gapchup, A., Wani, A., Wadghule, A., Jadhav, S., Emerging trends of green IoT for smart world. *Int. J. Innov. Res. Comput. Commun. Eng.*, 5, 2, 2139–2148, 2017.
25. Huang, J., Meng, Y., Gong, X., Liu, Y., Duan, Q., A novel deployment scheme for green internet of things. *IEEE Internet Things J.*, 1, 2, 196–205, Apr. 2014.
26. Rani, S., Talwar, R., Malhotra, J., Ahmed, S.H., Sarkar, M., Song, H., A novel scheme for an energy efficient Internet of Things based on wireless sensor networks. *Sensors*, 15, 11, 28603–28626, 2015.
27. Selvi, P.T., Sri, T.S., Rao, M.N. *et al.*, Toward efficient security-based authentication for the internet of drones in defence wireless communication. *Soft Comput.*, 26, 4905–4913, 2022.
28. Mardacany, E., Smart cities characteristics: importance of built environments components, in: *Proceedings of IET Conference on Future Intelligent Cities*, pp. 1–6, 2014.
29. Gharaibeh, A., Salahuddin, M.A., Hussini, S.J., Khreishah, A., Khalil, I., Guizani, M., Al-Fuqaha, A., Smart cities: A survey on data management, security, and enabling technologies. *IEEE Commun. Surv. Tutor.*, 19, 4, 2456–2501, 4th Quart., 2017.
30. Merwaday, A. and Guvenc, I., UAV assisted heterogeneous networks for public safety communications, in: *Proc. IEEE Wireless Commun. Netw. Conf. Workshops (WCNCW)*, pp. 329–334, Mar. 2015.
31. Hartmann, K. and Steup, C., The vulnerability of UAVs to cyber attacks—An approach to the risk assessment, in: *Proc. 5th Int. Conf. Cyber Conflict (CYCON)*, pp. 1–23, 2013.
32. Liu, A., Zhang, Q., Li, Z., Choi, Y.-J., Li, J., Komuro, N., A green and reliable communication modeling for industrial internet of things. *Comput. Elect. Eng.*, 58, 361–381, Feb. 2017.
33. Schaffers, H., Komninos, N., Pallot, M., Trousse, B., Nilsson, M., and Oliveira, A., Smart cities and the future internet: Towards cooperation frameworks for open innovation. *The Future Internet, Lect. Notes Comput. Sci.*, 6656, 431–446, 2011.
34. Acampora, G., Cook, D.J., Rashidi, P., Vasilakos, A.V., A survey on ambient intelligence in healthcare. *Proc. IEEE*, 101, 12, 2470–2494, 2013.
35. Chourabi, H., Nam, T., Walker, S., Gil-Garcia, J.R., Mellouli, S., Nahon, K., Pardo, T.A., Scholl, H.J., Understanding smart cities: An integrative

framework, in: *Proc. of the 45th Hawaii International Conference on System Science (HICSS)*, pp. 2289–2297, 2012.

36. Demirkan, H., A smart healthcare systems framework. *IT Prof.*, 15, 5, 38–45, 2013.

37. Trindade, E.P., Hinnig, M.P.F., da Costa, E.M. *et al.* Sustainable development of smart cities: A systematic review of the literature. *J. Open Innov.*, 3, 11, 2017.

    Bettencourt, L.M.A., The uses of big data in cities, SFI Working Papers, September, 2013.

38. Vermesan, O. and Friess, P., *Internet of things: Converging technologies for smart environments and integrated ecosystems*, River Publishers, Denmark, 2013, http://www.internet-of-thingsresearch.eu/pdf/Converging_Technologies_for_Smart_Environments_and_Integrated_Ecosystems_IERC_Book_Open_Access_2013.pdf, last accessed on 18 Feb 2016.

39. Corcoran, P., The internet of things. *IEEE Consum. Electron. Mag.*, 5, 1, 63–68, January 2016.

40. Harris, S., Securing big data in our future intelligent cities, in: *Proceedings of IET Conference on Future Intelligent Cities*, pp. 1–4, 2014.

41. Errichiello, L. and Micera, R., Leveraging smart open innovation for achieving cultural sustainability: Learning from a new city museum project. *Sustainability*, 10, 6, 1964, 2018.

42. Mohammed, F., Idries, A., Mohamed, N., Al-Jaroodi, J., Jawhar, I., UAVs for smart cities: Opportunities and challenges, in: *Proc. Int. Conf. Unmanned Aircr. Syst. (ICUAS)*, pp. 267–273, 2014.

43. Bucaille, I., Héthuin, S., Rasheed, T., Munari, A., Hermenier, R., Allsopp, S., Rapidly deployable network for tactical applications: Aerial base station with opportunistic links for unattended and temporary events absolute example, in: *Proc. IEEE Military Commun. Conf. (MILCOM)*, pp. 1116–1120, Nov. 2013.

44. Han, G., Shen, J., Liu, L., Shu, L., BRTCO: A novel boundary recognition and tracking algorithm for continuous objects in wireless sensor networks. *IEEE Syst. J.*, 12, 3, 2056–2065, Sep. 2018.

45. Lagkas, T., Argyriou, V., Bibi, S., Sarigiannidis, P., UAV IoT framework views and challenges: Towards protecting drones as 'things'. *Sensors*, 18, 11, 4015, 2018.

46. Gapchup, A., Wani, A., Wadghule, A., Jadhav, S., Emerging trends of green IoT for smart world. *Int. J. Innov. Res. Comput. Commun. Eng.*, 5, 2, 2139–2148, 2017.

47. Dameri, R.P., Using ICT in smart city, in: *Smart City Implementation*, pp. 45–65. Springer, Cham, 2017.

48. Huang, J., Meng, Y., Gong, X., Liu, Y., Duan, Q., A novel deployment scheme for green internet of things. *IEEE Internet Things J.*, 1, 2, 196–205, Apr. 2014.

49. Rani, S., Talwar, R., Malhotra, J., Ahmed, S.H., Sarkar, M., Song, H., A novel scheme for an energy efficient Internet of Things based on wireless sensor networks. *Sensors*, 15, 11, 28603–28626, 2015.

50. Al-Hourani, A., Kandeepan, S., Jamalipour, A., Modeling air-toground path loss for low altitude platforms in urban environments, in: *Proc. IEEE Global Commun. Conf. (GLOBECOM)*, pp. 2898–2904, Dec. 2014.

51. Lien, S.Y., Chen, K.C., Lin, Y., Toward ubiquitous massive accesses in 3GPP machine-to-machine communications. *IEEE Commun. Mag.*, 49, 4, 66–74, Apr. 2011.

52. Washburn, D., Sindhu, U., Balaouras, S., Dines, R.A., Hayes, N.M., Nelson, L.E., *Helping CIOs Understand "Smart City" Initiatives: Defining the Smart City, Its Drivers, and the Role of the CIO*, Forrester Research, Inc., Cambridge, MA, 2010, Available from http://public.dhe.ibm.com/partnerworld/pub/smb/smarterplanet/forr_help_cios_und_smart_city_initiatives.pdf.

53. Dawes, S.S., Cresswell, A.M., Pardo, T.A., From "need to know" to "need to share": Tangled problems, information boundaries, and the building of public sector knowledge networks. *Public Adm. Rev.*, 69, 3, 392–402, 2009.

54. Merwaday, A. and Guvenc, I., UAV assisted heterogeneous networks for public safety communications, in: *Proc. IEEE Wireless Commun. Netw. Conf. Workshops (WCNCW)*, pp. 329–334, Mar. 2015.

55. Alsamhi, S.H., *Quality of Service (QoS) Enhancement Techniques in High Altitude Platform (HAP) Based Communication Networks*, Dept. Electron. Eng., Ph.D. Dissertation, Banaras Hindu Univ., Uttar Pradesh, 2015.

56. Elmaghraby, A.S. and Losavio, M.M., Cyber security challenges in smart cities: Safety, security and privacy. *J. Adv. Res.*, 5, 4, 491–497, 2014.

57. Choi, D.H., Kim, S.H., Sung, D.K., Energy-efficient maneuvering and communication of a single UAV-based relay. *IEEE Trans. Aerosp. Electron. Syst.*, 50, 3, 2320–2327, Jul. 2014.

58. Granjal, J., Monteiro, E., Silva, J.S., Security for the Internet of Things: A survey of existing protocols and open research issues. *IEEE Commun. Surv. Tutor.*, 17, 3, 1294–1312, 3rd Quart., 2015.

59. Koulali, S., Sabir, E., Taleb, T., Azizi, M., A green strategic activity scheduling for UAV networks: A sub-modular game perspective. *IEEE Commun. Mag.*, 54, 5, 58–64, May 2016.

60. Mozaffari, M., Saad, W., Bennis, M., Debbah, M., Drone small cells in the clouds: Design, deployment and performance analysis, in: *Proc. IEEE Global Commun. Conf. (GLOBECOM)*, pp. 1–6, Dec. 2015.

61. Zorbas, D., Razafindralambo, T., Luigi, D.P.P., Guerriero, F., Energy efficient mobile target tracking using flying drones. *Proc. Comput. Sci.*, 19, 80–87, Jun. 2013.

62. Alvear, O., Zema, N.R., Natalizio, E., Calafate, C.T., Using UAV based systems to monitor air pollution in areas with poor accessibility. *J. Adv. Transp.*, 1–14, 2017, Art. no. 8204353, Aug. 2017.

63. Šmídl, V. and Hofman, R., Tracking of atmospheric release of pollution using unmanned aerial vehicles. *Atmos. Environ.*, 67, 425–436, Mar. 2013.

64. Koo, V.C., Chan, Y.K., Vetharatnam, G., Chua, M.Y., Lim, C.H., Lim, C.-S., Thum, C.C., Lim, T.S., Z. bin Ahmad, K.A., Bin Shahid, M.H., Ang, C.Y., Tan, W.Q., Tan, P.N., Yee, K.S., Cheaw, W.G., Boey, H.S., Choo, A.L., Sew, B.C., A new unmanned aerial vehicle synthetic aperture radar for environmental monitoring. *Prog. Electromagn. Res.*, 122, 245–268, 2012. [Online]. Available: http://www.jpier.org/PIER/pier.php?paper=11092604.

65. Mohamed, N., Al-Jaroodi, J., Lazarova-Molnar, S., Jawhar, I., Mahmoud, S., A service-oriented middleware for cloud of things and fog computing supporting smart city applications, in: *Proc. IEEE Smart World, Ubiquitous Intell. Comput., Adv. Trusted Comput., Scalable Comput. Commun., Cloud Big Data Comput., Internet People Smart City Innov. (SmartWorld/SCALCOM/UIC/ATC/CBDCom/IOP/SCI)*, pp. 1–7, Aug. 2017.

66. Sakhardande, P., Hanagal, S., Kulkarni, S., Design of disaster management system using IoT based interconnected network with smart city monitoring, in: *Proc. Int. Conf. Internet Things Appl. (IOTA)*, pp. 185–190, 2016.

67. Yannuzzi, M., van Lingen, F., Jain, A., Parellada, O.L., Flores, M.M., Carrera, D., Pérez, J.L., Montero, D., Chacin, P., Corsaro, A., A new era for cities with fog computing. *IEEE Internet Comput.*, 21, 2, 54–67, Mar. 2017.

68. Ray, P.P., Mukherjee, M., Shu, L., Internet of Things for disaster management: State-of-the-art and prospects. *IEEE Access*, 5, 18818–18835, 2017.

69. Cao, H., Liu, Y., Yue, X., Zhu, W., Cloud-assisted UAV data collection for multiple emerging events in distributed WSNs. *Sensors*, 17, 8, 1818, 2017.

70. Villa, T.F., Gonzalez, F., Miljievic, B., Ristovski, Z.D., Morawska, L., An overview of small unmanned aerial vehicles for air quality measurements: Present applications and future prospectives. *Sensors*, 16, 7, 1072, 2016.

71. Alvear, O., Calafate, C.T., Hernández, E., Cano, J.-C., Manzoni, P., Mobile pollution data sensing using UAVs, in: *Proc. 13th Int. Conf. Adv. Mobile Comput. Multimedia*, pp. 393–397, 2015.

72. Rashed, S. and Soyturk, M., Effects of UAV mobility patterns on data collection in wireless sensor networks, in: *Proc. IEEE Int. Conf. Commun., Netw. Satell. (COMNESTAT)*, pp. 74–79, Dec. 2015.

73. Peacock, M., Detection and Control of Small Civilian UAVs, Ph.D. Dissertation, *School Comput. Secur. Sci.*, Edith Cowan Univ., Joondalup, WA, Australia, 2014.

74. Hawbani, A., Wang, X., Al-Sharabi, Y.A., Ghannami, A., Kuhlani, H., Karmoshi, S., LORA: Load-balanced opportunistic routing for asynchronous duty-cycled WSN. *IEEE Trans. Mobile Comput.*, 18, 7, 1601–1615, Jul. 2019.

75. Motlagh, N.H., Bagaa, M., Taleb, T., UAV-based IoT platform: A crowd surveillance use case. *IEEE Commun. Mag.*, 55, 2, 128–134, Feb. 2017.

76. Bibri, S.E. and Krogstie, J., Smart sustainable cities of the future: An extensive interdisciplinary literature review. *Sustain. Cities Soc.*, 31, 183–212, May 2017.

77. Choudhary, G., Sharma, V., You, I., Sustainable and secure trajectories for the military Internet of Drones (IoD) through an efficient Medium Access Control (MAC) protocol. *Comput. Elect. Eng.*, 74, 59–73, Mar. 2019.

78. Michailidis, E.T., Vouyioukas, D.A., Review on software-based and hardware-based authentication mechanisms for the Internet of Drones. *Drones*, 6, 41, 1–26, 2022.

79. Hassanalieragh, M., Page, A., Soyata, T., Sharma, G., Aktas, M., Mateos, G., Kantarci, B., Andreescu, S., Health monitoring and management using Internet-of-Things (IoT) sensing with cloud-based processing: Opportunities and challenges, in: *Proc. IEEE Int. Conf. Services Comput. (SCC)*, pp. 285–292, Jun./Jul. 2015.

80. Khan, I.H.I., Kahan, I., Khan, S., Challenges of IoT implementation in smart city development, in: *Smart Cities-Oppurtunities and Challenges*, Ahmed, S., *et al.* (Eds.).

81. Villa, T.F., Gonzalez, F., Miljievic, B., Ristovski, Z.D., Morawska, L., An overview of small unmanned aerial vehicles for air quality measurements: Present applications and future prospectives. *Sensors*, 16, 7, 1072, 2016.

82. Granjal, J., Monteiro, E., Silva, J.S., Security for the Internet of Things: A survey of existing protocols and open research issues. *IEEE Commun. Surv. Tutor.*, 17, 3, 1294–1312, 3rd Quart., 2015.

83. Wang, J., Schluntz, E., Otis, B., Deyle, T., A new vision for smart objects and the internet of things: Mobile robots and long-range UHF RFID sensor tags, arXiv preprint arXiv:1507.02373, 2015. Available: https://arxiv.org/abs/1507.02373.

84. Mohammed, F., Idries, A., Mohamed, N., Al-Jaroodi, J., Jawhar, I., UAVs for smart cities: Opportunities and challenges. *2014 International Conference on Unmanned Aircraft Systems (ICUAS)*, 2014.

85. Shreih, R., *Intelligent Systems for Smarter Cities*, King Abdullah University Of Science & Technology, August 26, 2013, [Online]. Available: https://innovation.kaust.edu.sa/intelligent-systems-forsmarter-cities/. [Accessed 4 January 2014].

86. King Abdullah University Of Science & Technology, *Smart Cities*, King Abdullah University Of Science & Technology, 2013, [Online]. Available: https://innovation.kaust.edu.sa/industry/brows e-technology/smart-cities/. [Accessed 4 January 2014].

87. Elfrink, W., The smart-city solutions, McKinsey & Co, Transport. 27, 3, 335-343, 2012.
    Kharchenko, V. and Prusov, D., Analysis of unmanned aircraft systems applications in the civil field, Taylor and Francis Group, 2012.

88. Finn, R.L. and Wright, D., Unmanned aircraft systems: Surveillance, ethics and privacy in civil applications. *Comput. Law Secur. Rev.*, 28, 2, 184–194, 2012.

89. Franke, U.E., *The Five Most Common Media Misrepresentations of UAVs*, The Royal United Services Institute for Defence and Security Studies, London, 2013.

90. Chourabi, H., Nam, T., Walker, S., Ramon GilGarcia, S.M.J., Nahon, K., Pardo, T.A., Scholl, H.J., Understanding smart cities: An integrative framework, in: *Hawaii International Conference on System Sciences*, Hawaii, 2012.

91. Rodzi, A., Sharif, A.R., Ahmad, N., Al-Hader, M., Smart city components architecture, in: *International Conference on Computational Intelligence, Modelling and Simulation*, 2009.

92. Mitchell, S. and Falconer, G., *Smart City Framework-A Systematic Process for Enabling Smart+Connected Communities*, CISCO, San Jose, CA, 2012.

93. Bari, N. and Idowu, S., *A Development Framework for Smart City Services - Integrating Smart City Service Components*, Lulea University of Technology, Lulea, Sweden, 2012.

94. Asghari, P., Rahmani, A., Javadi, H., Internet of Things applications: A systematic review. *Comput. Netw.*, 148, 241–261, 2019. https://doi.org/10.1016/j.comnet.2018.12.008.

95. Atzori, L., Iera, A., Morabito, G., The internet of things: A survey. *Comput. Netw.*, 54, 15, 2787–2805, 2010. https://doi.org/10.1016/j.comnet.2010.05.010.

96. Tankard, C., The security issues of the internet of things. *Comput. Fraud Secur.*, 2015, 9, 11–14, 2015. https://doi.org/10.1016/s1361-3723(15)30084-1.

97. Jin, J., Gubbi, J., Marusic, S., Palaniswami, M., An information framework for creating a smart city through internet of things. *IEEE Internet Things J.*, 1, 2, 112–121, 2014. https://doi.org/10.1109/jiot.2013.2296516.

98. Bello, O. and Zeadally, S., Toward efficient smartification of the Internet of Things (IoT) services. *Future Gener. Comput. Syst.*, 92, 663–673, 2019.

99. Yaqoob, I., Hashem, I., Ahmed, A., Kazmi, S., Hong, C., Internet of things forensics: Recent advances, taxonomy, requirements, and open challenges. *Future Gener. Comput. Syst.*, 92, 265–275, 2019. https://doi.org/10.1016/j.future.2018.09.058.

100. Rathore, M., Paul, A., Ahmad, A., Jeon, G., IoT-based big data. *Int. J. Semant. Web Inf. Syst.*, 13, 1, 28–47, 2017. https://doi.org/10.4018/ijswis.2017010103.

101. Sagirlar, G., Carminati, B., Ferrari, E., Decentralizing privacy enforcement for internet of things smart objects. *Comput. Netw.*, 143, 112–125, 2018. https://doi.org/10.1016/j.comnet.2018.07.019.

102. Sicari, S., Rizzardi, A., Grieco, L., Coen-Porisini, A., Security, privacy and trust in internet of things: The road ahead. *Comput. Netw.*, 76, 146–164, 2015. https://doi.org/10.1016/j.comnet.2014. 11.008.

103. Weber, R., Internet of things: Privacy issues revisited. *Comput. Law Secur. Rev.*, 31, 5, 618–627, 2015. https://doi.org/10.1016/j.clsr.2015.07.002.

104. Serrano, W., Digital systems in smart city and infrastructure: Digital as a service. *Smart Cities*, 1, 1, 134–153, 2018. https://doi.org/10.3390/smartcities1010008.

105. Ahmed, E., Yaqoob, I., Hashem, I., Khan, I., Ahmed, A., Imran, M., Vasilakos, A., The role of big data analytics in internet of things. *Comput. Netw.*, 129, 459–471, 2017. https://doi.org/10.1016/j.comnet.2017.06.013.

106. RisteskaStojkoska, B. and Trivodaliev, K., A review of internet of things for smart home: Challenges and solutions. *J. Clean Prod.*, 140, 1454–1464, 2017. https://doi.org/10.1016/j.jclepro. 2016.10.006.

107. de Boer, P.S., van Deursen, A.J.A.M., van Rompay, T.J.L., Accepting the internet-of-things in our homes: The role of user skills. *Telemat. Inf.*, 36, 147–156, 2019. https://doi.org/10.1016/j.tele.2018.12.004.

108. Liu, L., IoT and a sustainable city. *Energy Proc.*, 153, 342–346, 2018. https://doi.org/10.1016/j.egypro.2018.10.080.

109. Albishi, S., Soh, B., Ullah, A., Algarni, F., Challenges and solutions for applications and technologies in the internet of things. *Proc. Comput. Sci.*, 124, 608–614, 2017. https://doi.org/10.1016/j.procs.2017.12.196.

110. Mital, M., Chang, V., Choudhary, P., Papa, A., Pani, A., Adoption of internet of things in India: A test of competing models using a structured equation modeling approach. *Technol. Forecast. Soc. Change*, 136, 339–346, 2018. https://doi.org/10.1016/j.techfore.2017.03.001.

111. Marques, P., Manfroi, D., Deitos, E., Cegoni, J., Castilhos, R., Rochol, J. et al., An IoT-based smart cities infrastructure architecture applied to a waste management scenario. *Ad Hoc Netw.*, 87, 200–208, 2019. https://doi.org/10.1016/j.adhoc.2018.12.009.

112. Rathore, M., Paul, A., Ahmad, A., Jeon, G., IoT-based big data. *Int. J. Semant. Web Inf. Syst.*, 13, 1, 28–47, 2017. https://doi.org/10.4018/ijswis.2017010103.

113. Riggins F.J, and Wamba, S.F., Research directions on the adoption, usage, and impact of the Internet of Things through the use of big data analytics. *2015, 48th Hawaii International Conference on System Sciences*, pp. 1531–1540, 2015.

114. Wang, H., Zhao, H., Zhang, J., Ma, D., Li, J., Wei, J., Survey on unmanned aerial vehicle networks: A cyber physical system perspective. *IEEE Commun. Surv. Tutor.*, 22, 2, 1027–1070, 2020.

115. Altawy, R. and Youssef, A.M., Security, privacy, and safety aspects of civilian drones: A survey. *ACM Trans. Cyber-Phys. Syst.*, 1, 2, 1–25, 2017.

116. Nassi, B., Bitton, R., Masuoka, R., Shabtai, A., Elovici, Y., SoK: Security and privacy in the age of commercial drones. *2021 IEEE Symposium on Security and Privacy (SP)*, 2021, pp. Section IV, pp. 73–90, 2021.

117. Ganz, F., Puschmann, D., Barnaghi, P., Carrez, F., A practical evaluation of information processing and abstraction techniques for the internet of things. *IEEE Internet Things J.*, 2, 4, 340–354, 20152015.

118. Merzouk, S., Cherkaoui, A., Marzak, A., Nawal, S., IoT methodologies: Comparative study. *Proc. Comput. Sci.*, 175, 585–590, 2020.

119. Alam, F., Mehmood, R., Katib, I., Albogami, N.N., Albeshri, A., Data fusion and IoT for smart ubiquitous environments: A survey. *IEEE Access*, 5, 9533–9554, 2017.

120. Hui, K., A secure IoT-based healthcare system with body sensor networks. *IEEE Access*, 4, 10288–10299, 2016.
121. Rathore, M.M., Ahmad, A., Paul, A., Jeon, G., Efficient graph-oriented smart transportation using internet of things generated big data, in: *11th International Conference on Signal-Image Technology & Internet-Based Systems (SITIS)*, Bankok, 2015.
122. Khanna, A. and Anand, R., IoT based smart parking system, in: *International Conference on Internet of Things and Applications (IOTA)*, Pune, 2016.
123. Chaudhuri, N. and Bose, I., Application of image analytics for disaster response in smart cities, in: *Proc. 52nd Hawaii Int. Conf. Syst. Sci*, pp. 3036–3045, 2019.
124. Sakhardande, P., Hanagal, S., Kulkarni, S., Design of disaster management system using IoT based interconnected network with smart city monitoring, in: *Proc. Int. Conf. Internet Things Appl. (IOTA)*, pp. 185–190, 2016.
125. Alazawi, Z., Alani, O., Abdljabar, M.B., Altowaijri, S., Mehmood, R., A smart disaster management system for future cities, in: *Proc. ACM Int. Workshop Wireless Mobile Technol. Smart Cities*, pp. 1–10, 2014.
126. Boukerche, A. and Coutinho, R.W.L., Smart disaster detection and response system for smart cities, in: *Proc. IEEE Symp. Comput. Commun. (ISCC)*, pp. 1102–1107, Jun. 2018.
127. Ray, P.P., Mukherjee, M., Shu, L., Internet of Things for disaster management: State-of-the-art and prospects. *IEEE Access*, 5, 18818–18835, 2017.
128. Reina, D.G., Camp, T., Munjal, A., Toral, S.L., Tawfik, H., Evolutionary deployment and hill climbing-based movements of multi-UAV networks in disaster scenarios, in: *Applications of Big Data Analytics*, pp. 63–95, 2018.
129. Mozaffari, M., Saad, W., Bennis, M., Debbah, M., Unmanned aerial vehicle with underlaid device-to-device communications: Performance and tradeoffs. *IEEE Trans. Wirel. Commun.*, 15, 6, 3949–3963, Jun. 2016.
130. Lien, S.Y., Chen, K.C., Lin, Y., Toward ubiquitous massive accesses in 3GPP machine-to-machine communications. *IEEE Commun. Mag.*, 49, 4, 66–74, Apr. 2011.
131. Mozaffari, M., Saad, W., Bennis, M., Debbah, M., Mobile unmanned aerial vehicles (UAVs) for energy-efficient Internet of Things communications. in: *IEEE Transactions on Wireless Communications*, 16, 11, 7574–7589, Nov 2017.
132. Mozaffari, M., Saad, W., Bennis, M., Debbah, M., Mobile internet of things: Can UAVs provide an energy-efficient mobile architecture?, in: *Proc. IEEE Global Commun. Conf. (GLOBECOM)*, pp. 1–6, Dec. 2016.
133. Tuyishimire, E., Bagula, A., Rekhis, S., Boudriga, N., Cooperative data muling from ground sensors to base stations using UAVs, in: *Proc. IEEE Symp. Comput. Commun. (ISCC)*, pp. 35–41, Jul. 2017.
134. Quaritsch, M., Kruggl, K., Wischounig-Strucl, D., Bhattacharya, S., Shah, M., Rinner, B., Networked UAVs as aerial sensor network for disaster

management applications. *Elektrotechnik und Informationstechnik*, 127, 3, 56–63, 2010.

135. Dawy, Z., Saad, W., Ghosh, A., Andrews, J.G., Yaacoub, E., Toward massive machine type cellular communications. *IEEE Wirel. Commun.*, 24, 1, 120–128, Feb. 2017.

136. Hassanalieragh, M., Page, A., Soyata, T., Sharma, G., Aktas, M., Mateos, G., Kantarci, B., Andreescu, S., Health monitoring and management using Internet-of-Things (IoT) sensing with cloud-based processing: Opportunities and challenges, in: *Proc. IEEE Int. Conf. Services Comput. (SCC)*, pp. 285–292, Jun./Jul. 2015.

137. Schaub, F. and Knierim, P., Drone-based privacy interfaces: Opportunities and challenges, in: *Proc. WSF@ SOUPS*, pp. 1–4, 2016.

138. Erman, A.T., Hoesel, L.V., Havinga, P., Wu, J., Enabling mobility in heterogeneous wireless sensor networks cooperating with UAVs for mission-critical management. *IEEE Wirel. Commun.*, 15, 6, 38–46, Dec. 2008.

139. Klimkowska, A., Lee, I., Choi, K., Possibilities of UAS for maritime monitoring. *Int. Arch. Photogramm. Remote Sens. Spat. Inf. Sci.*, 41-B1, 885–891, Jul. 2016.

140. Villa, T.F., Gonzalez, F., Miljievic, B., Ristovski, Z.D., Morawska, L., An overview of small unmanned aerial vehicles for air quality measurements: Present applications and future prospectives. *Sensors*, 16, 7, 1072, 2016.

141. Villa, T.F., Salimi, F., Morton, K., Morawska, L., Gonzalez, F., Development and validation of a UAV based system for air pollution measurements. *Sensors*, 16, 12, 2202, 2016.

142. Malaver, A., Motta, N., Corke, P., Gonzalez, F.J.S., Development and integration of a solar powered unmanned aerial vehicle and a wireless sensor network to monitor greenhouse gases. *Sensors*, 15, 2, 4072–4096, 2015.

143. Telesetsky, A., Navigating the legal landscape for environmental monitoring by unarmed aerial vehicles. *Geo. Wash. J. Energy Envtl. L.*, 7, 140, 2016.

144. Alvear, O., Calafate, C.T., Hernández, E., Cano, J.-C., Manzoni, P., Mobile pollution data sensing using UAVs, in: *Proc. 13th Int. Conf. Adv. Mobile Comput. Multimedia*, pp. 393–397, 2015.

145. Alvear, O.A., Zema, N.R., Natalizio, E., Calafate, C.T., A chemotactic pollution-homing uav guidance system, in: *Proc. 13th Int. Wireless Commun. Mobile Comput. Conf. (IWCMC)*, pp. 2115–2120, 2017.

146. Villa, T.F., Salimi, F., Morton, K., Morawska, L., Gonzalez, F., Development and validation of a UAV based system for air pollution measurements. *Sensors*, 2016.

147. Mozaffari, M., Saad, W., Bennis, M., Debbah, M., Drone small cells in the clouds: Design, deployment and performance analysis, in: *Proc. IEEE Global Commun. Conf. (GLOBECOM)*, pp. 1–6, Dec. 2015.

148. Cao, H.-R., Yang, Z., Yue, X.-J., Liu, Y.-X., An optimization method to improve the performance of unmanned aerial vehicle wireless sensor networks. *Int. J. Distrib. Sens. Netw.*, 13, 4, 1–10, 2017.

149. Cao, H., Liu, Y., Yue, X., Zhu, W., Cloud-assisted UAV data collection for multiple emerging events in distributed WSNs. *Sensors*, 17, 8, 1818, 2017.

150. Dong, M., Ota, K., Lin, M., Tang, Z., Du, S., Zhu, H., UAV-assisted data gathering in wireless sensor networks. *J. Supercomput.*, 70, 3, 1142–1155, 2014.

151. Zorbas, D., Razafindralambo, T., Luigi, D.P.P., Guerriero, F., Energy efficient mobile target tracking using flying drones. *Proc. Comput. Sci.*, 19, 80–87, Jun. 2013.

152. Sharma, V., You, I., Kumar, R., Energy efficient data dissemination in multi-UAV coordinated wireless sensor networks. *Mob. Inf. Syst.*, 2016, Art. no. 8475820, 1–13, May 2016.

153. Fujii, K., Higuchi, K., Rekimoto, J., Endless flyer: A continuous flying drone with automatic battery replacement, in: *Proc. IEEE IEEE 10th Int. Conf. Ubiquitous Intell. Comput., 10th Int. Conf. Auto. Trusted Comput. (UIC/ATC)*, pp. 216–223, Dec. 2013.

154. Hentati, A., II and Fourati, L.C., Comprehensive survey of UAVs communication networks. *Comput. Stand. Interfaces*, 72, 103451, 2020.

155. Shakhatreh, H., Sawalmeh, A.H., Al-Fuqaha, A., Dou, Z., Almaita, E., Khalil, I., Othman, N.S., Guizani, M., Unmanned aerial vehicles (UAVs): A survey on civil applications and key research challenges. *IEEE Access*, 7, 48572–48634, 2019.

156. Krishna, C.G. and Murphy, R.R., A review on cybersecurity vulnerabilities for unmanned aerial vehicles. *SSRR 2017-15th IEEE International Symposium on Safety, Security and Rescue Robotics, Conference*, pp. 194–199, 2017.

157. Shafique, A., Mehmood, A., Elhadef, M., Survey of security protocols and vulnerabilities in unmanned aerial vehicles. *IEEE Access*, 9, 46927–46948, 2021.

158. Syed, F., Gupta, S.K., HamoodAlsamhi, S., Rashid, M., Liu, X., A survey on recent optimal techniques for securing unmanned aerial vehicles applications. *Trans.Emerg. Telecommun. Technol.*, 32, 7, 1–34, 2021.

159. Altawy, R. and Youssef, A.M., Security, privacy, and safety aspects of civilian drones: A survey. *ACM Trans. Cyber-Phys. Syst.*, 1, 2, 1–25, 2017.

160. Yaacoub, J.-P., Noura, H., Salman, O., Chehab, A., Security analysis of drones systems: Attacks, limitations, and recommendations. *Internet Things*, 11, 100218, 2020.

161. Zhi, Y., Fu, Z., Sun, X., Yu, J., Security and privacy issues of UAV: A survey. *Mob. Netw. Appl.*, 25, 1, 95–101, 2020.

162. Fotouhi, A., Qiang, H., Ding, M., Hassan, M., Giordano, L.G., Garcia-Rodriguez, A., Yuan, J., Survey on UAV cellular communications: Practical aspects, standardization advancements, regulation, and security challenges. *IEEE Commun. Surv. Tutor.*, 21, 4, 3417–3442, 2019.

163. Mishra, D. and Natalizio, E., A survey on cellular-connected UAVs: Design challenges, enabling 5G/B5G innovations, and experimental advancements. *Comput. Netw.*, 182, 107451, August 2020.

164. Hayat, S., Yanmaz, E., Muzaffar, R., Survey on unmanned aerial vehicle networks for civil applications: A communications viewpoint. *IEEE Commun. Surv. Tutor.*, 18, 4, 2624–2661, 2016.

165. Sharma, A., Vanjani, P., Paliwal, N., Basnayaka, C.M., Jayakody, D.N.K., Wang, H.C., Muthuchidambaranathan, P., Communication and networking technologies for UAVs: A survey. *J. Netw. Comput. Appl.*, 168, 102739, June 2020.

166. Hassija, V., Chamola, V., Agrawal, A., Goyal, A., Luong, N.C., Niyato, D., Yu, F.R., Guizani, M., Fast, reliable, and secure drone communication: A comprehensive survey. *IEEE Commun. Surv. Tutor.*, 1, 23, 4, 2802–2832, 2021.

167. Boccadoro, P., Striccoli, D., Grieco, L.A., An extensive survey on the internet of drones. *Tech. Rep.*, 122, 102600, 2020. [Online]. Available: http://arxiv.org/abs/2007.12611.

168. Noor, F., Khan, M.A., Al-Zahrani, A., Ullah, I., Al-Dhlan, K.A., A review on communications perspective of flying AD-HOC networks: Key enabling wireless technologies, applications, challenges and open research topics. *Drones*, 4, 4, 1–14, 2020.

10

AI-Based Smart Surveillance for Drowning and Theft Detection on Beaches Using Drones

V. Sakthivel[1,2]*, Suriya, E.[2], Jae Woo Lee[1] and P. Prakash[2]

[1]Konkuk Aerospace Design-Airworthiness Institute, Konkuk University, Seoul, South Korea
[2]School of Computer Science and Engineering, Vellore Institute of Technology, Chennai, Tamil Nadu, India

Abstract

Some public areas like beaches are crowded during the weekends and are tedious to manage with a limited number of authorities. People in these places are prone to danger. Aside from cases of drowning, there have also been reports of suicide on beaches. Other mischievous or dangerous activities (e.g., theft) happen in such places. Most of these incidents happen when the crowd is bigger and those in charge of surveillance will find it challenging to monitor on the ground. The proposed solution is to build a deep learning-based model for the detection of drowning in the sea or any swimmers taken into sudden high waves. We will be using thermal images from the drone and extracting the features of humans, the model is trained to detect those large sets of images. So, once a person moves away from the safety swimming line on the beach, the system will alert the persons near the shoreline and the coast guards. We will also use the posture estimation model to detect miscellaneous activities and take pictures and alert the officials to take action. This drone will be self-automated based on its set boundary and it will map with GPS integration. With this, we can avoid untoward activities from happening with smart surveillance using drones.

Keywords: Theft detection, drowning detection, image processing, smart surveillance, deep learning, pose estimation, autonomous mapping

Corresponding author: mvsakthi@gmail.com

Sachi Nandan Mohanty, J.V.R. Ravindra, G. Surya Narayana, Chinmaya Ranjan Pattnaik and Y. Mohamed Sirajudeen (eds.) Drone Technology: Future Trends and Practical Applications, (243–256) © 2023 Scrivener Publishing LLC

10.1 Introduction

Drowning is one of the third leading causes of unintentional injury & death worldwide. Nearly 1.2 million people around the world die or are injured by drowning every year, more than 2 people per minute. Supervising beaches is the most difficult task for coastal guards and lifeguards as the effect of environmental factors like weather, high current, waves, etc. There is only around 20 to 60 seconds time budget to rescue people while they are being drowned during that time. Most people who are drowned are in panic mode about what action to be taken, they will get tense and die by drinking more seawater. With the tremendous development of machine learning applications that are used to find hidden patterns among the data collected by sensors. These new technologies can be built as a model, integrated, and deployed in various applications to enhance human needs. Machine learning enriches decision-making capabilities. Automating the beach drowning surveillance will greatly manpower and this model are robust and it even exceeds human performance in identifying, detecting, and classifying humans who are drowning or not. Computer vision approaches are used which allow machines to view the world helping ML models achieve greater performance. This paper includes the drown detection by computer vision and deep learning methods and alerts to nearby people using BLE beacons, abnormal event monitoring for theft detection.

10.2 Literature Survey

Sen-tag is a safety application that checks for the individual swimmers through a wristband which will monitor the person's depth in water, person's action, and time. Sensors are fixed to the wall of the pool to identify whether the swimmer wearing the wristband, it finds whether the swimmer spends a lot of time under the water. If it happens then an alert signal will be sent to the control unit. Lifeguards at the swimming arena will reach the victim immediately [1].

An automated video-based surveillance system provides real-time human behavior analysis for detecting any abnormal events in the surroundings. This records the person's activities through video frames by image processing. The challenge we face here is rapid light changing in environments [2].

This method is to detect fall incidents in humans. This proposed method has two steps first is to detect and track moving people and the next is to find a sequence of falling behavior. This method is robust in detecting human falls with intelligent video surveillance [3].

10.3   Proposed Model

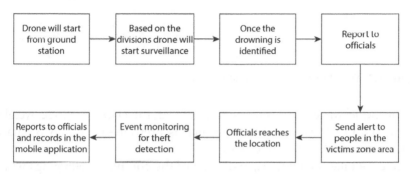

Figure 10.1 Proposed model.

10.3.1   Drown Detection by Deep Learning Methods

The proposed system (Figure 10.1) uses a frame of images that is sent from the edge server to the cloud server for deep learning and posture estimation model using the MoveNet model. Raspberry Pi 4b+ is fixed at the drone [4] and connected to the camera module. This sends the images via wireless communication to the server station. We will be using Wi-Fi over Zigbee since it has a higher data rate and also it covers the less physical area [5]. The data rate of Wi-Fi is up to 54Mb/s whereas Zigbee has 25Kb/s. We will be using ResNet50 architecture over the model built from scratch since datasets for this application are not available. ResNet50 is trained on the ImageNet dataset for huge applications. Before this actual approach, the initial trial work was conducted on the YOLOV3 object detection algorithm which uses features learned by a deep convolutional neural network to detect an object. YOLOv3 which is You Only Look Once is a popular real-time object detection to identify objects in the live feed, images, and videos. YOLO v3 is a more accurate version of the ML algorithm claimed by the developers. YOLO is basically a Convolutional Neural Network (CNN) model for performing object detection. YOLO seems to be faster than other models but still, it maintains better accuracy on the data being tested.

To use the YOLO v3 model (see Figure 10.2), first step is to identify the scope of the problem statement and identify the objects which are needed

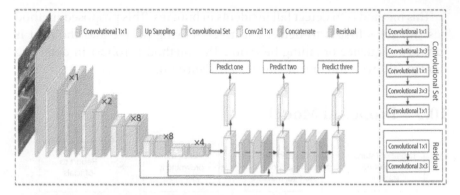

Figure 10.2 YOLO v3 architecture: Source [LINK].

for the model to keep focus in the frame. Next, the DarkNet model weights of the YOLO v3 are downloaded which is pretrained on a coco dataset which contains 80 different classes.

Here we take the humans, and some of the common belongings which people carry every day for the model to distinguish between people and objects. To display the results of the model's prediction after passing the learned feature by the model onto a classifier. The size of the prediction map will be exactly of the feature map size [6].

These features are interpreted by class confidence score with a bounding box. As each bounding box will have coordinates x, y, height, width, and box confidence scores. This confidence score shows the accuracy of the class object present inside the bounding box. The x and y values are offsets of the cell and class confidence score is obtained by multiplying box confidence score and conditional class probability. If the class confidence is higher than the threshold, here we kept the threshold value as 0.75 i.e. 75%. These bounding boxes are displayed as final predictions.

According to the problem statement, this is used to identify whether the person is drowning or not (see Figure 10.3). For example: if the person is underwater, the model will not be able to detect the person and so we won't be able to find out if the person is drowning or not (see Figure 10.4). This only detects the person's position in the sea, and the flowchart shows the model (see Figure 10.5). If the person went into the water (see Figure 10.6) or a little below the water, it is not able to classify them as humans. This might lead to the problematic situation for the model as it either raises a false alarm or doesn't raise an alarm on the drowning situation (see Figure 10.7).

Since yolov3 is not suitable for this problem, building the model from scratch with the custom dataset is the ideal solution. ResNet50 model

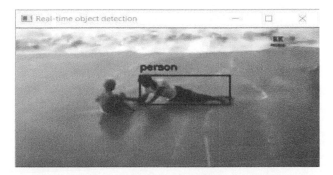

Figure 10.3 Yolo v3 for drowning detection – NORMAL.

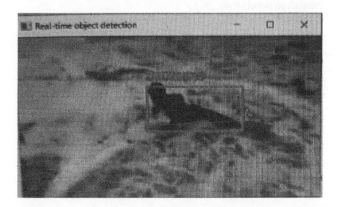

Figure 10.4 Yolo v3 for drowning detection – WARNING.

architecture consists of an input layer with a size of (225, 225, 3) image size. Input images are normalized with the standard scalar method which normalizes between zero and one. Following the input layer, we have five full layers with sizes of 1000, 500, 350, 100, 20 nodes. We have a dropout layer at the second and fourth layers with 0.25%. We will use stochastic gradient descent for optimizing cost function and minimizing the training errors. Here loss function is Categorical cross-entropy. Once after detecting the person with a probability score of zero to one. We also propose posture estimation (see Figure 10.8) by the movement model to confirm whether the person swimming on beaches knows proper swimming [8] or he/she moves his hands and body due to a panic attack. Once after confirming the person drowning, this system will try finding persons (see Figure 10.9) who are nearest to the shore and with BLE beacons, we will send alerts to them so that they can help the victims before the officials could reach there. Lifeguards will be intimated of the location of the victim through the mobile application.

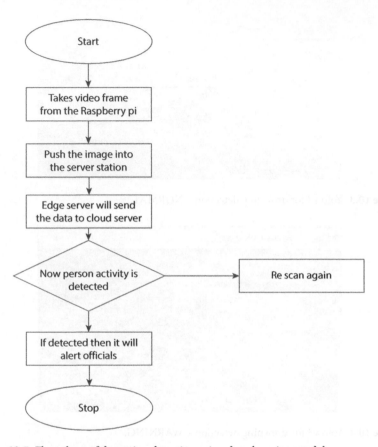

Figure 10.5 Flow chart of drowning detection using deep learning models.

Figure 10.6 Optical flow of persons on a beach – Method 1.

Figure 10.7 Optical flow of persons at a beach – Method 2.

Figure 10.8 Posture and swim pattern estimation above water.

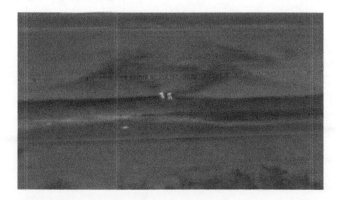

Figure 10.9 Thermal image from a drone.

10.3.2   People Alert System Using BLE Beacons

Once the people are entering the beaches they will be provided with beacon tags which is an entry passes for the beaches. We assume the condition that no persons are allowed inside the beaches without beacons. Beacon tags are very important since they are the medium for communication on beaches (see Figure 10.10). BLE is good for localization [9] applications as they are lower cost, lower energy consumption, and lower weight and it provides a moderate communication range of 100 m. BLE beacon devices are powered by a single-cell battery that has a shell life of up to two years.

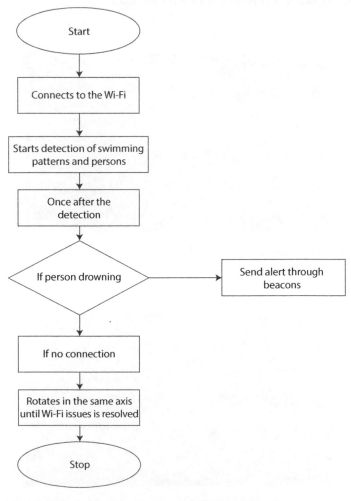

Figure 10.10 Flow chart of people alert system using BLE beacons.

This uses only three broadcasting channels and also uses a 2.4 GHz ISM band with 2 MHz channel spacing. It is governed by IEEE 802.11 and 802.15 protocols. Once the person is drowning, the drone will look for persons nearby the victims and once a set of all devices are discovered, the connection is established, and the drone sends the alert message through a mobile application with Bluetooth 5.0 technology. The easiest approach for positioning [10] is determining whether the target and the locator are within the specified range near the victim. To determine the distance between devices we have the equation

$$d = 20^{(\text{RSSI} + \alpha)} \tag{10.1}$$

where d is distance and RSSI is Received Signal Strength Indicator.

In order to maximize the accuracy of the system, the movement of the drone is locked around the specified area as we have segmented the beaches as various blocks by analyzing drone courage for over an hour among the individual blocks, while it continuously performs search operations in its proximity. If the connection drops suddenly, the drone is programmed to fly around the same line till it reconnects to the network. If reconnection time is the prolonged drone will return to the docking station to save power.

10.3.3   Abnormal Event Monitoring for Theft Detection

The proposed method will find humans from the video streams and estimate postures using a MoveNet model which works on RGB images and can predict 17 human key points of the full body. Basically human posture estimation is a computer vision technique used to predict a person's action or gestures [7].

Now MoveNet is a popular posture estimation model which was released by Google on mid of 2021. It is a convolutional neural network model that runs on RGB images and predicts human joint locations of a single person. For the key points apart from the 17 key points which are outside of the image frame, the model will emit low confidence scores. A confidence threshold can be used to filter out the very low confident prediction by the model. Compared to conventional pose estimation models, it performs better in video with intense motion and produce higher accuracy predictions [11].

MoveNet model is based on the 3D estimation which involves X, Y, Z coordinates of each key point in the human body. The operation takes place in a phase-wise manner like; first, the RGB image is fed to convolutional network as input, We will get the confidence threshold which is used to estimate whether the person's actions are suspicious or not with our dataset. To properly estimate the distance between the various points of the determined posture we must do an initial setup to get the Pixel to Meters Ratio. This is done by including a reference object of known width in the video feed. This model is tuned to be robust in detecting fast movement of difficult poses and with motion blur. Additionally, for Posture Evaluation, we require a reference image of the correct pose. Compared to conventional models, it performs with better accuracy even in videos with intense motion.

10.4   Deep Learning Model Safeties

We will train adversarial attacks on our own network and Mean Absolute error (MAE) a new network with the adversarial attributes. One of the adversarial attacks is the fast gradient sign method proposed by Ian Good fellow.

The term $J(x_{adv},y)$ represents the loss of the network for the classification of input image x as label y; ϵ is the intensity of the noise. We change the input image x in the direction of maximizing the loss $J(x_{adv},y)$. The input class to the network is the actual label of the input image. The target class is the label that we wish to change the input label form original class to different class. The image is modified by taking the gradient of the cost function with respect to the input. FGSM will add the noise to the image, this noise is not the random noise but that are whose direction is the same as the gradient of the cost function with respect to the data. Here direction of the gradient it moves is taken into consideration not the magnitude of the gradient.

$$\chi_0^{adv} = \chi, \quad \chi_{t+1}^{adv} = \chi_t^{adv} + \alpha \cdot \text{sign}\left(\nabla_\chi J\left(\chi_t^{adv}, y\right)\right) \qquad (10.2)$$

The next iterative methods take T gradient steps of magnitude $\alpha = \varepsilon/T$ instead of a single step t:

Starting with the original image to create an adversarial image. Instead of passing the function only, one can iteratively call the function to descend in the direction of the gradient of the loss function calculated with respect to that image and the adversarial target class obtained by subtracting magnitude and of the gradient with the direction

Input: A classifier f with loss function J; a real example x and ground-truth label y;

Input: The size of perturbation $\in$; iterations T and decay factor μ.

Ouput: An adversarial example x^* with $\|x^* - x\|_\infty \le \in$.

1: $\alpha = {}^{\epsilon}\!/_T$;

2: $g_0 = 0; x_0^* = x$;

3: **for** $t = 0$ to $T = 1$ **do**

4:     Input x_t^* to f and obtain the gradient $\nabla_x J(x_t^*, y)$;

5:     Update g_{t+1} by accumulating the velocity vector in the gradient direction as

$$g_{t+1} = \mu \cdot g_t + \frac{\nabla_x J(x_t^*, y)}{\left\|\nabla_x J(x_t^*, y)\right\|_1} \;; \tag{10.3}$$

6: Update x_{t+1}^* by applying the sign gradient as

$$x_{t+1}^* = x_t^* + \cdot\operatorname{sign}(g_{t+1}) \;; \tag{10.4}$$

7:     **end for**

8:     **return** $x^* = x_T^*$.

We will train adversarial attacks on our own network and mean absolute error (MAE) a new network with the adversarial attributes. As we have the last layer is the softmax layer and this is what we leveraged to create our defense the softmax layer works by applying an exponential function to the class course and dividing by the sum of the exponential values and there is a parameter in that function which is the temperature T which divides the class scores before the exponential function is applied and what this parameter does is that when you set a low temperature the probabilities will be extremely discreet so you will make confident predictions however if you increase the temperature the probability vector will Smooth and converge towards one over the number of classes so typically people will set the temperature to one because they want to make confident predictions what we saw is we can instead increase the temperature to increase the robustness of the model and We will train the data which is a set of input points and labels that we can train the first Network. This will produce the probability vectors that are very smooth next we use these probability vectors to label the data that is we replace the labels that simply indicate the correct class with the probability vectors.

Now we train the same architecture from scratch using this new labeled data set and we obtain a new model which we call the Distilled model to make predictions using this new model all you have to do is simply set the temperature back to one a test time this way you still have discrete predictions which are confident and it turns out that this is very efficient. In making adding virtual sample crafting much harder so why does it work what's the intuition behind this there are actually two of the first is that when models are trained they typically are trained to minimize a loss function which is something along the lines of the cross-entropy between the expected labels and the predictions and if you replace a label of the correct class by probabilities instead of constraining only the correct class to be the output Here, the attacks are white box as all the knowledge of network hyper-parameter setting with the network's architecture. A temperature of 100 was taken in our case. Results tell that FGSM attack reduces test accuracy from 90.33% to 88.01% with the same epsilon range, I-FGSM with the iteration of 10 reduced test accuracy from 90.80% to 88.16% similar to MI-FGSM with the same decay factor of 1.0 and iterations of 10, reduction in test accuracy from 90.26% to 87.97%.

10.5   Performance Evaluation

To analyze the performance of the proposed system, we go with evaluation metrics to measure the performance, based on the abovementioned metrics, the following metrics are computed: where TP is correctly identified poses, TN has correctly identified non-identified poses, FP is incorrectly classified non identified poses, and FN is incorrectly classified correct poses.

$$Accuracy\ (Acc) = (TP + TN)/(TP + TN + FP + FN) \qquad (10.5)$$

Sensitivity (Sen) = $TP/(TP + FN)$
Precision (Pre) = $TP/(TP + FP)$
Specificity (Spe) = $TN/(TN + FP)$

10.6   Conclusion

Safety in bodies of water is a major concern. Some of the latest advancements in technology helped us to arrive at an efficient drowning detection system. The biggest hectic challenge for the lifeguards is to monitor

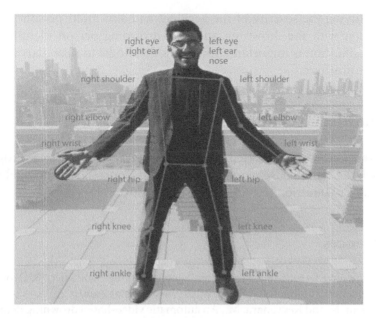

Figure 10.11 MoveNet model with 17 key points.

drowning on beaches in large crowded areas. But many existing solutions are very costly and require more infrastructures to deploy and maintain it. According to the recent survey, we found the highest number of details state most of the dead victims are of low and middle income. With the drowning model, we will take the input video frames from the camera and send them to the base server then the posture estimation and drowning classification model is run on a cloud server and processed data is sent to the base station and then to the drone. If they're a victim who drowned we can alert the officials and nearby people through BLE Beacons. At last, we will estimate the posture estimation model using the MoveNet (see Figure 10.11) developed by Google. This model is robust and has low latency also the MoveNet will perform better.

10.7   Conclusion and Future Work

This current method has some room for improvement as currently, it makes use of 17 key points of humans and recognizes whether he/she is drowning or not. This method can be enhanced with the face mesh landmarks of a total of 468 key points. This makes the model even more robust in identifying the severity of the accident and also multi-person posture estimation with the drowning and non-drowning person identification and classification is important. This helps the model to raise alarms without false positives.

Acknowledgements

This research was supported by the Basic Science Research Program through the National Research Foundation of Korea (NRF) funded by the Ministry of Education (No. 2020R1A6A 1A03046811).

References

1. Drowning Detection System from Sentag - Sentag Pool Safety & Drowning Detection. [Online]. Available: https://www.sentag.com/.
2. Salehi, N. and Keyvanara, M., An automatic video-based drowning detection system for swimming pools using active contours, 2014 MECS. *Int. J. Image, Graphics and Signal Process.*, 1, 1–3, 2014.
3. Tao, J., Turjo, M., Wong, M.-F., Wang, M., Tan, Y.P., *Fall Incidents Detection for Intelligent Video Surveillance*.
4. Maharana, S., Commercial drones. *Int. J. Adv. Sci. Eng. Technol.*, 5, 1 Suppl. Is. 3, 96–101, Mar. 2017.
5. Gonzalez-Aguilera, D. and Rodriguez-Gonzalvez, P., Editorial: Drones—An open access journal. *Drones*, 1, 1, 1–5, Dec. 2017.
6. Eng, H.-L., Toh, K.-A., Yau, W.-Y., Wang, J., Dews: A live visual surveillance system for early drowning detection at pool. *IEEE Trans. Circuits Syst. Video Technol.*, 18, 2, 196–210, 2008.
7. D.A. Grahn, H.C. Heller, R.M. Sapolsky, L. Share, sonar based drowning monitor. US Patent 7,330,123, Feb. 12, 2008.
8. Kharrat, M., Wakuda, A.Y., Kobayashi, A.S., Near drowning detection system based on swimmer's physiological information analysis. *Int. Life Saving Fed.*, 17, 201, 99635483, 2011.
9. Heukels, F.R., *Simultaneous Localization and Mapping (SLAM): Towards an Autonomous Search and Rescue Aiding Drone*, M. S. thesis, University of Twente, Enschede, Netherlands, 2015.
10. Półka, M., Ptak, S., Kuziora, Ł., The use of UAV's for search and rescue operations. *Procedia Eng.*, 192, 748–752, 2017. https://doi.org/10.1016/j.proeng.2017.06.129.
11. Wang, A. *et al.*, GuideLoc: UAV-assisted multitarget localization system for disaster rescue. *Mob. Inf. Syst.*, 2017, Art. no. e1267608, 1–13, Mar. 2017. https://doi.org/10.1155/2017/1267608.

11

Algorithms to Mitigate Cyber Security Threats by Employing Intelligent Machine Learning Models in the Design of IoT-Aided Drones

Devee Siva Prasad[1], Pyla Jyothi[2], G. Suryanarayana[3]* and Sachi Nandan Mohanty[4]

[1]Computer Science and Engineering, BABA Institute of Technology and Sciences (NR), Visakhapatnam, Andhra Pradesh, India
[2]Computer Science & Engineering, Vignan's Institute of Engineering for Women, Visakhapatnam, Andhra Pradesh, India
[3]Computer Science & Engineering, Vardhaman College of Engineering, Hyderabad, Telangana, India
[4]School of Computer Science & Engineering, VIT-AP University, Amaravathi, Andhra Pradesh, India

Abstract

Innovative and emerging technologies such as artificial intelligence, blockchain, business applications, cyber security, drones, and IoT have long helped organizations solve business problems and be more efficient, productive, and profitable. Developments in drones have opened new developments and possibilities in one-of-a-kind fields, especially in small drones [32, 33, 38]. Drones offer interlocution offerings for navigation, and this interlink is supplied via way of means of the Internet of Things (IoT). However, architectural problems make drone networks prone to privateness and protection threats. It is vital to offer a secure and stable community to collect preferred performance. Small drones are locating new paths for development within side the civil and protection industries [42, 44], however additionally posing new demanding situations for protection and privateness as well. The simple layout of the small drone calls for an amendment in its information transformation and information privateness mechanisms, and it isn't but [1, 8, 9] pleasant area requirements. This looks at additionally highlights the want for a secure and stable drone community this is unfastened from interceptions

*Corresponding author: surya.aits@gmail.com

Sachi Nandan Mohanty, J.V.R. Ravindra, G. Surya Narayana, Chinmaya Ranjan Pattnaik and Y. Mohamed Sirajudeen (eds.) Drone Technology: Future Trends and Practical Applications, (257–300) © 2023 Scrivener Publishing LLC

and intrusions. The proposed framework mitigates the cyber protection threats via way of means of using shrewd device studying fashions within the layout of IoT-aided drones via way of means of making them stable and adaptable.

Keywords: Cyber security, drones, blockchain, artificial intelligence

11.1   Introduction

Developments in drones have opened new trends and opportunities in different fields, particularly in small drones. Drones provide interlocation services for navigation, and this interlink is provided by the Internet of Things (IoT). However, architectural issues make drone networks vulnerable to privacy and security threats. It is critical to provide a safe and secure network to acquire desired performance. Small drones are finding new paths for progress in the civil and defense industries, but also posing new challenges for security and privacy as well. The basic design of the small drone requires a modification in its data transformation and data privacy mechanisms, and it is not yet fulfilling domain requirements. This paper aims to investigate recent privacy and security trends that are affecting the Internet of Drones (IoD) [46, 56]. This study also highlights the need for a safe and secure drone network that is free from interceptions and intrusions [2, 3]. The proposed framework mitigates the cyber security threats by employing intelligent machine learning models in the design of IoT-aided drones by making them secure and adaptable. Finally, the proposed model is evaluated on a benchmark dataset [12] and shows robust results. Internet usage is increasing as a powerful tool nowadays because of its unlimited benefits and applications. It is a global trend in which computers and devices are interconnected through some predefined rules and standards. Day-to-day information is carried by these communication networks over the internet. Nowadays, Internet of Things (IoT) is the most widely used network among all networks. IoT is an interconnection of devices that are using the internet to share their information. These devices can be small household objects or can be large industrial machines that are communicating to perform their operations. IoT devices can be used to monitor objects, the performance of machines, bank transactions, as well as industrial tasks [1]. There are 90 million IoT objects by the end of 2020 which can be 25 billion in 2021. These IoT objects can communicate intelligently with other devices by consuming low energy. This shows the impact of IoT devices on different areas of life. The manufacturing and health sector has the major share of IoT devices. IoT is most widely used in smart cities, surroundings observing, health, commerce, inventory, and business administration. Cyber safety incorporates of technologies, architecture, infrastructure, and software program programs which can be designed to guard computational assets [4] in opposition to cyber-attacks. Cyber safety concentrates on fundamental regions along with utility safety, catastrophe safety, facts safety, and community safety. Numerous cyber safety algorithms and computational techniques [58] are added via way of means of researchers to guard our on-line world from unwanted invaders and susceptibilities. But, the overall performance of conventional cyber safety algorithms suffers because of unique styles of offensive moves that focus on PC infrastructures, architectures, and PC networks. The implementation of sensible algorithms in encountering the extensive variety of cyber safety troubles is surveyed, namely, nature-stimulated computing (NIC) paradigms, system getting to know algorithms, and deep getting to know algorithms, primarily based totally on exploratory analyses to pick out the benefits of using in improving cyber safety techniques.

According to many security analysts [60], security incidents reached the highest number ever recorded in 2019. From phishing to ransomware, from dark web as a service economy to attacks on civil infrastructure, the cyber security landscape involved attacks that grew increasingly sophisticated during the year. This upward trend continued in 2020. The volume of malware threats observed averaged 419 threats per minute, an increase of threats per minute (12%) in the second quarter of 2020. Cyber criminals managed to exploit the COVID-19 pandemic and the growing online dependency of individuals and corporations, leveraging potential vulnerabilities of remote devices and bandwidth security. According to Interpol, 907,000 spam messages related to COVID-19 were detected between June and April 2020. Similarly, the 2020 Remote Workforce Cyber Security Report showed that nearly two thirds of respondents saw an increase in breach attempts, with 34% of those surveyed having experienced a breach during the shift to telework [5–7]. Exploiting the potential for high impact and financial benefit, threat actors deployed themed phishing emails impersonating government and health authorities to steal personal data and deployed malware against critical infrastructure and healthcare institutions. In 2021, the drive for ubiquitous connectivity and digitalization continues to support economic progress but also, simultaneously and 'unavoidably', creates a fertile ground for the rise in scale and volume of cyber-attacks: increasing ransomware and diversified tactics, increasing mobile cyber threats [11], ever more sophisticated phishing, cyber criminals and nation state attackers targeting the systems that run our day-to-day lives, and malicious actors attacking the cloud for every new low-hanging fruit.

11.2   Research Methodology

This book mainly focused to improve the basic design drones in order to ensure the security threats of drones such as data interception, data privacy and common cyber security threats. In the proposed approach, new layers are added in the layer architecture to help the implementation of security and data analysis mechanisms in the traditional drone's architecture. The improvement of layer architecture will support the easy regeneration for future enhancements. The addition of security [56] and privacy layer is with updating the data processing layer through the components of machine intelligence.

11.3   Motivation

The reliance on wireless communications makes drones vulnerable [59] to various attacks. These attacks can have drastic effects, including commercial and non-commercial losses. In this context, there is a lack of proper understanding on how hackers perform their attacks and hijack a drone [59], in order to intercept it or even crash it. In fact, drones can also be compromised for malicious purposes. Hence, there is a need to detect them and prevent them from causing any damage.

Drone Layer: The first layer of industrial drones [48, 50] is the drone layer where camera needs to attach with the main drone or quadcopter. In this layer, smart sensors are used such as altitude sensor, radar, GPS sensor, and camera. The purpose of this layer is to sense, record, and transmit the recorded information from drones to the next layer. DJ phantom drones are deployed that consist of communication link and custom remote controller [51].

Edge Processing Layer: This layer forwards the IoT raw data and drone data to the security and privacy layer where it verifies the data come from authenticated devices. The IoT gateways are used in this layer for wireless communication that provides a fast transmission of the information. This layer is responsible for data flooding, protecting, and cashing. For the purpose of cloud communication, it uses the Azure IoT gateway.

Security and Privacy Layer: This layer is responsible to provide the authentication to the devices and secure the access control through the machine learning algorithms. In this layer, some privacy threats are occurred such as physical, behavior and location privacy threat. Third party is secretly monitored and capture the drone information that effects the personal information of someone compromised. In behavior privacy, the unauthorized

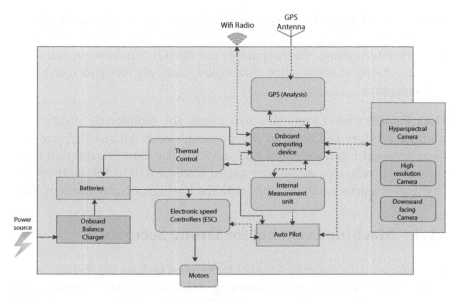

Figure 11.1 Proposed architecture for drones.

person can monitor someone's activities and behavior. Threats using location privacy involves to capture the location by authorized persons. These threats can be managed through the protocols and authentication schemes. Furthermore, machine learning algorithms are used by device authentication to alert and detect the security attacks (Figure 11.1).

Device Connection Layer: The security and privacy layers play a vital role to provide the communication link to a cloud based IoT Hub [47] at the base station and new module is added in this layer for security automation and orchestration which ensures the connection between the authenticated devices. In the IoT network, message passing between cloud system and IoT devices is allowed by the IoT hub. IoT security and devices are provided by blockchain mechanism in real-time.

Data Processing Layer: In order to analyze the drone data in stream, IoT hub data is passed to the data processing layer. In this layer, two new modules are developed such as machine intelligence that carried out the intelligent data analysis and data hub service that helps in simple and smooth cloud data storage. Naïve Bayes model are used in this layer that is intelligent machine learning algorithm. Flight data of drone is used for training and testing.

Data Storage Layer: Drones generate the data storage results for the cloud-based NoSQL database. It contains the IoT sensors data with the drone and network information. The purpose to use NoSQL database provides less storage schema of information that makes easily retrieve and access data in a short time.

Data Visualization Layer: This layer allows the monitoring of data with multiple services and tools. In our proposed work, Microsoft Azure services are used for storage and hub services. The data visualization layer shows the predictions made by proposed intelligent model about the security level of a drone and identified with the intelligent Naïve Bayes model.

11.4 Machine Learning for Drone Security

The basic types of machine learning techniques include supervised, unsupervised, and semi-supervised learning. A literature review revealed that machine learning models have been utilized by many researchers to deal with cyber-attacks in mobile-based networks, sensor-based wireless networks, cloud-computing, and IoT-based systems. Vedula *et al.* combined a supervised learning model with a self-learning model through RF and LSTM (autoencoder) to detect DDoS attacks using two features. Researchers proposed an approach to detect and control an actuation attack in a constrained cyber-physical system using a probabilistic approach in. No work has been found dealing with cyber-attacks using machine learning models in drone networks. We have also proposed an access control system in drone security [52, 53]. Our previous work showing the use of machine learning for wireless network security systems is presented. An extensive literature review ranging from 2010 to 2020 concerning privacy and safety concerns of drone data security shows a large number of research works have been published. Most of the studies discuss challenges, applications, and issues related to cyber security, data privacy concerns, spoofing, hijacking drones, and other threats. However, many researchers have highlighted the problem [35, 40] domain but have not provided a potential solution to resolve these problems. In, a solution based on blockchain is devised for data safety during transmission using 5G and IoT-enabled drones [55]. This system is based on the identification of threats manually. An authentication system based on keys for devices was proposed that was not suitable in an IoT-based network of drones. There is a clear research gap in the area of developing a safe and secure drone network by proposing a solution that deals with cyber security threats and makes drones adaptable in industry and the commercial sector. A smart and intelligent system is required for the security of drones that can investigate data of attacks and ensure the security of drones by taking proactive measures. In the past, machine learning models were proposed in the field of cyber security for wireless sensor-based networks and mobile-based networks but not for drone-based security. A machine learning-based solution is proposed in this study for drone security authentication and control access methods (Figure 11.2).

Small drones are opening new possibilities in the defense and civil industries. However, small drones are vulnerable to privacy and security threats due to a lack of appropriate architecture. Evolutions in the IoD [46] and IoT provide new directions and also pose additional challenges related to data privacy and data security. The available framework is not yet secure and reliable in terms of data privacy concerns.

Layers: A layered architecture, as shown in Figure 11.2, which is typically used for smart drones, is updated by adding a security and privacy layer and updating the data processing layer with machine intelligence components.

Drone Layer: In the proposed layered architecture of industrial drones, the first layer is the drone layer, where a quadcopter or other min drone has a camera attached to it. This layer is updated by IoT sensor data. Smart sensors are used, including a GPS sensor, altitude sensor, radar, and camera. This is the first step in the proposed architecture. This layer can perform sensing, recording, and sending the information recorded by drones

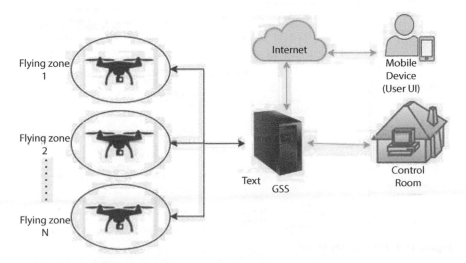

Figure 11.2 Proposed architecture for smart drone security.

to the next layer. In this layer, an unmanned aircraft system (UAS) drone is involved [39], which is responsible for drone flight operations, information recording by sensors, etc. The UAS consists of two parts: a ground controller and a communication connection. In the proposed architecture, a DJI phantom 3 drone is used [23, 25], which consists of a custom remote controller and communication link. In the proposed architecture, sensors are attached to the drone.

Edge Processing Layer: In the second layer—the edge processing layer— the drone and IoT raw data are forwarded to the security and privacy layer [34], where it is verified that the data originate from authenticated devices. This layer deals with the communication and transmission of data to the next station; i.e., the cloud layer. Several gateway device mechanisms are available which provide wireless communication. Wi-Fi communication transmits information at a fast speed. The edge processing layer provides device to cloud communication efficiently. This layer is responsible for data protection, cashing, and flooding. The proposed research uses the Azure IoT gateway for cloud communication. The architecture of the IoT gateway is shown in Figure 11.3.

Security and Privacy Layer: This layer plays an important role in providing device authentication and secure access control by using machine learning models. At this level, the safety and security of data are implemented,

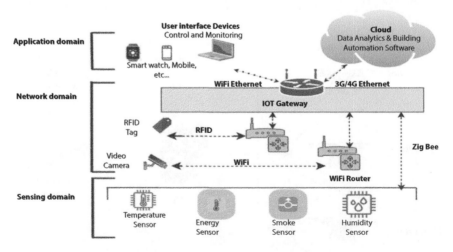

Figure 11.3 IoT gateway model.

which is the key element of this IoT framework. There are several types of privacy threats that can occur at this stage.

Physical privacy is related to capturing someone's property. If a third party is secretly monitoring the drone information, then the private information of someone's property can be compromised. A location privacy threat refers to someone's location being captured by an unauthorized person. A behavior privacy threat is related to the monitoring of someone's activities and behavior by an unauthorized person. Such types of security risks must be tackled by using authentication schemes and protocols. Several types of security breaches are used by unauthorized persons to make such security threats. Intrusion [8, 9], spoofing, jamming, and DoS attacks are the most common kinds of threat. In the proposed architecture, device authentication is maintained using a machine learning algorithm to detect and alert users of such security attacks.

AI Systems' Support to Cyber Security

Against this backdrop, organizations have started using AI to help manage a growing range of cyber security risks, technical challenges, and resource constraints by enhancing their systems' robustness, resilience, and response. Police dogs provide a useful analogy to understand why companies are using A to increase cyber security. Police officers use police dogs' specific abilities to hunt threats; likewise, AI systems work with security analysts to change the speed at which operations can be performed. In this regard, the relationship between AI systems and security operators should be understood as a synergetic integration, in which the unique added value of both humans and AI systems are preserved and enhanced, rather than as a competition between the two. Estimates suggest that the market for AI in cyber security will grow from $3.92 billion in 2017 to $34.81 billion by 2025, at a compound annual growth rate (CAGR) of 31.38% during the forecast period. According to a recent Capgemin survey, the pace of adoption of AI solutions for cyber security is skyrocketing. The number of companies implementing these systems has risen from one fifth of the overall sample in 2019, to two thirds of companies planning to deploy them in 2020. 73% of the sample tested A applications in cyber security. The most common applications are network security, followed by data security, and endpoint security. Three main categories can be identified in AI use in cyber security: detection (51%), prediction (34%), and response (18%).

11.5   Use of AI in Cyber Security

Speed of Impact: In some of the major attacks, the average time of impact on organizations is 4 minutes. Furthermore, today's attacks are not just ransomware, or just targeting certain systems or certain vulnerabilities; they can move and adjust based on what the targets are doing. These kinds of attacks impact incredibly quickly and there are not many human interactions that can happen in the meantime.

Operational Complexity: Today, the proliferation of cloud computing platforms and the fact that those platforms can be operationalized and deliver services very quickly—in the millisecond range—means that you cannot have a lot of humans in that loop, and you have to think about a more analytics-driven capability.

Skills Gaps in Cyber Security Remain an Ongoing Challenge: According to Frost & Sullivan, there is a global shortage of about a million and a half cyber security experts. This level of scarcity pushes the industry to automate processes at a faster rate.

1. AI can help security teams in three ways: by improving systems' robustness, response, and resilience. The report defines this as the 3R model. First, A can improve systems' robustness, that is, the ability of a system to maintain its initial assumed stable configuration even when it processes erroneous inputs, thanks to self-testing and self-healing software. This means that AI systems can be used to improve testing for robustness, delegating to the machines the process of verification and validation. Second, AI can strengthen systems' resilience, i.e., the ability of a system to resist and tolerate an attack by facilitating threat and anomaly detection. Third, AI can be used to enhance system response, i.e., the capacity of a system to respond autonomously to attacks, to identify vulnerabilities in other machines and to operate strategically by deciding which vulnerability to attack and at which point, and to launch more aggressive counterattacks. Identifying when to delegate decision-making and response actions to AI and the need of an individual organization to perform a risk-impact assessment are related. In many cases A will augment, without replacing, the decision-making of human security analysts and will be integrated into processes that accelerate response actions.

11.6   Use of AI in System to Achieve Robustness, Resilience and Response

Whenever AI is applied to cyber-incident detection and response the problem solving can be roughly divided into three parts, as shown in Figure 11.1. Data is collected from customer environments and processed by a system that is managed by a security vendor. The detection system flags malicious activity and can be used to activate an action in response Figure 11.4.

Companies today recognize that the attack surface is growing massively because of the adoption of the Internet of Things (IoT) and the diffusion of mobile devices, compounded by a diverse and ever-changing threat landscape. Against this backdrop, there are two measures that can be implemented: speed up defenders, and slow down attackers. With respect to speeding up defenders, companies adopt AI solutions to automate these detection and response to attacks already active inside the organization's defenses. Security teams traditionally spend a lot of time dealing with alerts, investigating if they are benign or malicious, reporting on them, containing them, and validating the containment actions. AI can help with some of the tasks that security operations teams spend most of their time on. Notably, this is also one of the primary and most common uses of AI in general.

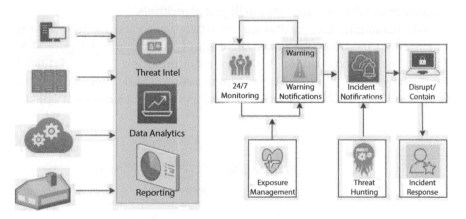

Figure 11.4 AI cyber incidents detection and response.

Intelligent algorithms have capability to discover hidden patterns and detect threats in the computer information systems. The development of hybrid cyber security methods and building computational systems that integrates with intelligent algorithms is needful to analyze big data [4, 58], mitigate threats, and protect against the new invaders. The implementation of optimization techniques is a continuous evolution process in improving the performance of cyber security algorithms in order to yield promising results. Optimization techniques are utilized [56, 60] in diverse perspectives, such as, parameter tuning, satisfy constraints, maximize or minimize objective function, feature selection, weight values optimization, meet multiple criteria, search strategy and finds trade-off solutions [27]. The objective of this chapter is to review the implementation of NIC paradigms, machine learning algorithms, and deep learning algorithms that covers a broad spectrum of cyber security problems. The overall aim of this chapter is to identify and summarize the need of aforementioned algorithms in solving cyber security applications. The proposed research work clearly states the need of intelligent algorithms in solving different kinds of cyber security problems and also, it infers the conceptual ideas, significance, and implications of these algorithms that improve efficiency and effectiveness for obtaining quantitative and qualitative experimental outcomes. The proposed chapter would be greatly beneficial to different peer groups of people, namely, research scholars, academicians, scientists, industrial experts and post graduate students who are working in the cyber security research area based on intelligent algorithms.

NIC paradigms are partitioned into two categories, namely, Swarm Intelligence (SI) and Evolutionary Algorithm (EA) as shown in Figure 11.5. Swarm Intelligence includes Ant Colony Optimization (ACO), Artificial Bee Colony (ABC), Particle Swarm Optimization (PSO), Firefly Algorithm (FA), Cuckoo Search (CS), and Bacterial Foraging Optimization (BFO).

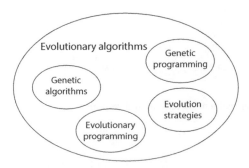

Figure 11.5 Evolutionary algorithm.

Evolutionary algorithm includes Genetic Algorithm (GA), Evolutionary Programming (EP), Memetic algorithm, Genetic Programming (GP), Evolutionary strategies, Differential Evolution (DE), and Cultural algorithms. Machine learning techniques are also partitioned into three categories, such as supervised adaptation, reinforcement adaptation [22] and unsupervised adaptation which are described below. Linear regression, Logistic regression, Linear discriminant analysis, classification and regression trees [14, 15], Naïve bayes [16], K-Nearest Neighbor (K-NN), Kmeans Clustering, Learning Vector Quantization (LVQ), Support Vector Machines (SVM), Random Forest, Monte Carlo, Neural networks, and Q-learning [23, 25] are traditional examples of machine learning algorithms. The abstract view of computational intelligence concept [4] is depicted in Figure 11.3. It comprises adaptation and self-organization using processed data and embedded knowledge as input and produces predictions, decisions, generalizations, and reason as output. Computational intelligence techniques [4, 58] partitioned adaptation characteristics into three categories, such as supervised adaptation, reinforcement adaptation, and unsupervised adaptation portrayed in Figure 11.5.

Supervised Adaptation: The adaptation is carried out in the execution of system at every iteration. The fine-tuned variables/parameters are subjected to generalize the behavior of a computational model in the dynamic environment, and the performance of the system is consistently improved.

Reinforcement Adaptation: The number of variables/parameters involved in the system is interacted to achieve best fitness solution through heuristic reinforcement approach. It deals with a time series of input vector space, evaluates the fitness of the system, and produces possible outcomes for each input (Figure 11.6).

Unsupervised Adaptation: It follows trial and error method. The number of variables/parameters involved in the system performs the task. Based on the obtained fitness value, computational model is generalized to achieve better results in an iterative approach.

Deep learning algorithms are [27, 30] a part of machine learning algorithms which involves multiple layers of deep learning data architectures, representations, and transformations. Convolutional Neural Network (CNN) and Recurrent Neural Network (RNN) are typical examples of deep learning algorithms. Aforementioned three categories of algorithms [26] are utilized to solve a wide range of cyber security problems which are discussed here. Machine learning algorithms are utilized to design and development of a learning-based security model. NIC algorithms are applied for

Environment

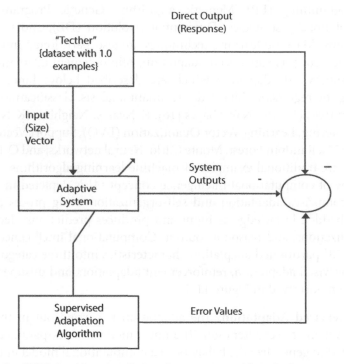

Figure 11.6 Role of NIC algorithms in solving cyber security problems [60].

fine-tuning parameters involved in the security model in order to improve the efficiency and performance. Deep learning algorithms are generally implemented for solving the complicated cyber security problems that involves huge volume of varied dataset [12]. The proposed chapter focuses on the implementation of NIC algorithms, machine learning algorithms and deep learning algorithms [30] in cyber security which spans across different applications, such as, network security, information security, secure communication in wireless sensor networks, cryptographic algorithm to reduce threats in data transmission over the network, intrusion detection [8], phishing detection, machine monitoring, signature verification, virus detection, insider attack detection, profiling network traffic, anomaly detection, malware detection, IoT security, security in web service, ad hoc security, cyber warfare, security in electronic services, biometric security, honeypot security, vulnerability assessment, social applications security, botnet detection, attack detection and sensor network security.

11.7   NIC Algorithms in Cyber Security

The characteristics of NIC algorithms are partitioned into two segments such as swarm intelligence and evolutionary algorithm. The Swarm Intelligence-based Algorithms (SIA) are developed based on the idea of collective behaviors of insects such as ants, bees, wasps, and termites living in colonies. Researchers are interested in the new way of achieving a form of collective intelligence called swarm intelligence. SIAs are also advanced as a computational intelligence technique based around the study of collective behavior in decentralized and self- organized systems. The development of evolutionary computation techniques is derived from three main observations. First, the selection of most appropriate individuals (parents) is determined by combination (reproduction). Second, randomness (mutation) expands the search space of the diversity. Third, the fittest individuals have a higher probability of surviving to the next generation. The combination of natural selection and self-organization is denoted in Equation 11.1 as follows.

$$\text{Evolution } n = (\text{natural selections}) + (\text{self-organization}) \quad (11.1)$$

Machine Learning in Cyber Security

Machine learning algorithm is defined as a methodology involving computing that provides a system with an ability to learn and deal with new situations, such that the system is perceived to possess one or more attributes of reason, such as generalization, discovery, association, and abstraction. The output of a machine learning model often includes perceptions and/or decisions. It consists of practical adaptation and self-organization concepts, paradigms, algorithms, and implementations that enable or facilitate appropriate actions (intelligent behavior) in complex and changing environments. Adaptation and self-organization are the two most important characteristics exhibited by intelligent paradigms which are widely applied to develop intelligent system and computational model that provide promising solution for solving the large scale of complicated optimization problems. The implementation of machine learning algorithms for solving cyber security problems are presented in Figure 11.7.

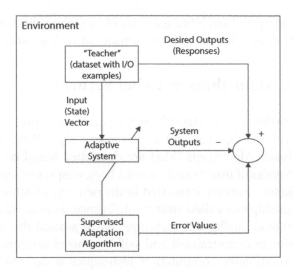

Figure 11.7 Role of NIC algorithms in solving cyber security problems.

11.8   Example Systems for AI and ML Applications for Cyber Security Diagnose

ML trained on user interaction provides a way of understanding local context and knowing what data to focus on; models trained to identify those more likely to be malicious improve the efficiency of a system by triaging the information to process in real time. In this way, using ML is cost saving but also allows for faster reaction in the most critical situations.

ML can be useful in detecting new anomalies by learning robust models from the data they have been fed with. ML is particularly good at identifying patterns and extracting algorithms in large sets of data where humans are lost.

ML can be useful for asynchronous user profiling and for measuring deviation from common behaviors as well as going back to much larger data volumes to understand behavior.

ML trained on immutable attacker 'Tactics, Techniques, and Procedures' (TTP) behaviors (those identified in the Mire Attack framework) can support durable and broad attacker detection.

To better illustrate the use of AI and ML for cyber security detection and response, Figure 11.8 presents an intrusion detection and prevention system that combines software and hardware devices inside the network. The system "can detect possible intrusions and attempt to prevent them. Intrusion detection and prevention systems provide four vital security

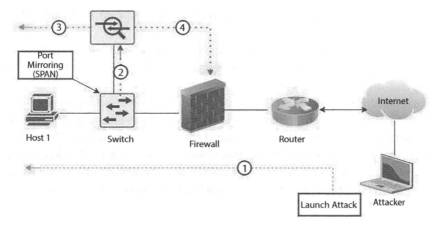

Figure 11.8 Intrusion detection and prevention system.

Table 11.1 Examples of A techniques for intrusion, prevention, detection, and response.

Technology	Advantages
Artificial neural networks	Parallelism in information processing, Learning by example, Nonlinearity – handling complex nonlinear functions Resilience to noise and incomplete data versatility and flexibility with learning models
Intelligent Agents	Mobility Rationality – in achieving their objectives Adaptability – to the environment and user preferences Collaboration – awareness that a human user can make mistakes and provide uncertain or omit important information; thus, they should not accept instructions without consideration and checking the inconsistencies with the user
Genetic Algorithms	Robustness Adaptability to the environment Optimization – providing optimal solutions even for complex computing problems Parallelism – allowing evaluation of multiple schemas at once Flexible and robust global search
Fuzzy Sets	Robustness of their interpolative reasoning mechanism Interoperability – human friendliness

functions: monitoring, detecting, analyzing, and responding to unauthorized activities.

There are a variety of AI techniques that can be used for intrusion [9], prevention, detection, and response. Table 11.1 illustrates examples of the main advantages of some of these techniques.

All these intrusion detection AI-powered technologies help in reducing the dwell time – the length of time a cyber attacker has free reign in an environment from the time they get in until they are eradicated. In December 2019, the dwell time in Europe was about 177 days, and attackers were discovered in only 44% of cases because of data breach or other problems. Using AI techniques, the dwell time has been dramatically reduced.

11.9   Introduction of New Threats

As well as existing threats expanding in scale and scope, progress in AI means completely new threats could be introduced. The AI characteristics of being unbounded by human capabilities could allow actors to execute attacks that would not otherwise be feasible.

Deepfakes

The use of 'deepfakes' has been steadily rising since a Reddit user first coined the term in 2017. Deepfakes are a developing technology that uses deep learning to make images, videos, or texts of fake events [21]. There are two main methods to make deepfakes. The first is usually adopted for 'face-swapping' (i.e., placing one person's face onto someone else's), and requires thousands of face shots of the two people to be run through an AI algorithm called an encoder. The encoder then finds and learns similarities between the two faces, and reduces them to their shared common features, compressing the images in the process. A second AI algorithm called a decoder is then taught to recover the faces from the compressed images: one decoder recovers the first person's face, and another recovers the second person's face. Then, by giving encoded images to the 'wrong' decoder, the face-swap is performed on as many frames of a video as possible to make a convincing deepfake. The second and very important method to make deepfakes is called a generative adversarial network (GAN). A GAN pits two AI algorithms against each other to create brand new images (see Figure 11.9). One algorithm, the generator, is fed with random data and generates a new image. The second algorithm, the discriminator, checks the image and data to see if it corresponds with known data (i.e. known

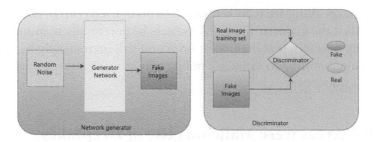

Figure 11.9 The functioning of a generative adversarial network.

images or faces). This battle between the two algorithms essentially winds up forcing the generator into creating extremely realistic images (e.g. of celebrities) that attempt to fool the discriminator.

These images have been used to create fake yet realistic images of people, with often harmful consequences. For example, a McAfee team used a GAN to fool a facial recognition system like those currently in use for passport verification at airports. McAfee relied on state-of-the-art, open-source facial-recognition algorithms, usually quite similar to one another, thereby raising important concerns about the security of facial-recognition systems. Deepfake applications also include text and voice manipulation as well as videos. As far as voice manipulation is concerned, Lyrebird claims that, using AI, it was able to recreate any voice using just one minute of sample audio, while Baidu's Deep Voice clones' speech with less than four seconds of training. In March 2019, AI-based software was used to impersonate a chief executive's voice and demand a fraudulent transfer of €220,000 ($243,000). In this case, the CEO thought he was talking to the chief executive of the firm's German parent company, who demanded the payment be made to a Hungarian subsidiary. Deepfakes used for text manipulation is also increasingly concerning. With GPT-3 generative writing it is possible to synthetically reproduce human-sounding sentences that are potentially even more difficult to distinguish from human-generated ones than video content. Even with state-of-the-art technology, it is still possible to tell video content has been synthetically produced, for example from a person's facial movements being slightly off. But with GPT-3 output there is no unaltered original that could be used for comparison or as evidence for a fact check. Text manipulation has been used extensively for AI-generated comments and tweets. Diresta highlights how "seeing a lot of people express the same point of view, often at the same time or in the same place, can convince observers that everyone feels a certain way, regardless of whether the people speaking are truly representative—or even real. In psychology, this is called the majority illusion." As such, by potentially manufacturing a majority opinion, text manipulation is and will increasingly be applied to campaigns aiming to influence public opinion. The strategic and political consequences are clear.

11.10   Areas were Malicious Use of Deepfakes is Trending

Pornographic

The number of deepfake videos online amounted to 14,678 in September 2019, according to Deeptrace Labs, a 100% increase since December 2018. The majority of these (96%) are pornographic in content, although other forms have also gained popularity. Deepfake technology can put women or men in a sex act without their consent, while also removing the original actor, creating a powerful weapon for harm or abuse. According to a Data and Society report, deepfakes and other audio and visual manipulation can be used with pornography to enact vendettas, blackmail people, or trick them into participating in personalized financial scams. The increasing accessibility of this technology makes this even more problematic. One recent example is the conjunction of 3D-generated porn and deepfakes, which allows a user to put a real person's face on another person's body, and do whatever violent or sexual act they want with them. Notably, audiovisual manipulation and other less sophisticated methods such as basic video and photo-editing software (part of what Paris and Donovan call 'Cheap Fakes' or 'Shallowfakes'), can also change audiovisual manipulation in malicious ways, much more easily and cheaply.

11.11   Model-Aided Deep Reinforcement Learning for Sample-Efficient UAV Trajectory Design in IoT Networks

Deep reinforcement learning (DRL) is gaining attention as a potential approach to design trajectories for autonomous unmanned aerial vehicles (UAV) used as flying access points in the context of cellular [20, 21] or Internet of Things (IoT) connectivity [44]. DRL solutions offer the advantage of on-the-go learning hence relying on very little prior contextual information. A corresponding drawback however lies in the need for many learning episodes which severely restricts the applicability of such approach in real-world time- and energy-constrained missions. Here, we propose a model-aided deep Q-learning approach that, in contrast to previous work, considerably reduces the need for extensive training data samples, while still achieving the overarching goal of DRL, i.e. to guide a battery-limited UAV on an efficient data harvesting trajectory, without prior knowledge of wireless channel characteristics and limited knowledge of wireless node locations. The key idea consists in using a small subset of nodes as anchors (i.e. with known location) and learning a model of the propagation environment while implicitly estimating the positions of regular nodes. Interaction with the model allows us to train a deep Q-network (DQN) to approximate the optimal UAV control policy [24, 28, 29]. We show that in comparison with standard DRL approaches, the proposed model-aided approach requires at least one order of magnitude less training data samples to reach identical data collection performance, hence offering a first step towards making DRL a viable solution to the problem. Index Terms—Deep reinforcement learning, UAV, IoT.

Rapid innovation in producing low-cost commercial unmanned aerial vehicles (UAVs) [31] has opened up numerous opportunities in the UAV market which is projected to reach Billion USD $ by 2025 [30, 11]. One key application scenario is the future Internet of Things (IoT), in which harvesting data from wireless nodes that are spread out over wide areas far away from base stations (BSs) generally requires higher transmission power to communicate the information, reducing the network's operating duration by draining the sensor battery faster. A UAV [57] that acts as a flying BS can describe a flight pattern that brings it in close range to the ground nodes, hence reducing battery consumption and increasing the energy efficiency of the data harvesting system. However, delivering this gain hinges on the availability of efficient methods to design a trajectory for the UAV, deciding when and where to collect data from ground nodes [50, 57]. The popularity of deep reinforcement learning (DRL) in this context can be explained by the fact that full information about the scenario environment (e.g., IoT sensor positions) is not a prerequisite. Further reasons include the computational efficiency of DRL inference, as well as the inherent complexity of UAV path planning, which is in general non-convex and often NP-hard [61]. However, one of the greatest obstacles to deploying DRL-based path planning to real-world autonomous UAVs is the prohibitively extensive training data required, equivalent to thousands of training flights. In this work, we address this issue by proposing a so-called model-aided DRL approach that only requires a minimum of training data to control a UAV data harvester under a limited flight-time. The training data demand of DRL methods for UAV path planning depends in large part on the scenario complexity and the availability of prior information about the environment. On the one hand, works such as, where a deep Q-network (DQN) is trained to control an energy-limited UAV BS, assume absolutely no prior knowledge of the environment, requiring large amounts of training even in a simple environment as the DRL agent has to deduce the scenario conditions purely by trial and error. On the other hand, near perfect state information in works such as, where cooperative UAVs are tasked with collecting data from IoT devices in a relatively simple unobstructed environment, enables faster convergence and requires less training data. In this work, prior knowledge available to the UAV agent is in between the two extremes: while some reference IoT node positions are known (referred to as anchors), other node positions, and the challenging wireless channel characteristics in a dense urban environment that causes alternation between line-of-sight (LoS) and non-line-of-sight (NLoS) links, must be estimated. In the context of sample-efficient RL, model-accelerated solutions have been proposed previously for a variety of applications. A method called imagination rollouts to increase sample-efficiency for a continuous Q-learning variant has been suggested for simulated robotic tasks in. Their approach is based on using iteratively refitted time-varying linear models, in contrast to a neural network (NN) model that we propose here. Learning a NN model in the context of stochastic value gradient learning methods has been proposed in. Other works in the area of RL trajectory optimization for UAV communications have suggested other ways of reducing training data demand [37]. A DRL method was proposed for sum-rate maximization from moving users based on transfer learning to reduce training time.

11.12   Model-Aided Deep Q-Learning

Employing standard deep Q-learning is often not practical due to the tremendous amount of training data points required and the cost associated with obtaining these data points, i.e., through real-world UAV experiments. To ameliorate this problem, we propose an algorithm where the agent learns an environment model continuously while collecting real-world measurements. This model is then used by the agent to simulate experiments and supplement the real-world data. More specifically in our scenario, the next state sn+1 given the current state sn and action a can be computed from. The reward function consists of two parts: the safety penalty, which is known from, and the instantaneous collected data from the IoT node devices. Therefore, we only need to estimate the instantaneous collected data from devices which according to (5), (4), is a function of ground node locations and the radio channel model. Hence, the approximation of the reward function boils down to ground node localization and radio channel learning from collected radio measurements. The problem of simultaneous wireless node localization and channel learning has been studied in previously. In this section, we propose a new approach of model-free node localization by leveraging the 3D map of the environment. Akin to, a LoS/NLoS segmented radio channel is assumed. However, in contrast to, our goal here is to estimate the radio channel using a model-free method while localizing the ground nodes. To learn the radio channel, we use a neural network (NN). This network is utilized along with a particle swarm optimization (PSO) technique and a 3D map of the city to localize the wireless nodes with unknown positions.

Simultaneous Node Localization and Channel Learning We assume the UAV follows an arbitrary trajectory denoted by $\chi = \{v_n, n \in [1, N]\}$ for collecting received signal strength (RSS) measurements, where v_n represents the UAV's position in the n-th time interval. We also assume that the UAV collects radio measurements form all K nodes at each location. Let g_n, k represent the RSS measurements (in dB scale) obtained from the k-th node by the UAV in the n-th interval. Assuming a LoS/NLoS segmented path loss model that is suitable for air-to-ground channels in urban environments with buildings.

Here $d_{n,k} = \|v_n - u_k\|2$ and ψ_n is the UAV heading angle at time step n. The function $\gamma(.)$ is the antenna gain which is unknown which needs to be learned. The $\omega_{n,k} \in \{0, 1\}$ is the classification [13–15] binary variable (yet unknown) indicating whether a measurement falls into the LoS or NLoS category. The function $\psi\theta(.)$ is the channel model parameterized by θ. Note that, neither function $\psi(.)$ nor parameters θ are known and need to be estimated. η_n, k,z stands for the shadowing effect with zero-mean normal distribution with known variance $\sigma2$ z. The probability distribution of a single measurement in (12) is modeled as

$$p(g_{n,k}) = (f_{n,k,LoS})_i \; w_{n,k} \; (f_{n,k,NLoS})^{(1-wn,k)}$$

where $f_{n,k,z}$ has a Gaussian distribution with $N (\phi_z(d_{n,k}) + \gamma(v_n, u_k, \psi_n), \sigma_z^2)$.

11.13   Algorithm Model-Aided Deep Q-Learning Trajectory Design

1: Initialize replay buffer (B),(B')
2: Initialize Q-network and target network parameters s
3: Initialize t = 0
4: for e = 0 to E_{max} do
5: t = t + 1
6: 1) Real-world experiment:
7: Initialize $s_0 = (vI, b_{max})$, n = 0
8: while $b_n \geq 0$ do
9: $a_n = \arg\max_a Q^\pi (s_n, a, \theta)$
10: Validate a_n using the safety controller (9)
11: Observe $r_n, s_{n+1}, \gamma_{1,n}, \cdots, \gamma_{K,n}$
12: Store (s_n, a_n, r_n, s_{n+1}) on (B)
13: Memorize $(v_n, \gamma_{1,n}, \cdots, \gamma_{K,n})$
14: n = n + 1
15: end while
16: 2) Learning the environment:
17: Learn the radio channel as described in Section IV-A1
18: Localize unknown nodes as described in Section IV-A2
19: 3) Simulated-world experiment:
20: for i = 0 to I do
21: t = t + 1
22: Initialize $s'_0 = (vI, b_{max})$, n = 0
23: while $b_n \geq 0$ do

24: $a_n' = \begin{cases} randomly \text{ select from A with probability } \varepsilon \\ \arg\max_a Q^\pi (s'_n, a, \theta) \text{ else} \end{cases}$

25: Validate an using the safety controller (9)
26: Compute $\tilde{r}_n$ from (7), and $\tilde{s}_{n+1}$ from (1), (3)
27: store $(\tilde{s}_n, \tilde{a}_n, \tilde{r}_n, \tilde{s}_{n+1})$ on $\tilde{B}$
28: for m = 0 to M do
29: Sample (s_m, a_m, r_m, s_{m+1}) uniformly from $\{BUB\}^{\sim}$

30: $y_m = \begin{cases} r_m & \text{if terminal} \\ r_m + \gamma\max_a Q^\pi (s_{m+1}, a, \theta) & else \end{cases}$

31: $\ell_{in}(\theta) = E[(y_m - Q^\pi (s_m, a_m, \theta))^2]$
32: end for
33: $\theta = \theta + \beta \dfrac{1}{M} \nabla_\theta \sum_{m=0}^{M} l_m(\theta)$

34: n=n+1

35: end while

36: $\varepsilon = \varepsilon_{final} + (\varepsilon_{start} - \varepsilon_{final}) exo(-kt)$

37: $f(t \bmod N_{target} = 0) \, then \, \theta = \theta$

38: end for

39: end for

11.13.1   Numerical Results

We consider a dense urban city neighborhood comprising buildings and regular streets as shown in Figure 11.2. The height of the buildings is Rayleigh distributed in the range of 5–40 m and the true propagation parameters are chosen similar to. The UAV collects radio measurements from the ground nodes every 5 m and we assume that the altitude of the UAV is fixed to 60 m during the course of its trajectory. The mission time of each episode is fixed to N = 20-time steps with a fixed step size of c = 50 m. We assume there are six ground nodes. Only the locations of anchor nodes u1 and u2 are known to the UAV in advance. The UAV starts from vI = [100, 100, 60]T and needs to reach the destination point vF = [300, 400, 60]T by the end of the mission. To learn the channel, we use a NN with two hidden layers where the first layer has 60 neurons with tanh activation function, and the second layer 30 neurons with relu activation function. The Q-network comprises hidden layers each with 120 neurons and relu activation function. In Figure 11.1, we compare the performance of the baseline Q-learning algorithm as explained in Section 11.2 and akin to, with the proposed model-aided Q-learning algorithm. Moreover, we show the result of an algorithm similar to, where the mixed-radio map of the nodes is embedded in the state vector. To compute the mixed-radio map, the individual radio maps of all nodes are combined. Individual radio maps are computed using the 3D map of the city and assuming perfect knowledge node positions and the radio channel. The model-aided algorithm outperforms the other approaches since it merely requires 10 real-world experiment episodes to converge to the same performance level as other algorithms. The algorithm introduced in is superior to the baseline since it uses more information, i.e., the map and perfect knowledge of node positions and the radio channel model. Figure 11.10 shows the final trajectory after convergence. The UAV starts flying towards the closest node and hovers above for several time steps in order to maximize the amount of collected data, and then reaches the destination vF. Moreover, the estimate of unknown node locations obtained at the last episode of the training phase of Algorithm 1 are shown and confirmed to be very close the true positions.

		smooth	gradient dominated
	Alg. 1, this paper (2-point + DGD)	$O\left(\sqrt{\frac{d}{m}}\log m\right)$	$O\left(\frac{d}{m}\right)$
distributed zero-order (nonconvex)	Alg. 2, this paper (2d-point + gradient tracking)	$O\left(\frac{d}{m}\right)$	$O\left(\left[1-c(1-\rho^2)^2\left(\frac{\mu}{L}\right)^{\frac{4}{3}}\right]^{m/d}\right)$
	ZONE [1]	$O\left(\frac{\gamma(d)}{\sqrt{M}}\right)$	—
distributed first-order	DGD	$O\left(\frac{\log t}{\sqrt{t}}\right)$ [2, 3] (convex) $O\left(\frac{1}{\sqrt{T}}\right)$ [5] (nonconvex)	$O\left(\frac{1}{t}\right)$ [4] (strongly convex)
	gradient tracking	$O\left(\frac{1}{t}\right)$ [6] (nonconvex)	$O\left(\left[1-c(1-\rho)^2\left(\frac{\mu}{L}\right)^{\frac{3}{2}}\right]^t\right)$ [7] (strongly convex)
centralized zero-order	[8] (2-point estimator)	$O\left(\frac{d}{m}\right)$ (nonconvex)	$O\left(\left[1-\frac{c}{d}\frac{\mu}{L}\right]^m\right)$ (strongly convex)

Figure 11.10 Comparison of different algorithms [61].

Figure 11.10 shows the comparison of different algorithms, showing accumulated collected data versus training their flexibility and the possibility of use in a wide range of applications, such as the security, control, monitoring, and exploration of terrestrial areas otherwise difficult to reach quickly. Furthermore, this is transformative technology, enhancing how first responders can reach and carry out rescue missions at the sites of natural disasters. In addition, it can provide support for delivering medical supplies, as well as for emergency management cases such as forest firefighting, critical infrastructure protection and inspection, coastal monitoring, and police augmentation, in addition to helping smart cities meet their public safety requirements. The relationship between public safety and the Internet of Things (IoT) was discussed in, while a taxonomy of the IoT-based smart city was provided in. Applying IoT technologies to smart cities could lead to changes and improvements in the economy, safety, management of public utilization, and transportation in smart cities. The most attractive application of drones [54, 62] is the collection of data from IoT using wearable devices in smart cities or at events. A drone can communicate with heterogeneous devices on the ground.

11.14   Machine Learning for Drone Security

The basic types of machine learning techniques include supervised, unsupervised, and semi-supervised learning. A literature review revealed that machine learning models have been utilized by many researchers to deal with cyber attacks in mobile-based networks, sensor-based wireless networks, cloud-computing, and IoT-based systems. Vedula *et al.* combined a supervised learning model with a self-learning model through RF and LSTM (autoencoder) to detect DDoS attacks using two features. Researchers proposed an approach to detect and control an actuation attack in a constrained cyber-physical system [49] using a probabilistic approach in. No work has been found dealing with cyber attacks using machine learning models in drone networks. We have also proposed an access control system in drone security.

An extensive literature review ranging from 2010 to 2020 concerning privacy and safety concerns of drone data security shows a large number of research works have been published. Most of the studies discuss challenges, applications, and issues related to cyber security, data privacy concerns, spoofing, hijacking drones [59], and other threats. However, many researchers have highlighted the problem domain but have not provided a potential solution to resolve these problems. A solution based on blockchain is devised for data safety during transmission using 5G and IoT-enabled drones. This system is based on the identification of threats manually. An authentication system based on keys for devices was proposed that was not suitable in an IoT-based network of drones. There is a clear research gap in the area of developing a safe and secure drone network by proposing a solution that deals with cyber security threats and makes drones adaptable in industry and the commercial sector. A smart and intelligent system is required for the security of drones that can investigate data of attacks and ensure the security of drones by taking proactive measures. In the past, machine learning models were proposed in the field of cyber security for wireless sensor-based networks and mobile-based networks but not for drone-based security. A machine learning-based solution is proposed in this study for drone security authentication and control access methods.

11.15   Surveillance

Drones can be equipped with various types of surveillance equipment that can collect HD video and still images day and night. Drones can be equipped with technology allowing them to intercept cell phone calls, determine GPS locations, and gather license plate information. The high payload compatibility allows the use of different surveying systems such as lidar scanners, multi- and hyperspectral devices and much more – around the clock, with low staffing requirements and low costs.

Drone surveillance is the use of unmanned aerial vehicles (UAV) to capture of still images and video to gather information about specific targets, which might be individuals, groups, or environments. Drone surveillance enables surreptitiously gathering information about a target as captured from a distance or altitude. Drone surveillance enables the surreptitious gathering of information about a target as captured from a distance or altitude of course, drone use in this wide-ranging industry (or area) extends well beyond these fundamental and simplistic boundaries. The government officials, police, and other security personnel use this drone technology. Automated surveillance is going to become increasingly common as companies and researchers find new ways to use machine learning to analyze live video footage.

11.16   Technologies Driving Drones' Success

UAV (Unmanned Aerial Vehicles) have been in existence for a long time now, however there seems to be a sudden hype for drones. In this section let us understand what is driving the need for drones. Like any disruptive innovation, Drones also reply on advancements in many fields that have enabled drones to come to reality. I have briefly outlined the various technology advancements here:

Digital Cameras: The most popular commercial use cases for drones, be it aerial photography or agriculture or aerial surveillance, rely on drones to provide high-definition footages or images of objects of interest. Cameras have become much cheaper these days. There are a handful of truly affordable quadcopters, as cheap as $60, ready to fly and equipped with built-in cameras. For surveillance applications, HD cameras are typically used. These cameras are also equipped with on-board analytics like object detection and tracking. An example could be DARPAs recently launched 1.8-gigapixel surveillance drone that can spot a terrorist from 20,000 feet.

Wireless Communication: Wireless communication has changed the world in many ways; many businesses today cannot survive without them being connected to their customers wirelessly. However, Wi-Fi is not still available universally. There are places where Wi-Fi may not be available. Drones are being used to bridge this gap by acting as Wi-Fi hot spots in places with no connectivity. Google acquired Titan Aerospace, a maker of high-altitude unmanned aerial vehicles (UAVs) [36] and Facebook bought Ascenta. Both companies are looking at drones as an opportunity to deliver

Internet access to the roughly five billion people who lack reliable land-based access today.

Computing Complexity and Analytics: For various applications, drones are used to collect data and communicate real-time intelligence to end users. For this purpose, drones have to be equipped with real time complex computing and analytics capabilities. Because most of the data collected by drones is in form of videos or images, the use of GPU or advanced CPUs for data processing has helped drones compute and communicate real time intelligence.

Internet of Things: Drones are an important end application for Internet of Things. They can be summoned to deliver things to end customers accurately and in-time. Near future applications could also include time critical delivery of medicines and other life-saving supplies by drones and this can be possible only by including IoT capabilities in drones. Equipped with Wi-Fi, they can provide broadband connectivity on demand to places where it would otherwise be unavailable. Equipped with video cameras, they can show us what's going on in locations where we might otherwise

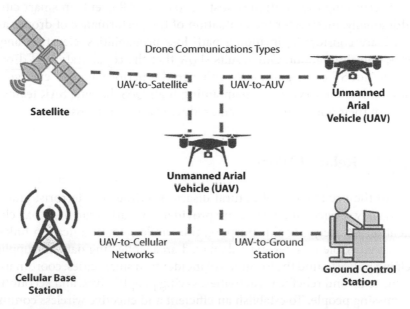

Figure 11.11 Typical cyber security and data privacy threats to smart drones.

have to guess. Equipped with RFID antennas or sensors, they can help us track objects or people, or alert us to changing environmental conditions (Figure 11.11).

This article mainly focuses on the network performance of drones and IoPST. Furthermore, we introduce the novel concept of collaboration between drones and IoPST in smart cities, in order to enhance the network performance and maintain the QoS. Furthermore, we discuss drones in public monitor a criminal on the run, find a missing person, survey a disaster scene, etc., especially in time-critical situations. Therefore, the best features of the integration of drones and IoPST are those that are supposed to achieve and provide maximum benefits, such as their reprogramma-bility, good sensing capability, ability to interconnect and identify things, ubiquity, communication capability, etc. This article mainly focuses on the network performance of drones and IoPST. Furthermore, we introduce the novel concept of collaboration between drones and IoPST in smart cities, in order to enhance the network performance and maintain the QoS. Furthermore, we discuss drones in public safety applications. With respect to public safety, we discuss the capability of drones and IoPST to collect data in real time, track crimes, and guide police in finding crime locations, as well as guiding Search and Rescue (SAR) teams effectively and efficiently, leading to improved orientation. In this case, the collaboration between drone technology and IoPST represents a key technology for determining the location of lost people in SAR events in smart cities. Additionally, metrics for the evaluation of the performance of drones and IoPST are considered, including path loss probabilities, elevation angle, delay, and throughput. Our results show that the collaboration of drones and the IoPST can significantly enhance the performance of emergency communication services by maintaining high QoS, helping SAR teams to perform their duties efficiently, and reducing economic losses.

11.17   Related Work

Due to the occurrences of natural disasters, crime, and terrorist attacks, emergency services and healthcare providers, in particular, must pay close attention to safety. For this purpose, the wireless communication linkage plays a vital role in assessing a damaged area, collecting data on supplies, helping police to find the locations of incidents in smart cities, coordinating rescue team and relief team activities, saving people's lives, and accounting for missing people. To establish an efficient and effective wireless communication network for delivering data in a disaster area spectrum width is highly recommended [17]. However, terrestrial wireless communication technologies could be missing, unavailable through congestion, or damaged. The critical factors for emergency communication network solutions and disaster recovery are rapid deployment, immediate availability, and reliability. Therefore, space technologies represent the best solution for disaster recovery, public safety, SAR, and emergency services. Space technologies are used for collecting information needed to protect human and reduce economic losses. In the aftermath of a disaster, a satellite is a reliable communication solution, but the weakness is a time delay and launching cost. Therefore, using an aerial platform can be a better solution, because it has the merits of both space and terrestrial wireless communication systems. Mohorcic *et al.* argued the use of aerial platforms for disaster and emergency situations and also showed the significance of rescue teams during the disaster. The ability of an aerial platform to deliver communication services such as E911 to facilitate SAR operations is discussed in. The advantages of aerial platforms are the capability and stable coverage area, survivability, mitigated interference that occurs in the wireless communication, and ability to manage traffic. They offer a valuable alternative to support emergency communications after a disaster. Disaster prediction coverage for mitigation of disaster impact a from low-altitude platform (LAP) is discussed in.

Drones belong to the LAP family, and are considered to be space robots. The significant advantages of drones are deployment cost, line of sight (LoS), low propagation delay, rapid deployment, fixed station, and use in disasters. The propagation models, mobility, and positioning of drones for a communication network are discussed in. Drones are attractive for emergency communication because of the possibility of rapid deployment and users operating them from their existing mobile handsets in disaster zones. Therefore, drones represent the best solution for disaster recovery and emergency services because they can be used to support relief and rescue teams in performing their tasks efficiently. Drones play a vital role in connected devices in smart cities. Hence, the authors of reviewed the various aspect of drones related to privacy, cyber security, and public safety in smart cities IoT devices, robots, and humans can communicate as a cooperative. In particular, when flying drones are used, they can support the connectivity of existing terrestrial wireless networks such as cellular and broadband networks. Compared to terrestrial base stations, the advantage of using drones as flying base stations is their ability to provide on-the-fly communications and to establish a LoS communication links to ground users. However, the coverage area for the deployment of drone base station was considered for minimizing energy cost, improving the coverage radius and optimal altitude of drones. Indeed, another important application of drones is in IoT scenarios, because devices often have small transmission power and may not be able to communicate over a long range.

Sharma et al. Summarized the various cooperation approaches for the formation of drones [48]. Also, Sharma et al. Developed a scenario for traffic management and cooperation of drones and nodes on the ground to provide continuous data transfers and network stabilization by using ad hoc technology. The target applications and technological implications of IoT-aided robotics were discussed. Furthermore, Dutta et al. addressed the network security enhancement of IoT-aided robotics in a complex environment. In addition to the work of the authors in, the authors of reviewed the convergence [43] in terms of network protocols, architectures, and embedded software for IoT robotics for smart cities. The interaction of robotic and IoT devices was investigated. Therefore, AI, robots, and IoT will provide the next generation of IoT applications. Moreover, renewable energy harvesting was discussed for the energy Internet of drone communication and networks. Therefore, drone technology and information and communication technologies pay a vital role in smart cities' reduction of resource consumption and costs. Greening information and communication technologies enabled the green IoT by reducing energy consumption, pollution, and hazardous emissions in smart cities. Most of the challenges regarding energy efficiency, interference, and communication networks are discussed in depth, along with intelligent techniques for processing data. Furthermore, the authors of explored the potential usefulness of the IoT to enhance public safety and discussed the challenges and opportunities of using the IoT to support public safety networks and SAR. The hybrid network architecture for public safety broadband communication was discussed in, in which the stationary base station supports heavy traffic after an incident. However, LTE network architecture towards 5G in order to support emerging public safety was discussed in. The authors of analyzed how the deployment of the drone as a base station could deliver communication services to a particular area. Furthermore, the coexistence between heterogonous devices and drones was discussed. The fundamental operation and techniques were introduced for enhancing the efficiency and accuracy of public safety network using IoT technologies. Moreover, the authors of reviewed the techniques and the availability of IoT for disaster management. Reina et al. Outlined the disaster management and the importance of using IoT and big data. The study focused on the use of drones as responders to deliver communication service to victims in a disaster. The main idea of the responder was to arrive at the disaster area before the first responders. The work was divided into two parts, the collection of data and the use of local searching to find the optimal position in which the drone could deliver communication services to the victims. Furthermore, the work in was supported by, in which the position of the responder was considered for delivering communication services to victims in disaster areas. However, keeping a connectivity link between the th responder and other responders was not discussed. Therefore, Alsamh *et al.* developed an artificial neural network to predict signal strength between the drone and wearable IoT devices or other responders on the ground to keep up connectivity for delivering services. The collaboration of drones and IoT plays a significant role in public safety. The proposed network architecture, that is, the integration of drones, IoT and smart wearable devices, offers numerous services like supporting disaster relief team to save human lives, long-distance communication, greening communication, etc. To the best of our knowledge, no study has been done to evaluate the collaboration of drone and IoPST performance network.

11.18   Drones for Public Safety

Drones can fly autonomously in the sky and are associated with different applications in civilian tasks such as transportation, communication, agriculture, disaster mitigation, emergency response, smart things, and environmental preservation. They are a promising technology because of their rapid and easy deployment, ability to dynamically change location in an emergency situation, quick reconfiguration, and flexible technology. Furthermore, they can provide effective communication for a public safety network, as shown in Figure 11.2. Moreover, drones can move around to provide large disaster coverage area faster, and achieve ubiquitous connectivity within a minimum time in a public safety network. Emergency communications are the most important and specialized field for giving high value to people's lives. During an earthquake or tsunami, there is no ground transportation, so drones can provide and maintain wireless emergency communication services during and after the disaster. Therefore, the SAR team can easily perform their tasks in a sequential manner. When a drone reaches the desired altitude, the affected area will be surveyed via a digital video camera. Then, collected images will be sent to a ground station for monitoring the area of operation and coordinating SAR teams with first responders' arrival. The authors of introduced a multi-drone network architecture and demonstrated the network benefits in SAR and disaster assistance. Drones are all about the payload, so the type of event determines what equipment should be in the payload. Therefore, the most common drones used for public safety are the Dà-Jiang Innovations (DJI) Inspire and DJI Phantom, as shown in Figure 11.3; both are from the DJ family. The DJI Phantom is easy to acquire, land and take off, flexible in terms of the camera tilting, and able to fly around objects, but even experienced pilots can get into tough spots. The features and advantages of the DJI drone are explained in. Furthermore, DJ drones provide better SAR, so they can be used or all sorts of activities such as environmental monitoring, disaster management, mapping, and 3D modeling. During the launch of a DJ drone, it starts looking for interference, and when found it will try to localize the interferer. It will also change the flight plan to get the next meaningful measurement point, reducing the flight time. The DJI drones are efficient at capturing imagery of sufficient quality for the 3D mapping required for accident reconstruction and crime scene visualization in smart cities (Figure 11.12).

Recently, a lot of studies have investigated various design challenges such as 3D deployment, energy efficiency, and time flight constraint. The optimal 3D deployment of drones for extending the coverage area and enhancing the QoS is discussed in. The idea is supported by a study that discusses the 3D deployment of a multiple drone base station to maximize the coverage area and maintain the link quality between drones and the ground station by using the practical swarm optimization technique. Furthermore, drones' 3D deployment represents a key technology that can assess drones to deliver network services for public safety and disaster management (Figure 11.13). Therefore, the authors of proposed a practical swarm optimization technique to enable the drones' 3D deployment to provide a large coverage area, maintain connectivity, and satisfy users' QoS

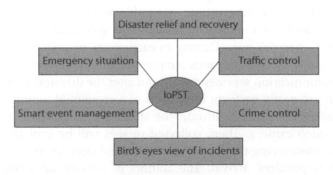

Figure 11.12 An illustration of devices connecting to a drone for the public safety network [62].

DJI Phantom   DJI Inspire

Figure 11.13 Public safety DJI drone.

requirements and drone capacity. Also, 3D deployment for swarm drones was introduced for maximizing the available lifetime of drones and the total throughput of all users.

IoT Devices for Public Safety

IoT devices can play a vital role in enhancing the performance of a public safety network in terms of accuracy, efficiency, and predictability. Furthermore, IoT devices have to analyze, aggregate information, and transmit without human intervention. This ensures the accuracy of received information and enhances the ability to anticipate crimes and other incidents. It is helpful to predict where crimes and natural disasters are more likely to occur. Figure 11.14 shows the classification [13–15] of IoT for disaster management, which includes natural, manmade, service orientation, and post-disaster. Natural disasters include floods, earthquakes, landslides, etc.; however, the post-disaster focus is only on saving victims in their locations. On the other hand, IoT for service-oriented disaster management has been discussed in Alsamhi *et al.* [62].

$$Plos = C1 - (C1 - C2)/(1 + ((\theta - C3)/C4))^{C5}, \qquad (11.7)$$

where $C1...,, C5$ are the environment parameters.

However, Houran *et al.* expressed the *Plos* by sigmoid term concerning elevation angle θ as:

$$Plos = 1/(1 + ae^{\wedge}(-b(\theta-a)) \qquad (11.8)$$

where a and b represent the S-curve parameters. Here, the Plos is easy to calculate and analytically flexible.

11.19   Securing Drones

A smart and intelligent system is needed for the security of drones that can investigate the data of attacks and ensure the security of drones by taking proactive measures. A secure IoD [58] relies on security, reliability, and consistency to develop a trustworthy system. In past, machine learning models have been proposed in the field of cyber-security for wireless sensor-based networks and mobile-based networks but not for drone-based security. A machine learning-based solution is proposed in this study for drone security authentication and control access methods. Various metrics are used for the evaluation of cyber-security systems. The purpose of using these metrics is that they are better for handling various performance indices in the cyber-security of a system. In this research, we propose the use of the following cyber-security metrics to evaluate the performance of the proposed system.

- Drone cyber-security threat exposure;
- Denial of service attacks;
- Malicious attacks;
- Jamming;
- Spoofing.

The machine-learning-based research solution for secure authentication and access control for drones and IoT devices is a key contribution of this research. This work aims to fill the research gap by making drones safe and reliable against major cyber-security issues and making them a useful monitoring tool, both commercially and for the industry. As explained in Section 11.3, the proposed architecture of the drone security system consists of seven layers. Data from the drone layer and edge processing layer are passed through the security and privacy layer before being forwarded to the device connection layer. At the security and privacy layer, data are protected from security threats by deploying machine learning models, where a mobile alert is sent when an attack is identified. Figure 11.14 shows the mobile alert of an identified attack.

Approach: In the previous section, the architecture and complete design details of the proposed system are explained. In this section, we discuss the implementation of the proposed system along with machine learning.

This experiment was performed on real-time data of drones. This drone dataset comprised geo-location-based features (latitude, longitude, and

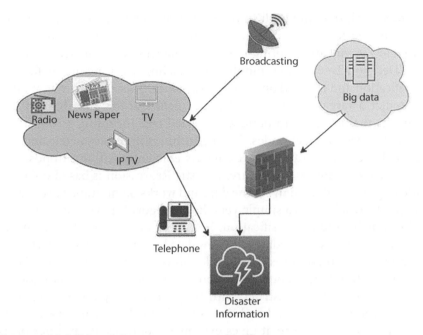

Figure 11.14 IoT-based disaster management classification.

altitude) [12], drone OBD data, and KDD intrusion detection features. The proposed model was trained and tested with the drone dataset as well as with two other benchmark datasets for intrusion detection [10] and cyber security attacks prediction.

11.19.1   Machine and Deep Learning Models

Machine learning has contributed considerably to improving the results of rating prediction based on reviews. There are many rich variants of machine learning classifiers that exist for performing rating classification [15]. A large number of machine learning classifiers can be found in the Python Scikit-learn library. It is an open-source library with a large user support base. In this study, the Scikit-Learn library was used for all the classifiers, including Random Forest, the Gradient Boosting model, Extra Tree classifier, Naïve Bayes, Logistic Regression, and Stochastic Gradient Classifier [13]. Random Forest is an ensemble learning classifier that works with the decision trees—also called estimators—for classification [30]. Bootstrap samples are used to train trees using the bagging technique. All trees are built in the same way to test the performance of the model on test data. A higher weight is assigned to the decision tree with a low error rate resulting in smaller chances of an incorrect prediction. Decision Tree is a commonly used machine learning model for text classification and is based on multiple variables. This algorithm is applied to predict a target variable. It classifies data features into branch-like segments that are used to build an inverted tree including a root node, internal nodes, and leaf nodes. The algorithm is non-parametric and can handle large-sized and complex datasets efficiently without imposing a complicated parametric structure. Logistic Regression is based on a statistical approach that analyzes the data and works on multiple variables to predict the results. It is a simple yet efficient algorithm with low variance and is mostly used in classification. Features can also be extracted using this model. It is easy to update with new data by using Stochastic Gradient Descent. Bayes' theorem is the foundation of the Naïve Bayesian classifier, in which independent assumptions are made between predictors. It is very convenient to construct, with simple iterative parameter estimation. Therefore, it is considered very suitable for large datasets. In spite of being simple in nature, it gives extremely good results and performs better than other sophisticated classifiers. The Support Vector Machine (SVM) is very common in text classification. It draws hyperplanes that separate classes by maximizing the marginal distance. The SVM [17, 18] hyperplane divides the text into two (non-overlapping) classes in the case of binary classification [1, 2]. It is simpler and less complex than deep learning methods and provides simple interpretability. SVM has also been widely used for intrusion detection [17, 18]. Multilayer Perceptron (MLP) is a simple deep learning model and has a reasonable classification ability. It is a layered model, where input layer neurons indicate the number of features and hidden layers work on the basis of weights to process input data and feed them to the output layer where neurons represent the output value. To obtain optimal results, numbers of neurons and numbers of hidden layers are selected according to requirements. To improve training efficiency for classification, the model is trained with the appropriate values of hyperparameters. To deal with the weights of MLP layers, backpropagation is generally used, which is based on Gradient Descent.

11.20   Future Work

This chapter proposes IoT-based cyber-security of drones using the Naïve Bayes algorithm. This model uses IoT sensors data, drones, and network information to generate patterns of security levels and identified the security attacks using these patterns. With this pattern, the model was able to identify attacks in the dataset. This model is tested with two datasets and achieves higher accuracy in real-time security attack detection. The accuracy achieved by the model is 96.3% which is higher and acceptable as compared to previous machine learning approaches. Precision recall and cost are calculated to estimate the performance. The Naïve Bayes model works by predicting items using two layers of processing in which independence between information items is assumed which shows a drawback in this suggested model. In the future, this problem will be addressed using a more efficient algorithm.

11.21   Contributions

In this work, we conduct a comprehensive review of the different aspects of drones' cyber-security including two main aspects: drones' security vulnerabilities, and the security concerns associated with compromised drones. Then, we review the countermeasures to secure drone systems, and to detect malicious ones. These contributions are summarized below:

- Identifying the main architecture of drones and their various communication types.
- Drone security and privacy concerns are discussed, mainly drone vulnerabilities, threats and attacks.
- Existing countermeasures of drones' security vulnerabilities and threats are reviewed, in addition to countermeasures in case of compromised (malicious) drones.
- Finally, the limitations of the existing works and recommendations for future research directions are included.

Conclusion

Drones have significant potential with many practical applications like aerial delivery systems, search and rescue operation, monitoring etc. Utilizing machine learning methods for the design and development of drones can make various drone operations better and more efficient. The recent development in the computational devices and the availability of data are enabling advancement in the field of machine learning especially, deep learning. Application of deep learning for visual perception can be seen in previous works related to self-driving car and drone navigating through forest trail while using the classical approach to model dynamics and low-level commands of these systems. Both studies were focused on the vision and need a path to navigate. This study attempts to consider the dynamics, low level control and perception in a forest environment without any path.

References

1. Wang, H., Gu, J., Wang, S., An effective intrusion detection framework based on SVM with feature augmentation. *Knowl. Based Syst.*, 136, 130–139, 2017.

2. Tao, P., Sun, Z., Sun, Z., An improved intrusion detection algorithm based on GA and SVM. *IEEE Access*, 6, 13624–13631, 2018.

3. Siddique, K., Akhtar, Z., Khan, F.A., Kim, Y., KDD cup 99 data sets: A perspective on the role of data sets in network intrusion detection research. *Computer*, 52, 41–51, 2019.

4. Tavallaee, M., Bagheri, E., Lu, W., Ghorbani, A.A., A detailed analysis of the KDD CUP 99 data set, in: *Proceedings of the 2009 IEEE Symposium on Computational Intelligence for Security and Defense Applications*, Ottawa, ON, Canada, 8–10 July 2009, IEEE, Piscataway, NJ, USA, pp. 1–6, 2009.

5. World Bank Group, Africa's pulse: An analysis of issues shaping Africa's economic future, Office of the Chief Economist for the Africa Region, vol. 15, pp. 35–43, 122, World Wide Web Foundation, April 2017.

6. Artificial intelligence: Starting the policy dialogue in Africa, December 2017. https://webfoundation.org/research/artificial-intelligence-starting-the-policy-dialogue-in-africa/.

7. For additional recommendations targeted to the agricultural sector, see Global Open Data for Agriculture and Nutrition (GODAN), "A Global Data Ecosystem for Agriculture and Food", August 2016. https://d1hiluowqo0t4b.cloudfront.net/posters/compressed/f1000research-215725.pdf.

8. Lian, W., Nie, G., Jia, B., Shi, D., Fan, Q., Liang, Y., An intrusion detection method based on decision tree-recursive feature elimination in ensemble learning. *Math. Probl. Eng.*, 2020, 2835023, 2020.

9. Hussain, J., Lalmuanawma, S., Chhakchhuak, L., A two-stage hybrid classification technique for network intrusion detection system. *Int. J. Comput. Intell. Syst.*, 9, 863–875, 2016.

10. Jiang, K., Wang, W., Wang, A., Wu, H., Network intrusion detection combined hybrid sampling with deep hierarchical network. *IEEE Access*, 8, 32464–32476, 2020.

11. Radanliev, P., De Roure, D.C., Nicolescu, R., Huth, M., Montalvo, R.M., Cannady, S., Burnap, P., Future developments in cyber risk assessment for the Internet of Things. *Comput. Ind.*, 102, 14–22, 2018.

12. MUmerSabir, Mumersabir/MDPIELECTRONICS: MDPIi Electronics Revision Dataset. *Electronics*, 10, 23, 10.3390/electronics10232926.
13. Svetnik, V., Liaw, A., Tong, C., Culberson, J.C., Sheridan, R.P., Feuston, B.P., Random forest: A classification and regression tool for compound classification and QSAR modeling. *J. Chem. Inf. Comput. Sci.*, 43, 1947–1958, 2003.
14. Song, Y.Y. and Ying, L., Decision tree methods: Applications for classification and prediction. *Shanghai Arch. Psychiatry*, 27, 130, 2015.
15. Korkmaz, M., Güney, S., Yiğiter, S., The importance of logistic regression implementations in the Turkish livestock sector and logistic regression implementations/fields. *Harran Tarımı Bilimleri Dergisi*, 16, 25–36, 2012.
16. Leung, K.M., Naivei bayesian classifier. *Polytech. Univ. Dep. Comput. Sci. Risk Eng.*, 2007, 123–156, 2007.
17. Ribeiro, A.A. and Sachine, M., On the optimal separating hyperplane for arbitrary sets: A generalization of the SVM formulation and a convex hull approach. *Optimization*, 1–14, 2020.
18. Xu, B., Shirani, A., Lo, D., Alipour, M.A., Prediction of relatedness in stack overflow: Deep learning vs. SVM: A reproducibility study, in: *Proceedings of thei 12th ACM/IEEE International Symposium on Empirical Software Engineering and Measurement*, Oulu, Finland, 11–12 October 2018, pp. 1–10.
19. Wood, L., $63.6 Bn drone service markets, 2025 - increasing use of drone services for industry-specific solutions - [news], Businesswire, ResearchAndMarkets. com, 17 Apr 2019, [Online]. Available: https://www.businesswire.com/news/home/20190417005302/en/.
20. Zeng, Y., Wu, Q., Zhang, R., Accessing from the sky: A tutorial on UAV communications for 5G and beyond. *Proc. IEEE*, 107, 12, 2327–2375, 2019.
21. Li, X., Wang, Q., Liu, J., Zhang, W., Trajectory design and generalization for UAV enabled networks: A deep reinforcement learning approach, in: *IEEE Wireless Communications and Networking Conference (WCNC)*, 2020.
22. Dulac-Arnold, G., Mankowitz, D., Hester, T., Challenges of realworld reinforcement learning, arXiv:1904.12901, 2019. https://doi.org/10.48550/arxiv.1904.12901.
23. Bayerlein, H., Gangula, R., Gesbert, D., Learning to rest: A Qlearning approach to flying base station trajectory design with landing spots, in: *52nd Asilomar Conference on Signals, Systems, and Computers*, pp. 724–728, 2018.
24. Zhang, Y., Mou, Z., Gao, F., Xing, L., Jiang, J., Han, Z., Hierarchical deep reinforcement learning for backscattering data collection with multiple UAVs. *IEEE Internet Things J.*, 8, 5, 3786–3800, 2021.
25. Gu, S., Lillicrap, T., Sutskever, I., Levine, S., Continuous deep qlearning with model-based acceleration, in: *International Conference on Machine Learning (ICML)*, 2016.
26. Heess, N., Wayne, G., Silver, D., Lillicrap, T., Tassa, Y., Erez, T., Learning continuous control policies by stochastic value gradients, in: *28th International Conference on Neural Information Processing Systems*, 2015.

27. Hu, Y., Chen, M., Saad, W., Poor, H.V., Cui, S., Distributed multiagent meta learning for trajectory design in wireless drone networks. *IEEE J. Sel. Areas Commun.*, 2021.

28. Bayerlein, H., Theile, M., Caccamo, M., Gesbert, D., UAV path planning for wireless data harvesting: A deep reinforcement learning approach, in: *IEEE Global Communications Conference*, 2020.

29. Chen, J., Yatnalli, U., Gesbert, D., Learning radio maps for UAV aided wireless networks: A segmented regression approach, in: *IEEE International Conference on Communications (ICC)*, 2017.

30. Mnih, V., Kavukcuoglu, K., Silver, D., Rusu, A.A., Veness, J., Bellemare, M.G. *et al.*, Human-level control through deep reinforcement learning. *Nature*, 518, 7540, 529–533, 2015.

31. Yin, Z., Song, Q., Han, G., Zhu, M., Unmanned optical warning system for drones, in: *Proceedings of the Global Intelligence Industry Conference (GIIC 2018)*, Beijing, China, 22–24 May 2018, International Society for Optics and Photonics: Bellingham, WA, USA, 2018; Volume 10835, p. 108350Q.

32. Ozmen, M.O. and Yavuz, A.A., Dronecrypt-an efficient cryptographic framework for small aerial drones, in: *Proceedings of the MILCOM 2018– 2018 IEEE Military Communications Conference (MILCOM)*, Angeles, CA, USA, 29–31 October 2018, IEEE, Piscataway, NJ, USA, pp. 1–6, 2018.

33. Ozmen, M.O., Behnia, R., Yavuz, A.A., IoD-crypt: A lightweight cryptographic framework for Internet of Drones. arXiv 2019, arXiv:1904.06829. https://doi.org/10.48550/arxiv.1904.06829.

34. Bertino, E., Data security and privacy in the IoT, in: *Proceedings of the 19th International Conference on Extending Database Technology (EDBT)*, Bordeaux, France, 15–18 March 2016, vol. 2016, pp. 1–3.

35. Rodday, N., Hacking a professional drone. Black Hat Asia 2016, pp. 1–2, 2016.

36. Highnam, K., Angstadt, K., Leach, K., Weimer, W., Paulos, A., Hurley, P., An uncrewed aerial vehicle attack scenario and trustworthy repair architecture, in: *Proceedings of the 2016 46th Annual IEEE/IFIP International Conference on Dependable Systems and Networks Workshop (DSN-W)*, Toulouse, France, 28 June–1 July 2016, IEEE, Piscataway, NJ, USA, pp. 222–225, 2016.

37. Shoufan, A., Continuous authentication of UAV flight command data using behaviometrics, in: *Proceedings of the 2017 IFIP/IEEE International Conference on Very Large Scale Integration (VLSI-SoC)*, Abu Dhabi, United Arab Emirates, 23–25 October 2017, IEEE, Piscataway, NJ, USA, pp. 1–6, 2017.

38. Luo, A., Drones hijacking; Tech. report, DEF CON, Paris, France, 2016.

39. Kerns, A.J., Shepard, D.P., Bhatti, J.A., Humphreys, T.E., Unmanned aircraft capture and control via GPS spoofing. *J. Field Robot.*, 31, 617–636, 2014.

40. Feng, Z., Guan, N., Lv, M., Liu, W., Deng, Q., Liu, X., Yi, W., Efficient drone hijacking detection using onboard motion sensors, in: *Proceedings of the Design, Automation & Test in Europe Conference & Exhibition (DATE)*,

Lausanne, Switzerland, 27–31 March 2017, IEEE, Piscataway, NJ, USA, pp. 1414–1419, 2017.

41. Feng, Z., Guan, N., Lv, M., Liu, W., Deng, Q., Liu, X., Yi, W., An efficient UAV hijacking detection method using onboard inertial measurement unit. *ACM Trans. Embed. Comput. Syst. (TECS)*, 17, 1–19, 2018.

42. Son, Y., Shin, H., Kim, D., Park, Y., Noh, J., Choi, K., Choi, J., Kim, Y., Rocking drones with intentional sound noise on gyroscopic sensors, in: *Proceedings of the 24th {USENIX} Security Symposium ({USENIX} Security 15)*, Washington, DC, USA, 12–14 August 2015, pp. 881–896.

43. Choi, H., Lee, W.C., Aafer, Y., Fei, F., Tu, Z., Zhang, X., Xu, D., Deng, X., Detecting attacks against robotic vehicles: A control invariant approach, in: *Proceedings of the 2018 ACM SIGSAC Conference on Computer and Communications Security*, Toronto, ON, Canada, 15–19 October 2018, pp. 801–816.

44. Gallacher, D., Drones to manage the urban environment: Risks, rewards, alternatives. *J. Unmanned Veh. Syst.*, 4, 115–124, 2016.

45. Lv, Z., The security of Internet of Drones. *Comput. Commun.*, 148, 208–214, 2019, [CrossRef].

46. Choudhary, G., Sharma, V., Gupta, T., Kim, J., You, I., Internet of Drones (IoD): Threats, vulnerability, and security perspectives. arXiv, 2018, arXiv:1808.00203. https://doi.org/10.48550/arxiv.1808.00203.

47. Zhou, J., Cao, Z., Dong, X., Vasilakos, A.V., Security and privacy for cloud-based IoT: Challenges. *IEEE Commun. Mag.*, 55, 26, 2017.

48. [CrossRef] 33. Nassi, B., Shabtai, A., Masuoka, R., Elovici, Y., SoK-security and privacy in the age of drones: Threats, challenges, solution mechanisms, and scientific gaps. arXiv 2019, arXiv:1903.05155. https://doi.org/10.48550/arxiv.1903.05155.

49. Giraldo, J., Sarkar, E., Cardenas, A.A., Maniatakos, M., Kantarcioglu, M., Security and privacy in cyber-physical systems: A survey of surveys. *IEEE Des. Test*, 34, 7–17, 2017.

50. Lagkas, T., Argyriou, V., Bibi, S., Sarigiannidis, P., UAV IoT frameworki views and challenges: Towards protecting drones as "Things". *Sensors*, 18, 4015, 2018, [CrossRef].

51. Tian, Y., Yuan, J., Song, H., Efficient privacy-preserving authentication framework for edge-assisted Internet of Drones. *J. Inf. Secur. Appl.*, 48, 102354, 2019.

52. Yaacoub, J.P., Noura, H., Salman, O., Chehab, A., Security analysis of drone's systems: Attacks, limitations, and recommendations. *Internet Things*, 11, 100218, 2020.

53. Albalawi, M. and Song, H., Data security and privacy issues in swarms of drones, in: *Proceedings of the 2019 Integrated Communications, Navigation and Surveillance Conference (ICNS)*, New York, NY, USA, 9–11 April 2019, IEEE, Piscataway, NJ, USA, pp. 1–11, 2019.

54. Bera, B., Saha, S., Das, A.K., Kumar, N., Lorenz, P., Alazab, M., Blockchain-envisioned secure data delivery and collection scheme for 5G-based IoT-enabled Internet of Drones' environment. *IEEE Trans. Veh. Technol.*, 69, 9097–9111, 2020.
55. Zhang, Y., He, D., Li, L., Chen, B., A lightweight authentication and key agreement scheme for Internet of Drones. *Comput. Commun.*, 154, 455–464, 2020.
56. Chriki, A., Touati, H., Snoussi, H., Kamoun, F., FANET: Communication, mobility models and security issues. *Comput. Netw.*, 163, 106877, 2019.
57. Mehta, P., Gupta, R., Tanwar, S., Blockchain envisioned UAV networks: Challenges, solutions, and comparisons. *Comput. Commun.*, 151, 518–538, 2020.
58. Nayyar, A., Nguyen, B.L., Nguyen, N.G., The Internet of Drone Things (IoDT): Future Envision of Smart Drones. *International Conference on Sustainable Technologies for Computational Intelligence*, Springer, Singapore, pp. 563–580, 2020.
59. Feng, Z., Guan, N., Lv, M., Liu, W., Deng, Q., Liu, X., Yi, W., Efficient drone hijacking detection using onboard motion sensors, in: *Design, Automation & Testin Europe Conference & Exhibition (DATE)*, IEEE, pp. 1414–1419, 2017.
60. Dhivyaprabha, T.T., Krishnaveni, M., Dhivyaprabha, T. T., Shanmugavalli, R., Review on intelligent algorithms for cyber security. *Handb. Res. Mach. Deep Learn. Appl. Cyber Sec.*, 7, 26, 2019.
61. Distributed Zero-Order Algorithms for Nonconvex Multi-Agent Optimization - Scientific Figure on ResearchGate. Available from: https://www.researchgate.net/figure/Comparison-of-different-algorithms-for-distributed-optimization-and-zero-order_tbl1_335565028 [accessed 20 Jan, 2023].
62. Alsamhi, S.H., Ma, O., Ansari, M.S., Gupta, S.K., Collaboration of drone and internet of public safety things in smart cities: An overview of QoS and network performance optimization. *Drones*, 3, 13, 2019. https://doi.org/10.3390/drones3010013.

12

IoT-Enabled Unmanned Aerial Vehicle: An Emerging Trend in Precision Farming

Gayatri Phade[1]*, A. T. Kishore[2], S. Omkar[3] and M. Suresh Kumar[4]

[1]*Sandip Institute of Technology & Research Center, Nashik, India*
[2]*Vidya Sangha Pvt Ltd, Bangalore, India*
[3]*Aerospace Engg IISc, Bangalore, India*
[4]*Aerospace Department, Sandip University, Nashik, India*

Abstract

Farmers of the twentieth century are adopting the use of leading technical facilities in the field of agriculture to sustain the competitiveness of the global market economy. With the help of modern technology, they are trying to reduce production costs and improve crop yield with better product quality. A new era in which many traditional agricultural practices are replaced by cutting-edge technologies like Precision Farming (PF), which entails applying the agronomic variables in the right place, at the right time, has been ushered in by technological advancements made in monitoring, supervision, management, and control systems. One of the methods that will enable quick and non-destructive analysis of agricultural data will turn out to be an IoT integrated smart agriculture drone. This includes taking pictures to analyze crop behavior, finding out how much water the soil can contain, managing irrigation systems, etc. Numerous engineering disciplines, including aerodynamics, electronics, computer programming, and economics are integrated in the design, development, and implementation of drone-based agricultural systems. In considering the above thought, a problem taken into consideration is for a grape field of approximately 10 acres located near Nashik (28 36'N, 77 12'E) in the state of Maharashtra to analyze the performance of IoT enabled agriculture drones for PF. Main objective is to spray the Insecticide with an agriculture drone only on the detection of the green zone of the crop or canopy for effective spraying and hence to reduce the wastage and cost of spraying. Drone image sensor will capture the field images, atmospheric

---

Corresponding author: gphade@gmail.com

Sachi Nandan Mohanty, J.V.R. Ravindra, G. Surya Narayana, Chinmaya Ranjan Pattnaik
and Y. Mohamed Sirajudeen (eds.) Drone Technology: Future Trends and Practical Applications,
(301–324) © 2023 Scrivener Publishing LLC

parameters like, temperature, humidity in the field and analyze it on the cloud platform for analyzing the crop health and related parameters. An agriculture drone of 16 L capacity is proposed for spraying 1 acre of land. It could take 7–10 minutes of flight time of the drone to spray the insecticide which can lead to reducing the labor cost by 65%, spraying time by 85% and amount of insecticide by 50%. Drone camera and spraying mechanism is to be synchronized to achieve the objective through IoT platform. For effective spraying from the bottom and the top of the leaf, synchronized UGV and UAV spraying is proposed. Further, time series analysis of the photographic images and video footage captured by the drone camera can be done for the prediction, modeling of crop yields and auditing them with computer vision capabilities of UAV, and developing suffi- cient datasets. These data sets can be used for machine learning algorithms for AI or DL and for future research related to crop disease forecasting, canopy cover, stress management, and be able to guide farmers for taking corrective actions in advance and evolve best practices for overall improvement in the yield.

Keywords: AI, computer vision, IoT, image sensors, UAV-UGV

12.1   Introduction to IoT Enabled UAV

According to the Indian Council ICAR vision 2030, it is expected that food requirement will be doubled by 2030 whereas there is significant impact of environmental imbalance and reduction in natural resources. Overall agri- culture production is declining due to different factors like, deterioration in the quality and health of soil, climate change due to global warming and the worst part, the effect of COVID19 on the agriculture ecosystem. The demand for food is growing because of increase in population along with rise in per capita income. In addition to the food grain, fruits and vegeta- bles are essential to maintain the health of an individual. The demand for fruit would increase from 43 million tons (MT) in 2020 to 110 MT in 2030. Similarly for vegetables from 76 MT to 182 MT and food grain from 192 MT to 345 MT [1].

For effective supply chain management, there is a requirement of han- dling different infrastructure, value addition processing and marketing. Climate change is a big challenge to meet this requirement. Due to global warming global earth temperature has been raised from 1.8°C to 3.0°C resulting in the rise in pests and diseases which leads to the instability in food production resulting in the question mark on the farmer's lifestyle. It is producing enough food for increased demand though they are fac- ing challenges due to unpredictable climate conditions. Despite the food and water problems, there is still a chance to increase farm incomes, create jobs, and involve more parties in the food supply chain through increasing farm incomes.

The production efficiency of next-generation precision agriculture would increase. For this, it would be necessary to collect pertinent parameters and simulate the most complicated systems using increasingly potent computers, sophisticated software, and cutting-edge sensors. To benefit from precision operations for crops and resource applications, improved long-range weather forecast technology would be necessary. An early information system is needed to be aware of crop health issues for efficient crop management and production for sustainable and profitable agriculture. In response, some corrective measures should be in place to mitigate the potential harm brought on by biotic and abiotic stress in order to prevent crop losses.

Agriculture plays a vital role in India's economy. One of the most productive agricultural ventures in India is grape production. India ranks among the top 10 grape-producing nations in the world. With a yield of 1.21 MT, this crop ranks fifth among fruit crops in India (or around 2% of the 57.40 MT of fruit produced worldwide). Maharashtra, Karnataka, and Tamil Nadu contribute 80% of the total production. Improvement of the agricultural fields is a very big challenge for countries like India, as weather becomes unpredictable nowadays. Farmers have much concern with environmental conditions as their crop yield depends on it. Soil moisture, temperature, and humidity are the major crop affecting parameters, especially for grapes. Human beings cannot control the environmental parameters but can monitor it, using sensor technology to take the corrective measures as per the unfavorable environmental changes, to avoid the yield loss. The farmer has started adopting modern farming techniques like IoT to deal with such challenges. The IoT contains sensing elements and embedded processing nodes that can collect and monitor the real time environment parameters, analyze collected data and can provide controlling parameters.

In the field of agriculture production, India ranks second in the world. Over all 96 million hectares of land is irrigated, which is the largest in the world. Agriculture is India's largest enterprise where 263 million people are involved in it [2]. Due to the pandemic, there are the challenges like a labor crunch and increase in the input cost of crop production which needs to be addressed immediately. Unmanned aerial vehicles (UAVs), commonly known as 'drones' can effectively be used for timely spraying of crop protection chemicals with minimum manpower requirements.

Drones can be used for timely diagnosis of insects and pests, crop health monitoring, targeted input application, and rapid assessment of crop yield and crop losses. Unlike ground spraying, spraying through drones can be carried out when field conditions prevent movement of wheeled vehicles. It enables the timeliness of spray treatments without inflicting soil compaction and reduces the human drudgery and health risks involved in manual spraying. Identification of infected crop areas is essential as it may damage the entire crop if not treated in time. With drone and imaging sensors it is possible to identify the infected crop and by spraying into that specific area may reduce the usage of the pesticides. It reduces the financial burden on the farmer, preserves the crop nutrients, and will be indirectly beneficial for the human being who will consume this healthy food. Fusion of these two emerging technologies will be revolutionary in the precision agriculture domain. With IoT, drone data collected with telemetry and drone sensors can be analyzed on cloud platforms.

Proposed system will monitor environmental parameters affecting the grape field, in real time, with the help of different sensors and a drone. Sensor data, through the gateway will be uploaded on cloud server where it will be analyzed. Same data can be monitored by the expert or farmers through the mobile application available on his/her mobile phone or on a personal computer, with the current environment condition displayed on the screen. Based on cloud data analysis, farmers or agro-expert can forecast the probable disease and its preventive measures. If the environmental parameters are not suitable for proper growth of the crop, farmers will be alerted through mobile phones and will be asked to take corrective action based on expert advice to avoid the probable yield loss. Environmental data recorded on cloud servers will be available for the policy makers to decide the policy for a particular crop like grapes.

IoT enabled drone-based spraying, monitoring and control is proposed for Nashik based SOMA Vineyard which is at present done in a manual operated way. At present, let us say, for one acre vineyard, 1000 L of pesticides or fungicides are sprayed for the effective growth of the crop. It took one hour and thirty minutes to spray the fungicides in a semiautomatic way by means of tractor spraying. For spraying, ceramic nozzles of 61 mm are connected to the liquid tank installed on the tractor. The drop size of the sprayer is pretty big and that too not focused on the crop. It leads to the wastage of approximately 30–35% of the sprayers that are costly enough for a medium scale vineyard like SOMA. The fertilizers or chemicals used for spraying are DAP, FYM and cow dung dissolved in the water which costs three to four thousand rupees per acre. The wastage of these sprayers due to unfocused spraying is an additional burden for the owner of the vineyard. For effective spraying either spot spraying or aerial spraying is recommended which will save the time required for the sprayings, number of sprayers and manpower as well. In addition, it will reduce the excess use of the chemicals which in turn is beneficial for the human who sometimes consumes the chemical along with the berries. The size, sweetness and the overall yield of the berry can be improved by means of the exact and timely application of chemical dose.

Yet, only surface monitoring and spraying at surface level is not sufficient as fungus attaches from the lower portion of the leaf. Rover or Unmanned Ground Vehicle (UGV), moving through the vineyard, close to the saplings can monitor the leaf from bottom side and spot spraying of manure like cow dung slurry can be achieved. Weeds growing in between the saplings consume the soil nutrients, water contents, as well as manure sprayed at the crop. Hazardous pathogens on the weed may attach to healthy plants or berries. In addition, it increases the humidity of the yard. Being very sensitive to the humidity, berries may lead to catching of the fungus which in turn may demand for additional spraying. Local application of weedicide with help of the rover will stop the weeds growing.

When there is a shift in the temperature and relative humidity, the fungus starts spreading on the leaf and the berry which need to be identified at an early stage. For this purpose, at present a conventional monitoring method is adopted. To the outer border of the yard, Rose bushes are grown as it catches the fungus as fast as that of the berry or grape leaves. Monitoring these bushes manually gives the decision of spraying. This conclusive method may mislead to excessive use of the chemicals if done earlier or if delayed, it may be prone to the fungus spreading at fast. So, it needs continuous monitoring. Presently, INR 25K–30K needed to spend on manual monitoring and spraying that is manpower charges per acre. Minimum 15 skilled persons are required to carry out these activities. Next unpredicted environment parameter that affects the yard is the fog that gets spread through the crop. Increased humidity in the air is the root cause of the fog spreading. This fog can be reduced by creating artificial smoke which needs to be spread evenly, across the crop. It can be achieved by moving the rover/UGV, with fog generating mechanism installed on it, through the crop.

Sufficient water is the next requirement of the vineyard as soil moisture is the sensitive parameter related to the healthy and juicy berries. To meet this requirement, a water pump of 7.5 HP is operated for 4 hours to water the yard of 8.5 acres. Manually operated motors may take more or less time to manage the on-off period and cause excess watering. Long-range (LoRa) operated pumps with effective time scheduling and by taking the timely on-off decision, based on water stress analysis with the help of drones, will provide and maintain exact soil moisture required for a healthy crop.

Skilled manpower, timely monitoring and spraying throughout the crop cycle, and maintaining the necessary soil moisture are the key requirements to meet the expected yield. Effective coordination between the drone (UAV), the rover (UGV) and the LoRa operated water pumps will prove as a catalyst to achieve the improved yield and quality of the berries.

12.2   Drones in Precision Farming

In 2015, 48,000 drones were purchased in the USA; 80% agriculture drones are to be used in the agriculture commercial market of the USA. Nearly 2500 drones are approved by USFAA for agriculture purposes as per the statement given by the authority of Association for Unmanned Vehicle System International. Drones can help farmers find weeds and pests, identify damaged plants or dry regions, and spray the right quantity of fertilizer and insecticide with less expense and environmental damage, increasing overall production. High-value produce like wine grapes have particularly steep returns. Since 2002, many countries like Australia, Japan, and South Korea, have adopted drone farming. Farmers started accepting drones as a standard farming tool [3].

In rice farming, a growth phase can be identified by using drone images captured for a massive agricultural area on the basis of collected information [4]. As rice farming is the biggest agriculture in Indonesia, to monitor the different phases of rice growth, they have captured the images of crop area with the drones and sensors with a height of 500 meters which can cover nearly 6Ha land view. Captured image histogram is computed and age group of rice is determined with help of Support Vector Machine (SVM). Image captured with drones gives 93.33% of accuracy as images get affected due to the sunlight and the shooting height. This can be taken as a reference height to capture the drone images.

A rigorous review is carried out to know about the work done for drone development, methodology adopted for development, different sensors used on the drone for data acquisition, payload that drone can carry, controllers used on drone board, and nozzles used on drone for spraying [5]. Further, different techniques used for crop monitoring using drones are explored. Yet it is required to review different issues addressed by the researchers for crop health monitoring and spraying.

A specific drone for a spraying application is developed with the objective of improving the orientation of UAV based on accelerometer and gyroscope. An eight bit microcontroller 8051 is used to develop a flight controller. It is a very basic prototype version of the drone and a lot of changes are to be adopted to commercialize it [6].

For further improvement, an attempt is made to design and develop an octocopter drone with inbuilt spraying mechanism with 6.7 kg of payload which reduces its weight [7]. Drone is constructed uniquely due to its four Y shaped arms of the octocopter. This drone can be used for spraying disinfectant liquid which can be used for agriculture as well as in the pandemic of COVID-19. It was a good attempt to develop such a drone.

The design and development of the spraying drone model, 'AeroDrone' is developed by a Ukraine based startup. An attempt is made to minimize the spraying error and to improve the autonomy by setting the policy parameters through their work. With this they have proposed a mission assignment scheme based on simulation which will minimize the spraying error, yet it is not full proof and needs to be reframed [8].

Different applications of UAV or drones in agriculture and livestock management are focused with bird eye which will be useful for the agriculture stakeholders [9]. It is seen that Skymet Company is providing agriculture surveys to the insurance company and the state government. So the Government of India is looking forward to adopting drone technology for precision farming.

The commercial drones with their technical parameters are investigated and summarized [10]. One can select the appropriate drone based on its specification for the intended application. Further, a drone application for agriculture and precision agriculture is developed.

Cameras used to capture the images and image processing techniques, aviation regulations, UAV data acquisition and processing system, UAV sensors and cost are some of the aspects that can be considered when an UAV is used for Precision Agriculture (PA). In PA, use of UAV based Remote Sensing (RS) over traditional platforms provides a fast survey platform. After the comparative study based on some specifications like cost, operating environ, automatic spraying mechanism, spatial resolutions, spatial accuracy, adaptivity, real time data processing operational complexity and many more, it is seen that among the platforms used for RS in PA like UAV, satellite, manned aircrafts, UAVs are best suited for PA. Further, resolution of the image capture with UAV is dependent on UAV architecture, camera resolution, image computing time, light intensity, image overlap, ground speed of the captured image and flight stability. RGB, multispectral, hyper spectral, thermal, and consumer grade camera is to be used for RS application using UAV for PA. Use of gimbal camera on UAV can give better image quality in motion. RGB & multispectral sensors are low on cost as that of the hyper spectral, thermal and LiDAR sensors [11].

By flying the drone at lower altitude, high resolution images of crop which are less affected by cloud cover and low light conditions than satellite imaging. Pest detection using drone images can provide differentiation, detection, and estimation of symptoms, whereas combination of hyper spectral and thermal images can offer promising results in disease detection in primary stages. Beneficial insects, predators, and parasites can be released over the crops to manage pests biologically. This can lower the amount of pesticides used and their adverse effects on the environment [12].

Drones are used to study the enrichment of the biosphere in Sumatra's lowlands. Estimating canopy cover using a drone and using conventional ground-based hemispheric photography has a strong connection. The investigation of canopy cover has found drone-based photogrammetry to be helpful. Numerous concerns relating to the management of agroforestry systems and ecological research can be addressed by this analysis [13]. It is seen that professors and students of the technical institute are making significant contributions to make use of drones in agriculture applications. Industry needs to come forward for the commercialization of drones on a large scale so that the end user, a common farmer, should get motivated to use it for agriculture application. Many of the researchers have made reviews on the proposed use of drones in the agriculture sector but very few came forward with actual implementation of an idea for a specific crop in India.

12.2.1   Types of Agriculture Drones for Precision Agriculture

Today's farmers have adopted technical farming alongside traditional farming. After adopting precision farming, farmers are now looking forward towards drone technology in addition to precision farming. Use of agriculture drones has been accepted by many farmers for different agriculture applications. Based on the agriculture applications, drones are categorized as follows,

    i.   Drones for aerial seeding and seedling surveillance;
    ii.   Drones for canopy and growth-stage scouting;
    iii. Drones for crop protection that spray chemicals to protect plants and detect illness or pests;

iv. Drones for crop-monitoring that collect data regarding NDVI, leaf chlorophyll and plant-N status. Such drones could also be utilized to apply liquid fertilizer-N, if the fuselage has such a facility with nozzles;

v. Harvesting drones detect the crop maturity. They map the entire field of over 10,000 ha per variation, seeding trends, seedling emergence, crop growth parameters, grain maturity, and so forth. Such drones add vast amounts of data to the 'big data bank'. Many of the computer-based decision-support systems rely on data banks;

vi. Drones are useful in imaging landslides, large-scale gully erosion of farm lands, avalanches, forest fires, dust storms, earthquakes etc. Such imagery from drones allows us to judge the extent of damage, locate the problem areas, note the topographic changes and the loss of vegetation/crop etc.;

vii. An octo-copter are used to derive multispectral imagery of the crop and provide accurate digital data/imagery for combine harvesters to operate. Sometimes, digital data could be applied to the robotic combine harvester. These machines later offer yield maps to farmers;

viii. Data drones are those vehicles that provide periodic data about the soil terrain, boreal vegetation, and moss growth. The extent of moss growth, its density, and spread rate was monitored using the drone imagery. The images were sharp, well focused and of high resolution, at 3 cm/pixel. Furthermore, drone-derived imagery could be applied to develop maps depicting the spread of moss species, particularly, to depict their fluctuations in response to temperature in the cold continent. In fact, impact of several other environmental parameters such as diurnal variations, irradiance, moisture, and temperature on boreal vegetation and moss growth could also be studied using drone imagery; and

ix. Drones to study natural vegetation, forest plantations, and agricultural cropping belts.

Major applications of drones with regard to natural resource monitoring, as quoted by many agencies are: (a) topographic mapping; (b) capture of images of natural features and other structures; (c) analysis of natural vegetation; (d) water level mapping in rivers and reservoirs; and (e) mapping environmental effects on soil resources. Drone imagery could help us in tracing and understanding climate change effects on several species of vegetation and fauna, in the remote continent.

Apart from the above listed categories, drones are evolving as per the requirements of agriculture applications.

12.2.2   Drone Architecture for Precision Farming

For building an agriculture drone, following primary drone components can be considered, (i) frame, (ii) brushless motors, (iii) Electronic Speed Control (ESC) modules, (iv) a control board, (v) an Inertial Navigation System (INS), and (vi) transmitter and receiver module [22].

According to the National drone policy, based on payload, drones are categorized as: Nano: up to 250 g; Micro: 250 g to 2 kg; Small: 2 kg to 25 kg; Medium: 25 kg to 150 kg; and Large: greater than 150 kg of payload. For pesticide applications using drones of small category with payload capacity 10–15 kg can be allowed by DGCA. All drones except those in the Nano category must be registered and issued a Unique Identification Number (UIN). The drone for agricultural purposes (Krishi drone) should have a payload capacity of 10–15 kg. The Krishi drone should have labeled with the information which includes, Year of manufacture, Designation of series or type, Serial number, if any, Empty Mass (kg): Mass of UAV when the tank is empty, maximum payload capacity (kg), Maximum flow rate, liter per minute (lpm).

If a drone is to fly above 200 feet for commercial use, a permit is required in India. The drone must be fitted with an accurate altitude sensor to ensure desired height above the crop is maintained throughout the spraying mission. Agriculture drones should have features like GPS, return-to-home (RTH), anti-collision light, ID plate, flight controller with flight data logging capability, RFID and a SIM. No Permission No take-off (NPNT) is applicable for flying all categories of drones except Nano drones. Before flying the drone, one has to ensure that the drone is registered on DGCA portal and permission is granted for flying the drone. By submitting a flight plan and getting a special Air Defence Clearance (ADC)/Flight Information Center (FIC) number, one can request permission to operate in controlled airspace. As India has a varied and complex topography including plains, plateau, coastal, forest, hills and mountains, the drone laws and regulations for pesticide application for different states can be framed keeping in view their topography particularly with respect to visual line of sight (VLOS).

Appropriate geofencing is needed to fly the agriculture drones. Drone spray systems must be leak-proof and shall support variable flow control to ensure uniform spraying. It should be well calibrated to have accuracy of ± 10% on the quantity of input sprayed. As the procurement of drones for small holding farmers would be expensive, the rental process of drones can be encouraged. In this case, DGCA approved Companies/organizations/training schools can offer the rental drones or Panchayat/Corporation offices can tie-up with them to provide drone services for farmers.

For any agriculture drone the basic competences needed are: (i) the drone need to fly in line with predetermined waypoints; (ii) manage its flying height; (iii) sense and avoid objects while in flight; (iv) land autonomously in accordance with the battery's condition; and (v) use a gimbal to stabilize the captured images. A multispectral Parrot Sequoia sensor, an RGB camera, and a luminance sensor are all integrated inside the device [23].

To minimize the spray drift, apart from wind limitation, some operational parameters need to be considered like, spray height above the crop canopy should be less than 3 m, speed of the drone should be less than 6 m/s, and droplet size should be greater than 100 µm. It is recommended to have following spray coverage guidelines to have satisfactory Bio-efficacy:

i.   20 droplets/cm² – for most insecticides and systemic fungicides.
ii.   5–10 droplets/cm² – for translocated herbicides
iii.   50–70 droplets/cm² – for non-systemic fungicides.

Considering weather condition, the drone-based spray should be permitted under following weather condition to get the best results:

a)   Wind Speed: ≤7 kmh
b)   Outside Air Temperature: ≤35°C
c)   Relative Humidity: ≥50%

It is recommended not to spray under fog, during cloudy and immediately after the rains to ensure input retention by the crop. Further, weather forecasting should also be considered before flying the drone. The conditions like cloudiness, light intensity, temperature, wind direction, and velocity should be recorded, and decisions should be taken accordingly. This can be achieved with IoT enabled drone sensors.

12.2.3   IoT-Enabled Drone in Precision Farming

Aerial automation modernizes the farming practices and aid which transforms traditional agriculture to Agriculture 4.0. In Aerial automation, for precision farming most of the service providers are using UAVs for spraying. UAVs act as key players in the agricultural IoT, which enables the collection of useful information to be automatically and seamlessly unified, letting the farmer to use this information for farm practicing. UAVs powered by AI scout for pests in the field or dry patches that need special care. Due to recent developments in sensor technology, drones can now analyze crops from the air and detect weeds and ill crops by using extra wavelengths of the visible light spectrum. Numerous agricultural applications, including crop health monitoring, agricultural aerial photography for site-specific development, variable rate applications, and pastoral management, can be carried out with AI-driven UAVs. Drones may collect a variety of data by using various sensors to scan a large area with remarkable accuracy and at a low cost. The main goal of Aerial Automation is to provide precise, economic and ergonomic, socially and environmentally responsible aviation services for Agriculture 4.0 applications.

With IoT in precision farming, productivity can be improved by means of yield forecasting and market connect, crop/soil health monitoring, farm weather monitoring [14], optimal use of natural resources, like water, soil and sunlight, early prediction and detection of crop diseases or pest infestation, livestock and crop health monitoring. To achieve these objectives with IoT, IoT infrastructure will sense weather conditions or climate data in real-time. Collected data using sensors will be uploaded on the remote server through a field gateway with internet access. The field data collected on the server then analyzed decision support models developed and executed. Based on the interpretation of the data analysis, experts can send personalized crop advice to the farmers in their regional language.

To validate the aerial IoT infrastructure and its effectiveness on the crop growth, crop parameters need to be monitored throughout the crop cycle. The parameters like Enhanced Vegetation Index (EVI), Green Atmospherically Resistant Index (GARI), Green Chlorophyll Index (GCI), Infrared Percentage Vegetation Index (IPVI), and Leaf Area Index (LAI) are to be analyzed with help of agriculture drones to get the threshold required for the crop decision making.

Focus has been given on different applications of UAV like UAVs for aerial photography, filming, security and logistics like first aid and postal delivery, GIS, land and water surveys [15]. For precision agriculture application, he suggested combining captured aerial photographs into one large using orthomosaic image and a reflectance map of the crops using Normalized Difference Vegetation Index (NDVI) algorithm. This map is the key to better yields and lower costs which leads to driving the agricultural business to the next level. It highlights the problematic area of the crop that needs to be examined and minimizes the scouting time so that more time and focus can be given on treatment of the plant.

UAV-based remote sensing provides extremely accurate biomass and quality parameter estimation. In nations like the Netherlands, where decisions regarding the timing of harvesting and the fertilizer rate are made using data on the quality and quantity of the grass, it can be seen as an ideal instrument for effective and precise management of silage grass production. To gauge the quantity and quality of grass, spectral imaging and photogrammetry using UAVs can be used. During the primary growing season, UAV-based remote sensing data sets are taken four times, and parameters like dry biomass and D-value related to quality, as an agricultural response, are measured concurrently. Furthermore, the data set for the image exterior orientation, camera orientation, 3D point cloud and orthostatic is generated using UAV. These data sets were trained, and from the trained data sets, spectral brands, vegetation indices, and 3D characteristics were retrieved. In order to extract 3D characteristics, the canopy height model is created using RGB data. Based on the analysis of the extracted features, to achieve a max yield, silage grass swards are harvested two to four times per season and fertilizer is applied once per year [16].

An extensive review of the developments in the use of drones to monitor and assess plant stressors such drought, diseases, nutritional deficits, pests, and weeds is provided [17]. A rigorous literature review is carried out with the objective to focus on the different approaches used for extracting the information contains the captured drone images. Costs, spatial resolution, area coverage, type of image to be used, and type of feature of index to be generated are the primary criteria that determine the best course of action. Various variables are estimated from the captured image information and its relevance towards the crop is identified. It is summarized with the help of the following Tables 12.1 and 12.2 [13, 18, 19]:

Table 12.1 Variables extracted from Image and its relevance to crop.

Variable	Significance
The leaf water potential (YL)	Water's propensity to go from one area of the leaf to another.
The stem water potential (YS)	Water's propensity to go from one area of the stem to another.
The stomatal conductance	The rate of carbon dioxide flow measured by open leaf stomates.

Table 12.2 Control variable and its source.

Sr. no.	Variable	Source of data
1	The vegetation indices (NDVI, GNDVI, etc.)	Spectral transformations aiming to highlight the specific vegetation properties
2	The photochemical reflectance index (PRI)	Sensitivity of reflectance measurement to variations in the carotenoid pigments found in leaves
3	The difference between the canopy and air temperatures (Tc -Ta)	Canopy temperature
4	The crop water stress index (CWSI)	The vapor pressure deficit (VPD) is used to normalize the difference between the canopy temperature and the air temperature (Tc - Ta)

The variable which decides the control strategies are listed as below:

Thermal sensors are typically utilized for applications that include water stress. This is due to the effects of water stress, which increase the likelihood of detecting changes in the canopy temperature by decreasing stomatal conductance and reducing heat dissipation in plants. It is advised that the optimal time to take thermal photos is around midday because at this time the impact of shadowed leaves and the variations in stomatal closure are minimal. Automatic co-registration of thermal and multispectral images along with a modified Scale Invariant Feature Transformation (SIFT) technique and K-means++ clustering can be used to mitigate this issue by eliminating the relevant pixels. When shadow pixels are removed, the correlation between remotely sensed and reference water stress indicators improves.

All possible applications of drone farming using drone imaging, drone based photogrammetric methods for forestry, agriculture and private urban gardens are explored [18]. In Finland, practical planning for foresting is done by using photogrammetric methods based on the data collected through drones. A drone is used to generate aerial image datasets, collected from multiple sites in Kanta-Häme, in southern Finland. To collect the data the radiometric cameras used are DJI ZenMuse XT2 and MicasenseAltum. Photogrammetry pipelines are proposed for higher quality data products.

Image fusion is the new trend nowadays for accurate prediction. A method based on fusion of drone image and optical satellite image data is recommended for precision agriculture monitoring [19]. It gives the possible classification, such as classification of spare and dense vegetation, which is an important aspect for precision agriculture monitoring. By contrasting the Landsat categorized image with the drone image, the suggested method is shown to be effective in differentiating the two types of vegetation for precision agricultural monitoring.

Once data is collected through the drone sensors, it is required to analyze and validate the parameters related to crop health monitoring throughout the crop cycle. Different software techniques are involved for data collected through sensors and its interpretation after analysis. Deep learning (DL) can be a better option over conventional Machine Learning (ML) for modeling the complexity in plant water stress conditions. Already UAV based remote sensing is proved as an advanced method to measure/analyze the water stress with high spatial, temporal resolution. DL with UAV based remote sensing is an emerging hand in plant water stress assessment which needs to be explored. Further, the fusion of 3D RGB data and DL will be the next assessment tool to improve the accuracy. Soil data, atmosphere parameters, and plant-based measurements can be integrated to create 3D input [20]. CNN is used to identify the environmental stress by taking the yield, soil and weather data as an input.

Mobile applications are being used to add smartness in precision farming. Today's farmer is equipped with a smartphone. Based on the agriculture data collected through IoT, a mobile application can be developed for crop monitoring, data management, disease forecasting, crop scouting activities, drone controlling and many more. Such applications can be used to improve productivity and profitability of the farmer. One of such mobile applications, Canopy cover, evaluates aerial field imaging and produces actionable data reports which will be helpful for farmers to metricate for potential causes [21]. It gives the amount of soil surface that is covered with plots. Higher the canopy cover, denser is the vegetation. Algorithm used here works on the vegetation index. This is used around the summer season when it is most crucial to gauge the growth of crops. Along with the amount of vegetation, it gives a robust agricultural data stack for soil, moisture and other measurements.

As Nashik is known as a 'Grape City', the crop undertaken for analysis is Grapes. For testing the proposed system we need the vineyard like field, specifically for drone spraying. A grape field of approximately 8-10 Acres located near Nashik (28 36'N, 77 12'E) in the state of Maharashtra named, 'SOMA Vinevillage', is chosen for monitoring and analysis. Understanding the Vine processing, two types of grapes are used namely, Shanin blow and Sovinie blow which decides the color of the vine, are taken for cultivation. However, the system can be tested for any crop situated at any location.

12.2.4 Safety and Security in IoT-Enabled Drones in Precision Farming

There should be a buffer zone of minimum five meter between drone treatment and the non-target crop. No human or animal movement shall be permitted within or in the close proximity of the farm (minimum 15 m) during and immediately after the spray operations (minimum 30 minutes). Drone based spray operations should be conducted away from (≥ 100 m) of water bodies, residential areas, fodder crops, public utilities, dairy, poultry etc. during pesticide application.

The drone and spraying unit should be properly calibrated before flying. The spraying unit should be checked for pump pressure, nozzle wear out for uniform spraying, and leakage in tank and lines. Proper pressure should be adopted for optimized droplet spectrum (100-150 μm).

12.2.5 IoT Architecture in Drone

Following schematic shows the conceptual view (Figure 12.1) of IoT architecture for drones. Users can monitor drone parameters and information collected with the help of a drone, on their smartphone. Multispectral camera, thermal camera, RGB camera and Light Detection and Ranging

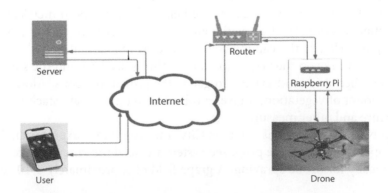

Figure 12.1 Conceptual view of IoT-enabled drone.

(LiDAR) systems are the sensors that can be used for aerial IoT. By using multispectral cameras, vegetation stage can be quantified in terms of: (i) chlorophyll content, (ii) leaf water content, (iii) ground cover and Leaf Area Index (LAI), and (iv) the Normalized Difference Vegetation Index (NDVI). Thermal cameras have demonstrated high potential for the detection of water stress in crops due to the increased temperature of the stressed vegetation [24].

Agriculture drone is enabled with the camera which will detect the green zone for spraying. With image processing algorithm, green part of the canopy will be detected which in turn will activate the sprayers installed on the drone. This automation process is illustrated with the flowchart below in Figure 12.2.

It is observed that spraying from top of the crop, specifically for Grapes is not sufficient. Plants get infected from the bottom of the leaves too [25]. In this case an IoT enabled drone will actuate the Unmanned Ground Vehicle (UGV) placed in the grape field which will spray from the bottom side of the grape leaves. Figure 12.3 depicts the UAV to UGV communication for spraying from both top and bottom side of the leaves.

UGV also monitors the ground parameters like, soil temperature, soil moisture, relative humidity, and temperature. Ground parameters and

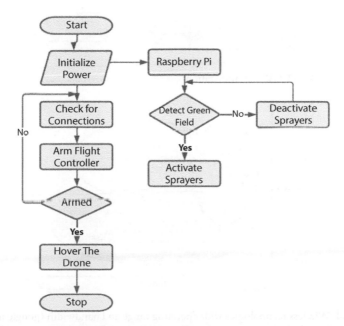

Figure 12.2 Flowchart for canopy detection and spraying.

aerial parameters can be analyzed on a common cloud platform of UAV and UGV for minute analysis of the crop.

LoRa module can be used as a part of IoT infrastructure. By analyzing soil moisture with UAV and UGV, water pumps can be operated with the LoRa module to maintain the soil moisture content. Typically, LoRa can achieve a communication distance of 15 to 20 km and can work on battery for a year. In India, the LoRa Technology licensing frequency spectrum ranges from 865 MHz to 867 MHz. It operates at low voltages in the range of 2.4V–3.6V, with maximum data rate of 250 kbps.

In wireless technology, Bluetooth Low Energy (BLE) works with low power, but has limitations for long-range transmission. With LoRa, long

Figure 12.3 UAV to UGV communication.

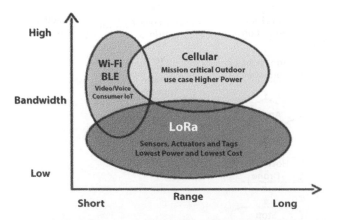

Figure 12.4 Wireless technologies with operating range and bandwidth (google image: https://www.kernelsphere.com/lora-technology.html).

distance communication can be achieved without an internet connection, which is a good option over the Wi-Fi and BLE communication. The following figure illustrates the low power technologies with respect to the bandwidth and their operating range.

Figure 12.4 shows the bandwidth limitation of the LoRa, yet it is suitable for agriculture applications where internet connectivity and battery power are the key parameters under consideration. For SOMA Vineyard, LoRa Trans-receiver Module LoRa SX1278, 433Mhz is used for water pump control operation.

12.3   Challenges and Future Scope in IoT-Enabled Drone

Major role of IoT-enabled drone in precession farming is crop monitoring, crop parameter analysis based on soil and atmospheric data like temperature, humidity, sunlight etc. and to generate actionable data. The assessment of the high resolution multitemporal data presents the primary challenge in the monitoring of plant health and yield potential. It is challenging to determine the proper vegetation index that will work best for mapping vegetation cover throughout the crop cycle. Therefore, it is essential to establish a link between crop yield and measured crop parameters.

AI-ML plus image processing and computer vision could enhance UAV-IoT Traits such as ability to clearly distinguish distributed weed and crop or herbicides that only harm weed and not affect the plants or friendly crop are only the state-of-the-art skills honed by multinational corporations and the intellectual properties still are not democratized into developing nations.

Aggro-tech or agri-science today are more focused on seeds or weeds and soil, moisture plus post yield management. Today's world does not have scalable service providers to cater to whole world to mechanize entire agricultural life cycles. E-waste and potential long battery life assumption of ten years of IoT is too ambitious an expectation on poor supply chains we have for availability to poor or marginal farmers across the third world nations.

In many developing countries, farmers from rural area are yet to get seamless internet connectivity. Further, There is no one-size-that-fits-all solution when it comes to application of sensors for smart technologies such as Zig Bee, Z-wave, RFID, NFC, LR-WPAN, 6LoWPAN, LoRaWAN, BLE, and Wi-Fi direct.

Though the technology is appreciated by modern farmer, very few of them have adopted the technology as a usual farming practice. Most of them still prefer the traditional farming practices, irrespective of the yield status. This may also due to lack of technology awareness and deployment issues. Rural or village farmers are reluctant to deploy the technology in their farm because of the dependency in adopting the technology. They may scare with over dose of technologies and digital divide. Potential solution can be literacy on technology based services through a common service centers in Rural Setup and motivation to village level entrepreneurism. Goapaddy.org is an experimental mechanization of paddy cultivation with innovative cluster and service provider model, and the implementation has plans to introduce UAV from JULY 2022 and spread the use of UAV-UGV for several other fields for sustainable farming strategy.

More Trained AgriTech or Scientific Community working at gross root levels is need of the hour.

12.4   Results and Discussion

To collect the data with drones or for effective spraying of the pesticides, a flying path needs to be finalized. A GPS installed on the drone to know its location and an IMU (Internal Measuring Unit) which helps the drone to follow the flight plan. Mission Planner is the open-source application that can be used for assigning flight path to a drone and a rover. After setting the home location and way points to fly, shown in Figure 12.5, an IoT enabled drone will follow the same path to collect the data. Takeoff altitude of ten meters needs to be set while planning the mission along with the landing waypoint at the end of the mission need to mark. Return to launch can be planned for the drone as per the last waypoint. QGroundControl application can be used to plan the flight which farmers can use easily from the field. He can change the way point at any time using this type of mobile application.

For live monitoring of the crop, GoPro Hero7 Black action camera is interfaced with Raspberry Pi (RP) installed on the drone which can capture the live images of high resolution of 1920x1080 Pixels. With this, the maximum video quality is 4/60 fps (frames per second) can be achieved to carry out time series analysis of the captured video data. Humidity and temperature sensors interfaced with RP collect the real time data and make it available on cloud. Sensors and RP module is powered up with battery supply of the drone. UGV or rover collect ground data related to soil moisture, temperature, sunlight, water contents in the atmosphere and send it to the same cloud as that of the drone data. Analyzing the data collected

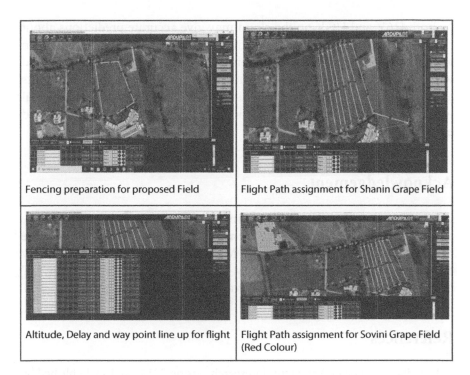

Fencing preparation for proposed Field	Flight Path assignment for Shanin Grape Field
Altitude, Delay and way point line up for flight	Flight Path assignment for Sovini Grape Field (Red Colour)

Figure 12.5 Flight assignment for IoT-enabled drone for SOMA vineyard.

 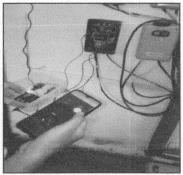

(a) Testing of IoT Module for
Soil Moisture

(b) Testing of LoRa Module for
Pump Control

Figure 12.6 (a) Testing of IoT module for soil moisture. (b) Testing of LoRa module for pump control.

by both UAV and UGV on the cloud may give accurate predictions about the crop health, growth, and harvesting and yield improvement. Based on analysis on the data of soil water content, a water pump can be operated with LoRa operation. Figure 12.6 shows testing of soil moisture contents and LoRa-based water pump operation (ON/OFF) with respect to soil water content. With this approach, both water and electricity can be used effectively.

Acknowledgement

Our sincere thanks to BIS team framing the standard for agriculture drone and it's SoP, for technical inputs on agriculture drone specifications. We thank SOMA Vineyard, Nashik, and Maharashtra for allowing us to carry out the experimentation and Ashoka Farm, Nashik Maharashtra for IoT testing. We thank SANDIP TBI for availing the Maker's Space.

References

1. Indian Council of Agricultural Research, Vision 2030, pp. 1–6. Published by the Project Director, Directorate of Knowledge Management in Agriculture (formerly DIPA), Indian Council of Agricultural Research, New Delhi, January 2011.
2. Annual report 2020-21, Department of Agriculture, Cooperation and Farmer's Welfare, Ministry of Agriculture and Farmer's Welfare, Government of India, New Delhi, www.agricoop.nic.in. 2021.
3. Agriculture drones are finally cleared for takeoff. *IEEE Spectr.*, 19th Oct 2016 date of access 15 August 2022 online magazine.
4. Marsujitullah, Zainuddin, Z. *et al.*, Rice farming age detection use drone based on SVM histogram image classification, SENTEN 2018 - Symposium of emerging nuclear technology and engineering novelty, IOP conf. series. *J. Phys.: Conf. Ser.*, 1198, 092001, 2019, 10.1088/1742-6596/1198/9/092001.
5. Study of the aerodynamic characteristics of an agricultural spraying UAV. *15th International Conference on Aerospace Sciences & Aviation Technology, ASAT - 15*, May 28-30, 2013, Military Technical College, KobryElkobbah, Cairo, Egypt, pp. 1–12.
6. Chavan, M.S., Automatic aerial vehicle based pesticides spraying system for crops. *Int. J. Innov. Technol. Exploring Eng. (IJITEE)*, 8, 11, 2278–3075, September 2019.
7. Shaw, K.K. and Vimalkumar, R., Design and development of a drone for spraying pesticides, fertilizers and disinfectants. *Int. J. Eng. Res. Technol. (IJERT)*, 9, 05, 1181–1185, May-2020.

8. Anand, K. and Goutam, R., An autonomous UAV for pesticide spraying. *Int. J. Trend Sci. Res. Dev. (IJTSRD)*, 3, 3, Mar-Apr 2019, Available Online: *Journal of Statistics and Management Systems*.

9. Rani, A., Chaudhary, A. *et al.*, Drone: The green technology for future agriculture. *Harit Dhara, Soil Health: Technological Interventions*, 2, 1, 3–6, January–June, 2019.

10. Puri, V., Nayyar, A., Raja, L., Agriculture drones: A modern breakthrough in precision agriculture. *J. Stat. Manage. Syst.*, 20, 4, 507–518, 2017.

11. Delavarpour, *et al.*, A technical study on UAV characteristics for precision agriculture applications and associated practical c-hallenges. *Remote Sens.*, 13, 1204, 2021, https://doi.org/10.3390/rs13061204.

12. Drone-based bio control services to protect your crops UAV-IQ, Available online: https://www.uaviq.com/en/biocontrol/(accessed on 19 January 2021).

13. Khokthong, W. *et al.*, Drone-based assessment of canopy cover for analyzing tree mortality in an oil palm agroforest. *Front. For. Glob. Change*, 2, Article volume 2, 12, 1–10, 2019.

14. Phade, G. *et al.*, Grape disease forecasting system. *Annual Technical Volume of Agricultural Engineering Division Board*, vol. 3, pp. 43–46, 2020.

15. Beloev, I.H., A review on current and emerging application possibilities for unmanned aerial vehicles. *Acta Technol. Agric.*, 19, 2016, 3, 70–76, September 2016. Published Online: 23 Sep 2016.

16. Alves, R. *et al.*, Assessment of RGB and hyperspectral UAV remote sensing for grass quantity and quality estimation. *Int. Arch. Photogramm. Remote Sens. Spat. Inf. Sci.*, XLII-2/W13, 489–494, 2019, ISPRS Geospatial Week 2019, 10–14 June 2019, Enschede, The Netherlands.

17. Barbedo, J.G.A., A review on the use of unmanned aerial vehicles and imaging sensors for monitoring and assessing plant stresses. *Drones*, 3, 40, 2019, doi: 10.3390/drones3020040; www.mdpi.com/journal/drones.

18. Niemitalo, O., Koskinen, E. *et al.*, A year acquiring and publishing drone aerial images in research on agriculture, forestry, and private urban gardens. *Technol. Innov. Manage. Rev.*, 11, 2, 5–16, February 2021.

19. Murugan, D., Garg, A., Ahmed, T., Singh, D., Fusion of drone and satellite data for precision agriculture monitoring. *2016 11th International Conference on Industrial and Information Systems (ICIIS)*, IEEE Xplore, pp. 910–914, 2016.

20. Ihuoma, S.O. and Madramootoo, C.A., Recent advances in crop water stress detection. *Comput. Electron. Agric.*, 141, 267–275, 2017.

21. Mendes, J. *et al.*, Smartphone applications targeting precision agriculture practices—A systematic review. *Agronomy*, 10, 855, 2020, https://doi.org/10.3390/agronomy10060855.

22. Mr. I. D. Pharne *et. al.*, Agriculture Drone Sprayer, *International Journal of Recent Trends in Engineering & Research (IJRTER)*,Volume 04, Issue 03; March- 2018.

23. Marica Franzin *et. al.*, Geometric and Radiometric Consistency of Parrot Sequoia Multispectral Imagery for Precision Agriculture Applications, *Appl. Sci.* 2019, 9, 5314; doi:10.3390/app9245314

24. Gayathri Devi, K., Sowmiya, N. *et al.*, Review on application of drones for crop health monitoring and spraying pesticides and fertilizer. *J. Crit. Rev.*, 7, 6, 667–672, 2020.

25. Daponte, P. *et al.*, A review on the use of drones for precision agriculture. *IOP Conf. Ser.: Earth Environ. Sci.*, 275, 012022, 1st Workshop on Metrology for Agriculture and Forestry (METROAGRIFOR), Ancona, Italy, 1–2 October 2018, 2019.

13

Unmanned Aerial Vehicle for Land Mine Detection and Illegal Migration Surveillance Support in Military Applications

C. Anil Kumar Reddy[1*] and B. Venkatesh[2†]

¹Department of Mechanical Engineering, KLEF, KL University, Vaddeswaram, India
²Mechanical Engineering, Vardhaman College of Engineering, Hyderabad, India

Abstract

UAVs or drones have found widespread use in both military and civilian purposes in recent years, and anti drone systems have been created to reduce risks using AI and GPS systems. Drones were once intended for military use and deployed in high-risk areas; but, as technology advances, drones are becoming more inexpensive. Due to the growing demand for their use in the military, drones are expected to play an increasingly significant role in future military operations on terrorism and illegal migration. Drones are currently used for land mine detection because landmines still go undetected, increasing the death rate and wreaking havoc on the environment. They can be replaced for a low price, which makes them acceptable for risky and politically sensitive operations. The goal of this project is to use a surveillance drone built to detect landmines to detect mines. A quad copter with a mine detector is placed to the prototype. This uses two separate detecting modalities, one of which is an induction metal detector. These are widely utilized to assist with the entire operation. The GPS was used to track the mine's location, and the location was recognized and communicated using the GSM module and Aurdino. A camera is also mounted to the quad copter to monitor suspected

Corresponding author: anildsnron5@gmail.com
†*Corresponding author*: bvtvardhaman@gmail.com

Sachi Nandan Mohanty, J.V.R. Ravindra, G. Surya Narayana, Chinmaya Ranjan Pattnaik and Y. Mohamed Sirajudeen (eds.) *Drone Technology: Future Trends and Practical Applications*, (325–350) © 2023 Scrivener Publishing LLC human presence at borders, with the data being sent with the operator on a regular basis. It can be improved further to strike targets with cruise missiles, allowing the UAV to manage (Ground Control Station) the borders and troop paths.

Keywords: UAV, GPS, mine detection, counter drone, GSM module, and metal detector

13.1   Introduction to Military Drones

Remotely or autonomously controlled drones armed with weapons will replace soldiers because of their strategic battlefield operations and surveillance. During World War II, a few nations used radio-controlled aircraft and anti-aircrafts for offensive operations in conflict zones, and in 1971, India faced a large-scale influx of illegal immigrants from east Pakistan. However, India was unable to stop them because of conventional surveillance. Therefore, advanced drones equipped with weapons, cameras, sensors, and reflectors to mimic the enemy drones' radar range will aid border soldiers in providing effective protection and surveillance. The Indian government has equipped indigenous and imported drones for border protection with cutting-edge military technology in order to compete with superpowers like the USA, China, UK, and others while preventing harm and reducing human fatigue for Indian forces. The Indian government has acquired a few drones such the Predator C Avenger, Heron TP, LMQ-9B Skygaurdian, and Yabhon combined with big cargo carrying capacity, effective speed, and guidance systems. Updating and innovating the border soldiers and naval fleets is crucial for national security.

13.1.1   Unmanned Aerial Vehicle (UAV)

Through its nomenclature, unmanned aircraft system (UAS) highlights the use of components other than airplanes. Without the aid of a human operator on the other end of the line, military unmanned aerial vehicles (UAVs) are capable of killing. They have the ability to fly to a certain area, pick their own targets, and execute. This has transformed the concept of a "killing robot" from science fiction to reality. Most people are likely ready to embrace "autonomous drones" as "smart technology," such as drones that operate based on a self-selected option like 100% autonomy. These drones are made to handle a variety of problems that could occur while doing their tasks. The limiting problem, not technology, is the political will to develop and adopt such politically "sensitive" technology that would allow lethal machines to operate without direct human oversight [10].

There is no legal definition for autonomous military operations drones. To handle increasing issues, modern drones are equipped with algorithms for a plethora of human-defined courses of action. Military forces were the first to employ drones. As a result, drone warfare is not a novel idea. Drones have been used by the Department of Defense to provide observation, surveillance, and intelligence for hostile forces in almost every combat operation and attacks. According to reports, there are currently about 100 nations using military drones. They are equipped with the newest generation cameras and are utilized in war and rescue missions to provide an exact topography of the area. Artificial intelligence systems contact with soldiers on a regular basis and give them with information about enemy movements. Drones are capable of transporting ever-heavier loads. They are equipped with anti-tank guided missiles and assist in the development of battle strategies [10].

Unmanned aerial vehicles (UAVs) have become more important in a variety of civilian and military applications because to their enhanced durability and stability in a variety of conditions and activities. Unmanned aerial vehicles (UAVs) are pilotless aircraft that can take off and land without a person on board, perform critical tasks without endangering human safety, and operate more effectively and inexpensively than manned systems of a similar size (UAV). They can be remotely piloted, in which case control operations are carried out by a ground control station, or autonomously, in which case the UAV is capable of conducting control operations onboard using the autopilot and a range of sensors, such as IMUs and GPS [1]. They are capable of carrying out missions like transportation, search and rescue, armed assaults, and remote sensing. Drones, remotely piloted aircraft (RPAs), unmanned aerial systems (UASs), and remotely piloted aerial systems (RPAS's) are other names for unmanned aerial vehicles (UAVs). The primary foundation for these utterances is the disparity between the standards and requirements of military and civilian institutions. These systems are referred to as unmanned aerial vehicles in this study (UAVs). The military was the first to employ UAVs for tasks including reconnaissance, surveillance, and weapon delivery. Primitive UAV technology was employed for military purposes, such as fighting and surveillance, in at least one out of every two wars before the first human aircraft flight [7].

The manned system is utilized in challenging circumstances where there is a lack of control, autonomy, or legal restrictions. The amount of time that some soldiers are exposed to risk has been decreased because to modern military manned vehicles like planes. Some defense analysts believe that modern manned military aircraft will soon be replaced by inexpensive unmanned aerial vehicles (UAVs). In hazardous areas where human aircraft run a high risk of being turned away, it is more economical to deploy UAVs. Due to the absence of an aircrew, an unmanned vehicle's (UV) range, endurance, and risk of death are all constrained. A human operator is not required for the electromechanical operation of an unmanned system, often known as an ultraviolet system (UV). Remote or autonomous control of unmanned vehicles is possible with the use of preprogrammed programming. Because of improvements in safety, a UV/US is employed in a variety of circumstances. Benefits including increased mission safety and decreased operational expenses are becoming available as the range of drone uses expands [11].

Military drones have been credited with changing combat and opening the door for modernization of the armed forces. Using cutting-edge technologies to address the technical innovation gap in Europe is another trend that experts have noticed. Drones hold great promise for air power and can improve both military and civilian personnel's safety. Military forces are known to deploy drones of all sizes and weights, despite the fact that the majority of armed drones are far larger and heavier than unarmed drones and weigh more than 600 kg. Small drones can be produced or modified to carry weapons or carry out ISR missions, unlike nano and micro drones, which are typically used for data collecting. Although radio line-of-sight can occasionally be utilized to convey imagery to the operator, large drones often fly outside the operator's line of sight. Medium-altitude long-endurance (MALE) and high-altitude long-endurance (HALE) drones are large drones that can be deployed for targeted attacks, tactical ISR, and battlefield support missions. Both military information gathering and search and rescue missions can greatly benefit from their use. As a result, these drones are utilized to collect vast volumes of data, which are subsequently processed and analyzed by professionals on the ground [12].

The military's use of unmanned aerial vehicles (UAVs) Unmanned aerial vehicles (UAVs) have made it possible for the military to conduct military operations more safely and effectively than they could in the past when pilots controlled the planes. The armed services are currently equipped with ISR, tactical air support, real-time RSTA, and new firing capabilities thanks to UAV systems. As seen in Figure 13.1, they can be utilized in a conventional combat, such as counter-insurgency operations, on the advance line of their own forces or far beyond it, on the flanks or in the back. In unconventional wars like counter-insurgency operations, they can also be deployed.

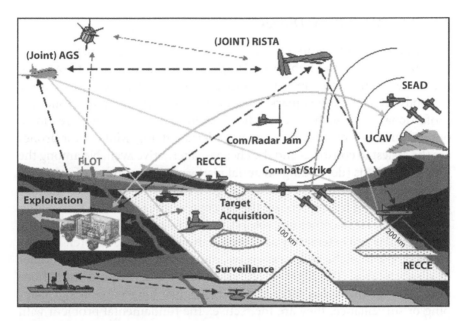

Figure 13.1 UAV role in battle field. [Source: NATO].

Target laser illumination, route and zone detection, combat damage assessment, and communications relay are all crucial roles. The integration of unmanned aircraft into combined weapon teams can be challenging, but it is necessary. From a military standpoint, Figure 13.1 displays a centralizer with the principal, but not all, military objectives that UAV systems can carry out [13]. Unmanned aerial vehicles (UAVs) typically include the following components: control system, ground station control, control connection, specific data link, Aurdino, and microcontroller.

13.1.2   UAV Types

There are many other methods to classify "drones," including "drones for photography," "drones for aerial mapping," "drones for surveillance," and more. But the most accurate classification of "drones" might be based on aerial platforms. Based on the kind of aerial platform used, drones are divided into four kinds.

a) Single-rotor drones
b) Drones with fixed wings
c) A solitary-rotor helicopter
d) The fixed-wing hybrid VTOL

13.1.2.1 Multi-Rotor Drones

Drones with several rotors are the most prevalent, and they are utilized by both professionals and hobbyists. They're employed for a variety of tasks, including aerial photography and video surveillance [8]. There are a variety of goods available in this market segment, such as multi-rotor drones for professional usage such as aerial photography and a variety of recreational options such as amateur drone racing or leisure flying. Multi-rotor drones are the easiest to construct and the cheapest alternative available among the four drone kinds (depending on aerial platform). The number of rotors on the platform can be used to further categorize multi-rotor drones. The four different types of helicopters are the Tri copter (three rotors), Quad copter (four rotors), Hexa copter (six rotors), and Octo copter (eight rotors). The most popular and extensively used type of drone is the quadcopter. Despite being simple to make and inexpensive, multi-rotor drones have a number of disadvantages. The flight duration, endurance, and speed limitations stand out the most. For large-scale projects like long-range aerial mapping or surveillance, they are ineffective. The fundamental problem with multi-copter aircraft is that they have to use a lot of energy (perhaps from a battery source) just to defy gravity and maintain altitude. Most multi-rotor drones available today have a flight time of 20 to 30 minutes.

13.1.3 Problem Statement

Explosive landmines in the route of troops and illegal immigration or entry are key security concerns for border security forces as well as residents of war zones. The military has been the first to deploy machines and mine detectors in an attempt to mitigate the risks associated with human land-mine detection and surveillance. There are currently fully autonomous systems that do not require a human operator for monitoring explosive land mine detection and deactivation as well as human surveillance, but these systems are expensive, difficult to integrate, and require experienced staff.

13.1.4 Objective

This chapter tries to mitigate the negative effects of explosive detection and human surveillance in border areas where security threats to the country and patrolling forces may exist. This device consists of a quad copter frame with a metal detection sensor, an Aurdino, a camera, and GPS tracking. This allows for the detection of land mines by remote sensing using GPS coordinates, as well as the recording of human trespassing at border crossings.

13.1.5   Previous Work

A bomb squad monitoring system necessitates the presence of soldiers whose job it is to locate and defuse a device in order to save human life [5]. This robot is being utilized to help strengthen our country's defense against terrorists, suicide bombers, and other similar actions. A command and control robot is the type of robot that will be built. This robot receives commands in the form of control signals from the operator and then conducts the needed action and locates the bomb. The disadvantage of such a system is that because the robot is running on the ground, the risk of a bomb exploding and causing harm to the robot is great. And, because it's difficult to watch a large area at once with human monitoring and trespassing, they used to split the troops. To address this, we're going with the recommended approach.

13.2   Literature Review

A drone is unmanned aircraft and formally recognized by the term unmanned aerial vehicles (UAVs) As we know that Aerial vehicles are well suited to provide an overview and aerial perspective over any natural climatic disaster, surveillance, or incident. When any natural climatic disaster happens in any populated region, only speedy and effective disaster management can help in reducing the impact of the destruction caused due to the disaster and also helps in the reduction in the number of victims. Whenever a disaster takes place the rescue and search unit requires and useful information and live data to predict the location and information about the lost person. An unmanned aerial vehicle which is AI- Drone resembles a vehicle that does not need a pilot to operate. When a natural calamity occurs the organization of calamity administration is to aid the inhabitants to lessen the figure of sufferers and limit the financial impact. Therefore, during such an event, it is necessary to keep up the statement link between the targets and all other responses occurring at the time of calamity by using Raspberry Pi w spec sheet in a given. The saving squads need quick and important data about the situations they have to face for better precautions. Since UAVs are self-governed therefore autonomous control comes into being. By drastically lowering the cost and danger of exploration, it can quickly give situational vigilance over a vast area and also contributes to reducing the time it takes the number of volunteers needed to find the injured, lost, or in need of rescue. To receive information to rescue the victim it is required to be trained. Therefore, drone needs to be trained using a supervised learning process. Live data are important to recognize the problems earlier and put a stop to them rapidly. To resolve the matter, we need a robust decision-making machine vision alone with a deep learning algorithm of neural networks. By doing so the system would be trained well to learn through complex data and multiple sources. In the case of an AI drone molded camera and inbuilt IOT sensors act as an input. This is where computer vision would help. Computer vision plays a vibrant role in sensing the various types of objects while hiking in the air. Computer vision trains in three steps.

1) Object tracking
2) Self navigation
3) Obstacle detection

On the other hand, neural networks sanction and train drones to perform tracking and object detection. With the help of neural networks, drones could recognize objects like trees, land animals, vehicles, etc. and they can count no of objects and can recognize various symbols effectively. Since these fields deep learning, machine learning, and computer vision share some common platform and applications. Computer vision so, however, we can say that Computer vision is bigger than the ML process put in it embraces and unified by a set of task dealing with an image to derive useful information from digital and real-time images, videos, and other visual inputs which helps in the better decision making for rescue operations after being served with the lots of training data, simplify and realize shapes by using various algorithm and model. Additionally, it requires numerous modeling tasks, including point cloud processing, motion estimation, stereo correspondence, structure-from-motion, multi-view camera geometry, and 3D scene modeling. With the combination of computer science and mathematics, using computer vision in AI drones can be stated to be a significant factor in automation. In this research paper, we are trying to develop a management system for an AI Drone to record, manage, and share the data to the base station so that it can be utilized for the detection algorithm [9].

Swarm robotics is a technique in which numerous robots work together in a synchronized manner to achieve a common goal, according to Aashish Raj *et al.* The goal is to separate the total operations into three categories: landmine detection, mapping, and defusing. We are applying the pure notion of swarm robotics with the help of auto bots in this project. Which are capable of completing the assignment with the assistance of other bots. According to the auto-bot concept, we are utilizing a UAV (drones) that can detect landmines in warzones. The concept of swarm robotics in defense not only lowers the cost of labor, but it also saves militants' lives by reducing trial and error methods [2].

According to Prem Mahadevan *et al.*, military drones play an increasingly essential role in future military operation, particularly in counter-terrorism and counter-insurgency operations. Because of their low cost, they are easily replaceable, making them suited for both dangerous and politically sensitive tasks. Technical limitations, as well as predicted advances in rival technology, such as air defense systems, will limit the military utility of drones. Unmanned aircraft, despite their importance in future conflict, are unlikely to totally replace manned aircraft and will instead function as a complement [3]. Yuvaraj Ganesh and colleagues have looked into a brand-new conceptual idea for mine detection employing a surveillance drone. During World War II, landmines were primarily employed as defensive and tactical barriers. In places like Afghanistan and Korea, they are still heavily utilized. Because so many land mines go undetected, more people die and the ecosystem is destroyed. In this mine detector is mounted to drone and IR Camara with metal detector is used to confirm metal presence and GSM Shares the location of identified object [4].

According to Jo Frew and Peter Burt, among others, the employment of unmanned aerial vehicles (UAVs), also referred to as drones, for military operations has generated a lot of discussion, controversy, and research during the past 20 years. The employment of military-style drones for security has attracted much less attention in the civilian world, though. Drone border control is becoming increasingly prevalent, which directly affects these challenges. According to this study, drones are now being deployed in a number of places throughout the world to patrol national borders. Because of their capacity for long-term monitoring, the ability to keep an eye on remote and inaccessible locations, and relative ease of deployment, border control authorities find them appealing. On the other hand, surveillance drones and their sensors are built on cutting-edge military technology, which governments frequently view as offering a straightforward answer to challenging social and political problems. Drones used for border control are a symbol of a society that views boundaries primarily in terms of "security" and regards border crossers as a threat to that security [14].

According to a study by Rey Koslowski *et al.* border control organizations, vigilante groups, and criminal organizations use drones to monitor activity over the US-Mexican border. This article discusses the complex security situation that is developing as countries use military drones for border security while non-state organizations with wildly different aims create their own drone surveillance capabilities. We contend that the political, governmental, and moral ramifications of border security drones are diametrically opposed. First, even though incorporating military technology into non-military security operations may compromise national security, drones may save the lives of migrants as they make their way through perilous oceans and deserts. Second, although offering new accountability mechanisms, drone surveillance threatens people's privacy. Last but not least, drones have the power to replace some physical security measures like fences while also constructing an invisible security system that transcends national boundaries. These puzzling results contribute to an understanding of the complex processes of policy creation that underpin drone border security programs in the US and Europe, as well as the challenges in determining if drone security is desirable [15].

Drone deployment along the southern European maritime borders by both state and non-state actors was examined by Panagiotis Loukinasa *et al.* along with the multiple surveillance uses of drones. The paper centers on the blending of security and humanitarian justifications in the use of this border technology as it develops around the idea of the multipurpose drone. The article claims that a number of variables, including the actions taken in response to the data collected by the multipurpose drone, affect its personality. Drones can endanger undocumented migrants while also making search and rescue efforts more efficient because to the data they offer. The use of drones in border operations raises concerns about transparency and accountability as well as the blending of civil and military, public and private spheres [18].

13.2.1   Need of Drones for Indian Borders

India shares a 15,106.7-km land border with its neighboring countries with typical eco system like forests, mountains, deserts etc. Due to the difficult and varied terrain, the specific conditions associated with each terrain, the climatic conditions, the relationships with particular neighbors, and other factors, among others, border security is difficult and vulnerable to insurgency, illegal migration, smuggling, and other anti-national activities. India also has a 7,517-km coastline, with 5,422 kilometers on the mainland. The shoreline of Lakshadweep stretches for 132 kilometers, while the coast of the Andaman and Nicobar Islands stretches for 1,962 km [19]. The Indian coastline is divided into nine coastal states and four union territories, with the tropics covering nearly the whole coastline. The territorial boundary is defined up to 12 nautical miles along the coastal border line. It is India's sovereign territory in this zone. The contiguous zone, often known as the hot pursuit zone, extends up to 24 nautical miles beyond this zone. In this zone, violations of customs, sanitary, immigration, and fiscal restrictions may result in penalty.

Border management is a security obligation for which numerous government organizations across the nation must work together and take coordinated action [16]. The purpose of border management is to protect our borders and the country from the dangers posed by the movement of people and goods between India and other countries, as well as the reverse. The term "border management" is inclusive and covers a wide range of activities, including managing both legal and illegal immigration, preserving the free and unrestricted movement of people and goods, and preventing human trafficking, infiltration, and smuggling. India's economy, which is largely dependent on the movement of people and goods, is becoming more and more globalized and service-oriented. Smuggling, trafficking, crime, terrorism, and illegal migration can all rise if these movements are not controlled, regulated, or overseen, creating a range of issues for the nation. In order to protect our borders from any attack, border management requires proactive intelligence, technological breakthroughs, coordinated action by bureaucrats, economic agencies, security personnel, and other national stakeholders. In the graphic below, the major participants in border management are depicted [17].

Table 13.1 UAV technical specifications based on modules.

Mechanical and electrical module	Sensor technology module	Embedded system and other electronic module	Software module
• Quad Rotor Frame • Landing Stand • 4x Motors • 4x Propellers • 2300 Mah Li-Po Battery With Power Distribution System	• IMU/3 Axis Digital Compass/ Digital Pressure Sensor • On-Board Camera • GPS • Metal Detector	• Flight Controller Using ARDUINO UNO Board • PI • ESC • Trans receiver • RX • RF	• Cadsoft Eagle/ Fritzing • Matlab/Simulink (Drake Tool Box) • Arduino Ide, Linux, Opencv • Network Analyzer app

13.2.2 UAV Technical Specifications

While the overarching aims, strategies, and objectives have been specified, the component specifications will be determined when they are identified for their project applicability. On the basis of the application and engineering involved, the technical specifications are separated into the following engineering modules. Table 13.1 shows how the modules are organized [6].

13.3 Methodology of UAV's in Military Applications

13.3.1 Proposed System

In this paper it is discussed about integrated system which uses flight controller board and multiple motors, where surveillance camera and mine detector are connected to drone and controlled with Aurdino program and network analyzer app as shown in the Figure 13.2, this is designed to reduce threats for patrolling troops where mine detection sensor connected to drone detects the mine and give the GPS location of mine so that detection team can diffuse the mine, and during wars illegal immigration is major problem, in view of this problem drone is attached with camera which gives live streaming to operator with the

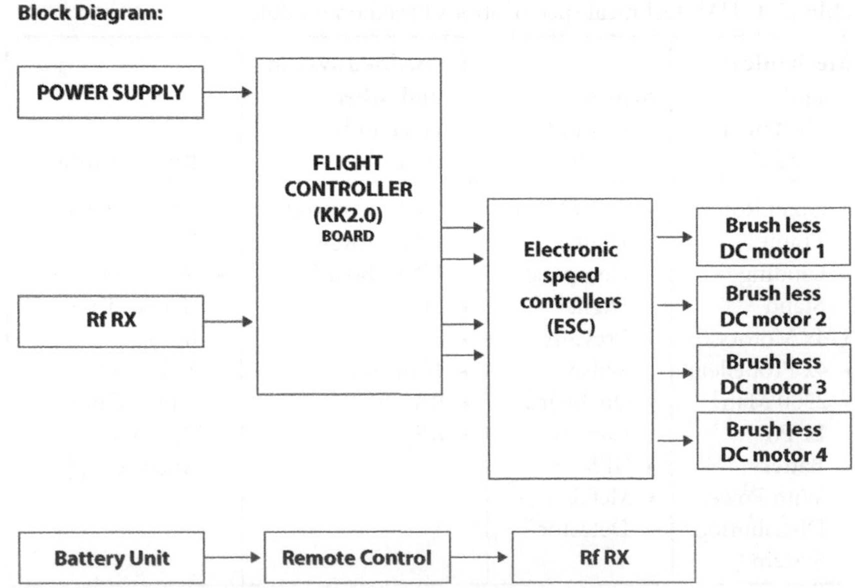

Block Diagram:

Figure 13.2 Block diagram of UAV.

help of network analyzer app and code for camera functioning and live streaming can be seen in mobile linked with receiver or in Interfacing Unit. So by this integration troops can diffuse the mines with accuracy and observe the long borders to resist illegal migration.

The system is intended for drone implementation that is both cost-effective and low-maintenance. According to estimates, the market for drones is expected to reach $6.52 billion by 2026, growing at a rate of 31.4 percent. As the cost of drones decreases and drone software for agriculture becomes more sophisticated, demand continues to rise [13].

13.3.2   Methodology

13.3.2.1   UAV Work Principle

A quad copter is a four-rotor helicopter, also referred to as a quad rotor heli-copter. In a square arrangement as shown in Figure 13.3, with an equal distance between each rotor and the quadrocopter's center of mass, the rotors are oriented upward. The rotors' angular velocities, which are propelled by electric motors, are changed to control the quadcopter. A quadcopter is a common option for small unmanned aerial vehicles due to its simple engineering (UAVs). Quad copters are employed for a variety of operations, including surveillance, search and rescue missions, and construction inspections. The information is read and transmitted to the central flight controller system when the joysticks are flipped in either direction. The central flight controller now receives data from the IMU, Gyroscope,

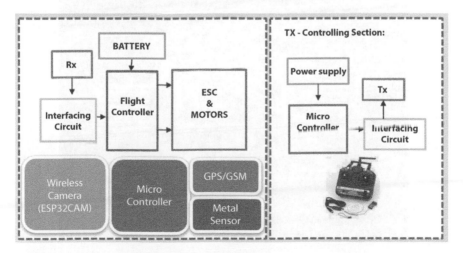

Figure 13.3 Proposed UAV layout.

GPS, and obstacle detection sensors, if the quadcopter is equipped with them. The outcomes of calculations made using pre-programmed flying parameters and algorithms are sent to the electronic speed controllers. The central flight controller receives data from the movement of the remote control sticks. This information is sent from the central flight controller to the motor's ESC, which regulates motor speed.

Steps: Remote Control Stick Movement → Central Flight Controller → Electronic Speed Control Circuits (ESCs) → Motors and Propellers → Quad copter Movement or Hover.

13.3.2.2   UAV Controls and Installation

a) Multi Controller

For two, three, and four-rotor remote-control multicopters, the KK.2 multi controller as shown in Figure 13.4 serves as the flight control board. Its task is to maintain flying stability for the aircraft. This is done by the multi controller providing the integrated circuit data from the three on-board gyros, roll, pitch, and yaw (Atmega IC). After the data has been analyzed by the KK.2 software, the Electronic Speed Controllers (ESCs) that are attached to the motors and wired into the board are then provided a control signal. The ESCs will either accelerate or decelerate the motors (and tilt the rear

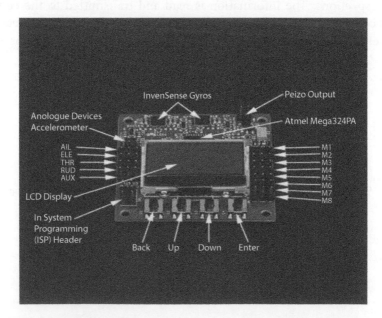

Figure 13.4 KK2.0 flight controller (6V).

rotor through a servo in a quadcopter) based on the signal from the IC to ensure level flying.

A flight control board for remote-controlled multi-copter models with two, three, four, or six rotors is called the KK.2 multi controller. Its responsibility is to maintain stability during takeoff and landing. Data from the three on-board gyros (roll, pitch, and yaw) are fed into the integrated circuit to do this (Atmega IC). The Electronic Speed Controllers (ESCs) on the board, which are connected to the motors, receive a control signal after the data has been analyzed by the KK software. To ensure level flying, the ESCs will either speed up or slow down the motors in response to the signal from the IC (and, in the case of a quadcopter, tilt the rear rotor using a servo). Additionally, the board receives a control signal from the RX, which it then transmits to the IC via the pins for the aileron, elevator, throttle, and rudder. In response to the RC Pilot's Transmitter command, the IC will send a signal to the motors (through the M1 to M6 pins on the board) to speed up or slow down in order to achieve controlled flight (up, down, backwards, forwards, left, right, and yaw) (TX). The Atmega168 chip and ISP header in the version.5.5 allow users to modify and upload their own controller code.

b) Electronic Speed Controllers (ESC)
The direction, speed, and braking power of an electric motor are managed by an electrical circuit called an electronic speed control, or ESC. On electrically driven radio-controlled models, ESCs are widely utilized, and versions with brushless motors essentially function as a low voltage three-phase electronic energy source.

c) RF Transmitter and Receiver
A compact electrical circuit known as an RF Module (Radio Frequency Module) can transmit and/or receive radio signals on a range of carrier frequencies. Because designing radio circuits can be challenging, electronic designers frequently use RF Modules. Good electronic radio design is notoriously challenging because of the sensitivity of radio circuits and the accuracy of components and layouts needed to achieve performance on a certain frequency. In this example, we're using a fly sky (fs) CT6b transmitter and receiver. The six-channel RF Remote Controller was created with Microchip's CC2500 RF Transceiver Modules and PIC16F1847 microcontroller. Six tact switches and four Address Jumpers are included on the transmitter to allow numerous units to be paired so that they do not interfere with one another. The board comes with a power LED and a valid transmission LED. The project employs a 5V DC power source with an integrated LM1117-3.3V regulator to power the CC2500 Module.

A transmitter and a receiver can both be used with the dual-purpose PCB. A 5V DC supply powers the receiver. Control signals are sent from the RX to the IC via the aileron, elevator, throttle as shown in the Figure 13.5, and rudder pins on the board. Included are a 9-pin connector for outputs, a valid signal LED, a power LED, and four jumpers for connecting the RX and TX devices. On the same PCB, both the transmitter as shown in Figure 13.6 and receiver as shown in Figure 13.7 are assembled. All outputs are latch type and TTL 5V signal for simple communication with other devices such as relay boards and solid state relays.

Control signals are sent from the RX to the IC via the aileron, elevator, throttle as shown in the Figure 13.5, and rudder pins on the board. Once this processing is complete, the IC will send a signal through the motors (through the M1 to M6 pins on the board) to speed up or slow down in response to the RC Pilot's Transmitter's command to achieve controlled flying (up, down, backwards, forwards, left, right, and yaw) (TX). On a quadcopter, a servo will be turned on using an M4 pin connector to gain yaw authority.

Figure 13.5 Electronic speed control.

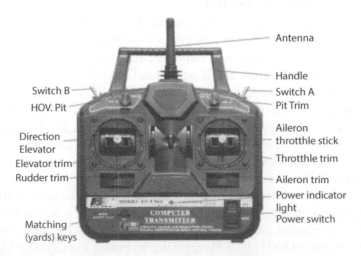

Figure 13.6 Drone transmitter.

Figure 13.7 Drone receiver (2.5 GHz).

d) Hardware Installation

Initially remove the propellers, turn on the transmitter and flight controller, set the throttle to 14%, and start the motors in the first phase, which is concerned with the transmitter channels. Pitch (elevator) stick: move it forward. Be sure to turn up the back motor. If not, shift your rolls (aileron) and elevator pitch to the left. The appropriate motor should speed up. If not, roll once more (aileron).

The second step is to adjust the throttle, which is the first step. Lower your trim if the LED does not turn on and stays on after turning on the transmitter and flight controller. If the problem persists, try reversing the throttle channel. Make sure your Transmitter docsn't have any mixing switches turned on.

The third step is to establish the ESC throttle range, which requires turning the yaw pot to zero, then turning on the transmitter and setting the throttle stick to full. Later, turn on the flight controller and wait for the ESCs to beep twice after the first beep (Throttle stick to off) (Plush and SS ESC's). The ESCs beep, the flight controller is turned off, and the yaw pot is restored.

We initially double-checked the gyro directions in the fourth stage. New propellers should be installed in place of the old ones. Turn on the flight controller and transmitter, then cut the throttle by roughly a quarter. Engines should start to run. Turn the multicopter forward. The forward motor needs to pick up speed. When it fails, try flipping the pitch gyro. Tipping the multicopter to the left is advised. The left motor needs to rev up. Roll the gyro rearward if not. Clockwise rotation of the multicopter. The motors in the front and back should both be turned up. Use the yaw gyro in reverse if not. Reversing gyros where the UAV is flying is what the fifth phase entails. Set the roll gain pot to zero. The LED flashes 10 times quickly as soon as the flight controller is turned on. Later, you can reverse it by moving the stick to the desired gyro. The LED will keep blinking. Deactivate the flight controller. Go to step 2 if there are still more gyros to be reversed; otherwise, reset the roll gain pot.

The UAV is tested at the last phase. Hold the quadcopter firmly in place over our heads while progressively increasing the throttle to roughly half. The multi controller calibrates its gyros when the throttle leaves zero, so maintain it steady as you increase the throttle because the gyros must be at rest after that to allow for calibration. If the multi-copter tries to twist away, check the propeller and engine directions, gyro positions, and trim settings. It's acceptable if there is a slight twist. When that fails, try bending the quad. It needs to be sturdy enough to endure your motions. Gyro gain is accompanied by an increase in resistance. If the gain starts to oscillate, lower it. The benefit shouldn't have to be reduced by more than 40%.

13.3.2.3   Drone Material and Frame

It is a critical component of the quad copter and should be composed of fiber or a light, robust material. We use aluminum discs as the main frame at the centers of our project, and steel arms. This aluminum quad frame is well-designed and constructed from high-quality components. The F450 Multi-Copter Quad copter Rack Frame is used here, and it is built of superior engineering material, making it incredibly sturdy and smooth. It also includes a new design PCB board for 50mm controller boards such as MK, KK, FF, and MWC. It has a 21.5*4*5cm arm size.

❖ **Dimensions:** Its dimensions are as follows:
  - Width: 363 mm
  - Height: 40 mm
  - Weight: 270g (w/out electronics)

Light weight and strong materials should be adopted for drone construction.

13.3.2.4   Program Used/Software Used (e.g., Aurdino) and Data Collection

a) Network Analyzer
The broad selection of tools offered by Network Analyzer can assist you in analyzing a wide range of issues with your wireless network setup, Internet connectivity, and identifying a wide range of errors on remote servers. It has a rapid Wi-Fi device discovery tool that provides a list of all LAN device addresses, names, and makers. The usual net diagnostic tools offered by Network Analyzer include ping, trace route, port scanner, DNS lookup,

and who is. In order to help choose the best channel for a wireless router, it also displays all local wireless networks along with other data such as signal strength, encryption, and router manufacturer. IPv4 and IPv6 are both compatible with everything.

b) Camera
A small, low-power camera module called the ESP32-CAM is built around the ESP32 microcontroller. It has an inbuilt TF card slot and an OV2640 camera. The ESP32-CAM can be used for smart IoT applications including Wi-Fi picture upload, QR identification, wireless video surveillance, and more. Utilizing the ESP32-CAM is identical to utilizing the ESP32 modules we previously examined, with one important exception. The ESP32-CAM board does not include a USB port, so you cannot just plug it in and begin loading programs. Instead, an external FTDI adaptor will be needed. A dual-mode WIFI+ Bluetooth development board with integrated antennae and ESP32 chip-based cores is called the ESP32-CAM. ESP incorporates Wi-Fi____33, traditional Bluetooth, and BLE Beacon.

c) Flight Controller Interface with Aurdino
The flight controller will have five input control pins that will allow the drone four movement actions: throttle, roll, yaw, and pitch. Because these pins require a PWM signal to function, we'll use the Arduino Uno to generate one. We may program the Arduino by executing functions with Capture and Detect using appropriate planning and sketching of a specific area. Appendix-A contains the program.

13.3.2.5   Illegal Migration Surveillance with Camera

The ESP32-CAM is a tiny camera module that costs about $10 and runs on the ESP32-S microcontroller. It has an OV2640 camera, many GPIOs for attaching peripherals, and a micro SD card slot for storing the camera's pictures. It must turn to face the camera The IP address can be entered into your preferred web browser, such as Chrome or Mozilla, after being copied from the LAN scanner. As a result, the web page containing all of the camera settings will show up on the left side. You can review each setting individually. At the bottom of the page, you can see a button that says "Start Streaming"; click it to watch the live video. Using the Camera option, increase the resolution. The video data is real-time, and you may always see it on the local Web server.

13.3.2.6   Data Collection from Mine Detector and Camera

When the metal sensor triggers it immediately triggers the signal to check for the GPS coordinates of the drone which is send as the location data through SMS to the registered mobile number.

a) Locating Mines with GPS
Metal sensor is interfaced with the microcontroller which sends the location upon sensing the metal at any given point of time during the operation. The metal sensor is controlled through the movement of the drone manually and based on the drone flight height the metal can be detected from certain distance. The locations are identified by using the GPS coordinates which triggers whenever metal sensor gets activates as it also includes live camera from the drone which provides us the access of remote monitoring to live view the camera, Metal sensor is interfaced with Microcontroller which waits for the sensor to get activated in case of metal being detected. If the sensor gets activated microcontroller checks for the GPS coordinates and send the latitude and longitude values to the operator through flight control interface.

b) Image Live Sharing/Surveillance and GPS Locating
With the implementation of the latest technology it's easy to access the images and GPS location by just using your smart phone or network analyzing apps. The code creates a virtual bridge between drone and the smart phone to share the information, as the live stream is available from the camera with sample IP address is needed to view the live camera.

13.3.2.7   Testing Conditions Applied for this Drone

Metal detection and Human Surveillance are the two conditions considered for testing. Initially in first phase after the setup of hardware is done accordingly and all the components are connected to each other then it is set to ARM position and makes it ready to fly. The quad copter will receive waypoints and goes in second phase and comes back to ground as destination arrives. The kit is having the components like controller, telemetry, frame, GPS, compass, battery, motors, ESC's.

13.4   Software Implementation

This chapter overviews the software use for Flight Controller. The software used is Arduino IDE is shown in the Figure 13.8.

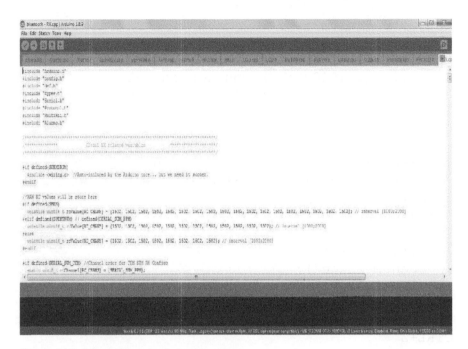

Figure 13.8 Sample from the drone program.

13.4.1   Arduino IDE

Java was used to construct the cross-platform application known as the Arduino Integrated Development Environment (IDE). With the aid of third-party cores, it is used to create and upload programs to Arduino-compatible boards as well as other vendor development boards.

13.4.2   UAV Program/Coding

Appendix A

```
#include <SoftwareSerial.h>
#include <TinyGPS.h>
float lat,lon; // create variable for latitude and
longitude object
SoftwareSerial gpsSerial(8,9);//rx,tx
TinyGPS gps; // create gps object
const int analogInPin = A0;
int sensorValue = 0;
```

```
int outputValue = 0;
unsigned int i=0;
void setup(){
Serial.begin(9600); // connect serial
gpsSerial.begin(9600); // connect gps sensor
}
void loop(){
  sensorValue = analogRead(analogInPin);
  outputValue = map(sensorValue, 0, 1023, 0, 12);
  Serial.println(outputValue);
  delay(200);
  if(outputValue>=7)
  {
    if(i==0){
    SendMessage();
  gpsSerial.print("LA:"+String(lat,6)+"   N,"    +
"LO:"+ String(lon,6");
    gpsSerial.print((char)26);// ASCII  code  of
CTRL+Z
    delay(1000);
    i++;
    }
  while(gpsSerial.available()){ // check for gps
data
  sensorValue = analogRead(analogInPin);
  outputValue = map(sensorValue, 0, 1023, 0, 12);
  Serial.print("Output:");
  Serial.println(outputValue);
  delay(200);
  if(gps.encode(gpsSerial.read()))// encode  gps
data
  {
  sensorValue = analogRead(analogInPin);
  outputValue = map(sensorValue, 0, 1023, 0, 12);
  Serial.println(outputValue);
  delay(200);
  if(outputValue>=7)
  {
    if(i%2==0){
    SendMessage();
```

```
      gpsSerial.print("MetalAT:LA:"+String(lat,6)+"
N,"+"LO:"+ String(lon,6));
      gpsSerial.print((char)26);//  ASCII   code   of
CTRL+Z
      delay(1000);
      i++;
      }
   gps.f_get_position(&lat,&lon);  // get latitude
and longitude
    //Serial.print("Position: ");
    //Serial.print("Latitude:");
    //Serial.print(lat,6);
    //Serial.print(";");
    //Serial.print("Longitude:");
    //Serial.println(lon,6);
  }
}
String latitude = String(lat,6);
String longitude = String(lon,6);
Serial.println(latitude+";"+longitude);
delay(1000);
}
   }
}
void SendMessage()
{
  Serial.println("Sending mesg");
  gpsSerial.print("AT\r\n");        //Sets  the  GSM
Module in Text Mode
  delay(2000);   // Delay of 1000 milli seconds or
1 second
  gpsSerial.print("AT+CREG?\r\n");
  delay(2000);
  gpsSerial.print("AT+CMGF=1\r\n");
  delay(2000);
  gpsSerial.print("AT+CNMI=1,2,0,0\r\n");
  delay(2000);
  gpsSerial.print("AT+CMGS=\"8919112563\"\r\n");
  delay(100);
}
```

13.5 Conclusion

By using this integrated testing system mine detection and human trespassing can be identified efficiently in safe manner and range of mine detection can be minimized and maximized according to requirement and it can be specific and accurate by using high performance thermal imaging camera and high frequency metal detectors. Because mine detection with human intervention is dangerous and during night times it may not visible, so to avoid this threats this integrated drone was developed which can avoid threats and able to work during day and night effectively. In this mine detection sensor senses mine and shares the GPS Location to the operator and the same time the camera will capture the illegal migration activities with the help of Aurdino program and network analyzer app integrated with flight control unit. By using UAV it can be easily monitored human trespassing and mine detection and the results can be interpreted by tracking GPS module and Aurdino program. the main advantage of this drone is it will be helpful for soldieries, Border security forces, illegal migrants surveillance and war environments, while being controlled by a single person operating from a safe and secure location. The prototype which is being created is one of its kinds and has not been tested ever. This can be a boon for the future of drones and robotics which can bring a variable change in development for the mankind. This is a trial and error method which has been tested again and again to get favorable results for the requirement of proper setting and functions of the drone. The drone has a lot of potential for development and enhancement of different parts and proper research can lead to a remarkable evolution in the field of robotics.

References

1. Dinesh Tharun, S. and Siva Kumar, P., Robust estimation by drone for surveillance and bomb detection using actuators. *Int. J. Eng. Adv. Technol. (IJEAT)*, 8, 6S3, 1505–1512, September 2019.
2. Raj, A., Rafiq, I., Abhishek Gowda, J., Krishna, L.S., Landmines detection using UAV. *Int. Res. J. Eng. Technol. (IRJET)*, 06, 04, 2716–2719, Apr 2019.
3. Mahadevan, P., The military utility of drones, CSS analysis in Security Policy, No. 78, © 2010 Center for Security Studies (CSS), ETH Zurich, July 2010.
4. Ganesh, Y., Raju, R., Hegde, R., Surveillance drone for landmine detection. *2015 International Conference on Advanced Computing and Communications*, IEEE, p. 33, © 2016, 978-1-4673-9777-3/16 $31.00.

5. Casey, B. and Rocks, T., *Landmine Detection Rover*, Worcester Polytechnic Institute, Worcester, USA, 2017.
6. Reasad, A.M., Bappy, A., Asfak-Ur-Rafi-Anonno, *Design and Development of Unmanned Aerial Vehicle (Drone) for Civil Applications*, BUET, Dhaka, 2014.
7. Elmeseiry, N., Alshaer, N., Ismail, T., A detailed survey and future directions of unmanned aerial vehicles (UAVs) with potential applications. *Aerospace*, 8, 363, 2021, https://doi.org/10.3390/aerospace8120363.
8. Hartley, C., Drone Project Report, Permit #TE-181713-3, Ormond Beach, California, September 26, 2018.
9. Rouf, M. and Yadav, A., RF controlled solar based robotic drone, GCET, 2021.
10. Konert, A. and Balcerzak, T., Military autonomous drones (UAVs) - From fantasy to reality. Legal and Ethical implications. *Transp. Res. Proc.*, 59, 2021, 292–299, 2352-1465, 2022.
11. Sivakumar, M. and Naga Malleswari, T.Y.J., A literature survey of unmanned aerial vehicle usage for civil applications. *J. Aerosp. Technol. Manag.*, São José dos Campos, 13, e4021, 2021.
12. Boulanin, V. and Verbruggen, M., *Mapping the development of autonomy in weapon systems*, Stockholm International Peace Research Institute (SIPRI, November 2017, European Parliamentary Research Service, Signalistgatan, Sweden, 2019.
13. Jeler, E.G., Military and civilian applications of UAV systems. *Strategies Xxi International Scientific Conference*, The Complex and Dynamic Nature of the Security Environment, p. 379, 2019, https://fas.org/irp/program/collect/nato-uav-99/r-I-5/sld008.htm.
14. Burt, P. and Frew, J., *Crossing a line | The use of drones to control borders*, Drone Wars UK, Peace House, London, 2020.
15. Koslowski, R. and Schulzke, M., Drones along borders: Border security UAVs in the United States and the European Union. *The International Studies Association Meeting (ISA)*, San Francisco, CA, April 4-7, 2018.
16. Mazzeo, A., Border surveillance, drones and militarisation of the Mediterranean, A-DIF, Sicily, Italy, November 2020.
17. Gupta, S., Smart Border Management, Tansen Marg, Sept 2018, Federation House.
18. Loukinas, P., Drones for border surveillance: Multipurpose use, uncertainty and challenges at EU borders. *Geopolitics*, Geopolitics, 27, 89–112, 2021.
19. Bier, D.J. and Feeney, M., Drones on the border, Cato Institute's Center for Global Liberty and Prosperity, Washington, D.C, May 2018.

14

Importance of Drone Technology in Agriculture

Karuppiah Natarajan[1]*, Karthikeyan R.[2] and Rajalingam S.[3]

[1]*Department of Electrical and Electronics Engineering,
Vardhaman College of Engineering, Hyderabad, India*
[2]*Department of Artificial Intelligence & Machine Learning (CSE),
Vardhaman College of Engineering, Hyderabad, India*
[3]*Sunyani Technical University, Bono Region, Ghana, West Africa*

Abstract

This chapter deals with the role of the Unmanned Ariel Vehicle (UAV) in managing natural resources and agricultural farms. The application of drones in agriculture reduces a lot of manpower and increases productivity by a huge amount. Smart drones help farmers to estimate crop production. The development of technology in drones supports the farmers to test the soil fertility. It provides effective and efficient results. Agricultural drones (AD) are a recent wonder that has been introduced into farming areas. These are small and fly above the cropland from a distance of 100–400 feet. It is able to capture and analyze images instantly. The results are amazing and more accurate. Most of the start-ups in North America and European countries are funding huge amounts of money for developing AD. The AD helps to identify the soil fertility, crop diseases, natural resources available in the soil, and crops yielding estimation. It also helps to find the intruders to crops such as insects, rats, and other animals that affect the crops.

Keywords: Unmanned ariel vehicle (UAV), agricultural drones, European countries

Corresponding author: natarajankaruppiah@gmail.com

Sachi Nandan Mohanty, J.V.R. Ravindra, G. Surya Narayana, Chinmaya Ranjan Pattnaik
and Y. Mohamed Sirajudeen (eds.) Drone Technology: Future Trends and Practical Applications,
(351–374) © 2023 Scrivener Publishing LLC

14.1   Introduction

More than 60% of the world's population relies on agriculture. It is also an important factor for ecological conservation. Modernization is essential in the field of agriculture as it has a lot of issues and challenges. In our country, the monsoon fails miserably. Pets and destructive insects affect the yield of crops. Moreover, the drastic increase in population demands more food production. So, Drones are the best suitable machines that can solve these issues [1]. Studying the natural resources, analyzing the fertility of soil, effective water, and irrigation management, crop disease identification, and pest control management are the areas where drones can be used [2].

Unmanned aerial vehicles or drones are best suitable for modernizing agriculture. In recent years people from villages have moved to cities and there is a shortage of human manpower to do agriculture-related activities. Also, the wages of the laborers have also been increased. On the other hand, productivity is decreasing day by day due to unknown diseases and the failure of monsoons. Smart agriculture and precision farming are the buzzwords in today's modern agriculture. Smart agriculture is data-driven. Starting from assessing the soil fertility to taking the yield to markets everything is data-driven [3].

Precision agriculture refers to the usage of GPS in agriculture farming along with sensors to maximize the output and minimize the input. It gives an idea to the farmers such that they make sure there is a continuous supply of water and fertilizer so that the production of the crops reaches maximum. In addition, it also surveys the fertility of the soil, pest control, crop disease identification, etc. The usage of drones is helpful in analyzing the health of the crop, suitability of fertilizer to be used, nutrients availability, and weed control.

This chapter is well organized with the following sections. A detailed overview of the agricultural drone is given in the introductory section. The important components of the agricultural drones are discussed in section 14.2. Sections 14.3 and 14.4 concentrate on how to use drones effectively in assessing the resources and fertility of the soil. Section 14.5 focuses on the important aspect of agriculture i.e., managing the usage of water for irrigation purposes. Sections 14.6, 14.7, and 14.8 deal with crop disease identification, pest control management, and crop yield management. Section 14.9 discusses the issues and challenges with agricultural drones.

14.2   Components of a Drone

The list of components that make an agricultural drone is shown in Figure 14.1. It consists of a sprayer.

Figure 14.1 Agricultural drone with sprayer.

Frame

It is the base on which the entire structure is built. The motors and other components are housed on the frame. The frame should be light and endurable so that the motor can withstand crashes and save power while flying.

Propellers

Propellers are used for lifting the drone. The inherent characteristics of a propeller are its pitch and diameter. The pitch of a propeller is the distance travelled by the drone in one revolution. A propeller with a low pitch value has high torque and it provides good stability. It saves power since the motor draws less current from the battery while flying. High-pitch propellers cause turbulence while flying. This increases the endurance time of the drones. Similarly, the diameter of the drone propeller determines the amount of air that it contacts. This determines the efficiency of the drone. A propeller with a large diameter will have more contact with the air.

Motor

The motor provides the necessary torque for the propeller to lift the drone. Hence motors with high torque capability are preferred. Brushless DC motors are generally used since it has good speed vs. torque characteristics, high efficiency, and noiseless operation. They are light-weighted motors with no restriction on the output waveform. These motors are electronically commutated motors capable of varying the speed of the drone by adjusting its power output.

Flight Controller

This part of the drone is considered its brain. It performs arithmetic and logical operations based on the signals received from the various sensors and user commands. Based on the input received, it sends signals to the Electronic Speed Controller which controls the torque and speed of the motors. This steadies the flight of the drone. Naza, Rabbit, and KK are some examples of flight controllers. For a small drone of quadcopter type, gyro sensors are used as flight controllers.

A flight controller is used to control the drone in mid-air by controlling its pitch, roll, and yaw axis. The IMU sensors send the orientation of the drone as signals to the microcontroller. The microcontroller processes the raw data to estimate the angles and hence provides error compensation so that the drone remains in its steady state. Drones are controlled by AVR-based ATmega32-bit microcontroller or Arduino-based systems.

Electronic Speed Controller

ESC is used to vary the speed and direction of the drone. It also acts as a dynamic brake. They are powered electrically and radio-controlled and used to provide signals to a brushless dc motor.

Power Source

Batteries are the main power source for drones. Lithium-ion batteries are extensively used as power sources. Batteries occupy less space and provide comparatively more energy than other sources.

Power Distribution Board

It is an assembly from which electric power is distributed to various points of the circuit. It has protective devices for each feed point and all are enclosed in a common frame. It has residual current breakers with over-current protection and the main switch for each feed point.

Flight Sensors

Accelerometer

It is an electromechanical device that is used to sense the acceleration of a moving body. The acceleration is measured by the static and dynamic forces acting on it. It calculates the amount of tilt of the object with respect to gravity and then it analyses the dynamic acceleration and calculates the position of the moving object.

Gyrometer

It is used to determine the orientation of the object. It uses the earth's gravity to determine its orientation. It consists of a rotor mounted on the spinning axis of a stable wheel.

Working of Drone

It is made up of light composite material to lessen the weight and increase its manoeuvrability. These drones are capable of flying in higher altitudes because of their above property. These drones are equipped with GPS, infra-red cameras, and laser. These drones are controlled by a remote-control system. It does not require long runways. It is also called an Unmanned Aerial Vehicle (UAV) since it is controlled by humans from the ground and does not house any humans.

14.3   Study of Natural Resources

It is necessary to know about natural resources like air, soil, water, sunlight, and vegetation pattern (forest, grassland, tundra, desert, and Iceland) for agricultural production and its applications. This information about natural resources also helps for the purpose of aesthetic value, cultural value, and scientific interests. The traditional way of collecting information on natural resources and vegetation pattern is either by ground-based survey or by using satellite images from the satellite. Satellite-based method of collecting information has challenges in the cloudy atmosphere, inadequate resolution of small spots, and insufficient close-up images. A ground-based survey is not applicable to all areas due to its complexity and dangerous environmental conditions. Both methods are expensive and time-consuming [4]. Close-up images are important for detailed and precise analysis of natural resources and vegetation patterns. Drones or unmanned aerial vehicles play a key role in collecting high-resolution close-up images that replaces satellite images with impressive high resolution. These high-resolution images from a drone may help to study natural resources, water management, soil fertility analysis, pest control, irrigation, and crop yield. It also helps to identify crop diseases from remote locations. Drones help to map water resources and the environment with low cost and less time using Multispectral sensors. Some of the applications of drones in natural resource monitoring include Topographical mapping, High-resolution imaging of natural structures, Natural vegetation analysis, Water level mapping, and Mapping of Environmental effects on soil [5].

Drones collect information by flying over dangerous areas that are inaccessible to humans. The collected information/data shall be stored at a protective remote location to avoid data loss. Then, the collected information will be processed and analyzed using computer software. The 2D and 3D drone imaging makes the analysis easier than the traditional method.

The aim of recent research is to study the loss of vegetation on greenhouse gas emissions and the loss of soil fertility. It is very expensive and challenging to collect such information using satellite imaging. Sometimes ground-based survey is adopted when satellite imaging fails due to poor resolution. A ground-based survey is time-consuming, expensive, and needs skilled manpower to record data accurately. Conservation drones are available for such studies and are helpful for the researchers doing research. The conservation drones can fly 100 m above the terrain and captures high-quality images and record videos for analysis. These conservation drones cover a 25 km distance in 15 minutes. These conservation drones are tested in Indonesia and the researchers Koh and Wich (2012) reported that this conservation drone is successful in showing soil degradation, water resources, river flow, and other changes [6]. Similar agricultural drones are used to

1. Study natural and manmade pastures
2. Monitor water resources, floods, and droughts
3. Study weather patterns and climate change
4. Monitor soil erosion
5. Adopt cloud seeding

14.3.1   Study of Natural and Manmade Pastures

Knowledge about pastures is needed for agriculture and cattle rearing. Collecting such information in dense forest and other risky areas are challenging. Identifying and monitoring such pastures become easy with the help of drones (see Figure 14.2). With digital cameras and sensors, remote sensing of pastures becomes effective and reliable. It helps to analyze spectral reflectance properties of mono species and mixed legume pastures.

14.3.2   Monitor Water Resources, Floods, and Droughts

Drones are an effective monitoring vehicle to monitor the growth of algal blooms in ponds, pools, and reservoirs. These algae shall be used as biofertilizers and soil stabilizers in agriculture. Drones with infrared and thermal imaging cameras could monitor water bodies like rivers, irrigation canals, dams, and other such resources used for agricultural purposes (see Figure 14.3). By detecting water movement in the resources, various research analyses can be done. It is also reported that Drones with aerial imagery help to monitor crop loss due to drought and floods.

14.3.3   Study of Weather Patterns

Weather condition is another important factor that affects the agricultural process. The common method of weather monitoring is done by

Figure 14.2 Natural pastures (Source: farm and dairy).

Figure 14.3 Drone monitoring water resources (Source: MIKE).

using radars, mobile instruments, and balloons. In case of a storm, it was reported 20 minutes ahead of the storm event. Weather drones reported by NASA could be better than the usual methods and they can report 60 minutes ahead of a storm event which is three times faster. It was reported that drones are handy in collecting information about atmospheric pressure, temperature, humidity, and wind speed. As this information is quicker, it is easy to forecast quickly and develop weather models. Drones are impressive in collecting air samples ,to analyze gaseous and water parameters above the agricultural field. Weather experts are trying to implement drones to analyze various parameters related to storms, rain, flood, drought, and other related weather patterns.

14.3.4   Monitoring of Soil Erosion

Soil erosion is a phenomenon of soil degradation that affects the natural vegetation and agriculture process. It is important to monitor soil erosion to maintain the natural vegetation. Erosion pegs are one of the common methods of monitoring and calculating soil erosion. This method is time-consuming and requires skilled manpower. In Australia, drones are used for calculating soil erosion [7]. The drone-derived images using

Figure 14.4 Cloud seeding drone (Source: News Deeply).

Multiview stereopsis techniques are accurate compared to other methods. Different types of soil erosion can be periodically monitored using drones with multispectral cameras by capturing 2D and 3D images of eroded areas with close-up pictures. This helps to get more details about the erosion than the satellite-mediated observation.

14.3.5   Cloud Seeding

Cloud seeding is an artificial process of weather modification method that produces rain by adding toxic or nontoxic chemicals like silver iodide, and calcium chloride for cloud precipitation (see Figure 14.4). In many drought regions, cloud seeding is adopted. This cloud seeding is performed in Australia, America, and China by using manned aircraft which could be replaced by drones with the advancement in technology.

Thus, drones are the best agricultural surveillance and remote sensing vehicle that even fly at low altitudes to study natural resources, geographic features, soil erosion, and vegetation patterns.

14.4   Soil Fertility Management

The agricultural drone is the best agricultural instrument for surveillance and managing soil fertility. Soil fertility management procedures done by farmers and skilled mankind will be replaced by drones at an affordable cost. Agricultural drones can perform tasks with great speed, accuracy, and instantaneous repeatability.

14.4.1 Management of Soils and Their Fertility

It is important to maintain and monitor the fertility of the soil for crop production in agriculture. Analysis of minerals, nutrition, and distribution of organic matter in soil is done by soil sampling. Agriculture drones with suitable cameras and sensors are capable of doing this sampling and analysis with less human intervention. This analysis helps the farmers to do suitable irrigation. The United States Department of Agriculture reports that the basic requirement of soil fertility management involves soil nutrient status assessment as follows

- Identify soil field variation and manage and collect information by using a drone imaging technique.
- Prepare nutrition enriching procedure including sources like organic manures, inorganic fertilizers, irrigation schedule, and residue re-cycle.
- Identify soil fertility diseases like salinity, acidity, and soil erosion that affect soil fertility.
- Identify deficiency and sufficiency of soil nutrients that may affect crop growth.
- Determine soil variability and supply amendments at variable rates using appropriate computer programs.

Soil sampling is a complex process that can be done with ease with Drones' imagery for further analysis (see Figure 14.5). Drone imagery with high-resolution cameras and sensors helps to analyze the topographic

Figure 14.5 Drones in soil sampling (Source: British Geological Survey).

and soil type variation for effective management. Grid sampling and zone soil sampling helps the farmers authenticate drone imagery for soil fertility variations. Grid sampling involves the rectangular formation of grids of different sizes to cover a large area in acres. The formation of grids, the number of grids to be sampled, and the number of soil samples per grid can be decided quickly using drones. The spots to be sampled will be pre-programmed in the drone for effective operation. Zone sampling involves the formation of zones that depends on the availability of labor, topography, and soil type. Sampling will be taken at particular spots of the zone. It also depends on the previous knowledge about the zone's productivity and losses. Drone imaging techniques and physical reality information help in soil sampling. Fertilizers and Nutrients are applied to the grid and zone based on the plan through drones [8]. Digital maps are prepared from drone imagery, and it shows the production of each grid or zone. Each grid cell can be captured with high resolution for further analysis that is complex for human scouts. Hence drone aided imagery avoids complications compared to manual observation and recording for large areas. By perfect soil sampling, drones help to detect and solve soil salinity and acidity problems to maintain the soil fertile.

14.4.2   Variable-Rate Technology for Soil Fertility Management

A well-nourished soil can improve agricultural yields. It has been established that approximately 70% of growth in farming is due to the effective usage of Nutrients, fertilizer, and manures.

Variable rate fertilizer application is one of the operations in precision agriculture. This method applies fertilizer at different rates at different parts based on pre-planned prescription. It uses sensors, computers, and GPS technology to monitor and control its operation. The tractors can be replaced by drones for better efficiency and operation (see Figure 14.6). It is reported that drone-based variable rate fertilizer application increases fertilizer efficiency and is less expensive [9].

Generally, farmers collect satellite maps, then implement ground-based soil sampling. The soil sampling will be processed and analyzed using computers to build a variable rate map. The same can be done using drones (example: Agribotix). It can refine maps and fix variable rates of fertilizers using drones with multispectral cameras. This can also reduce fertilizer supply, losses, and cost. Finally improves the crop yield. Thus, Drones are among the recent techniques in the farming world, with a promise to improve and impart uniformity to soil fertility, thereby managing large farms and attaining yield goals.

Figure 14.6 Drone in variable-rate fertilizer spray (Source: Fly Dragon).

14.5   Irrigation and Water Management

Irrigation depends on the water resources like ponds, lakes, dams, and rivers. Irrigation expands based on the availability of water. Hence it is necessary to manage water for irrigation by monitoring its usage. It is also important to measure the water content on the surface soil and the rooting zone for effective usage of water. Experiments were conducted using drones to calculate the Surface Soil Moisture (SSM) (see Figure 14.7). It was found that data collected from drones with IR correlated with the ground data. It is concluded that SSM data using drones is comparatively cheaper and more efficient. Using this IR sensor in a drone can also detect

Figure 14.7 Drone in irrigation (Source: business insider).

soil temperature and moisture content variations. Using the multispectral camera in a drone, high-resolution pictures can be recorded for mapping. Using the soil moisture maps, water resources can be effectively utilized and managed.

There are many methods to detect soil and crop water status. The proximity sensor-based method is used to collect the moisture content. Such data may help to decide the amount of water to be supplied for irrigation. The water required for each individual plant needs to be estimated for effective usage of water. For this estimation, drones are appropriate with IR sensors and multispectral cameras. The water status of a particular plant can be measured as many times as needed using such drones [10].

14.5.1   Crop Water Stress Index

The Crop Water Stress Index (CWSI) is defined as the difference between air temperature (T_a) and tree canopy temperature (T_c). It is used to find the water deficiencies in field crops.

In the year 2013, Gonzalez-Dugo *et al.* conducted experiments with five trees using drones to estimate CWSI. Agricultural drones fitted with IR sensors and multispectral cameras are used for this experiment. Air and tree canopy temperature was recorded using a drone flying three times a day. The difference between air and tree canopy was correlated with CWSI. Plots were clearly detected concerning the water status of plants/trees using a drone. The collected data helped to adjust the water stress index values with suitable techniques.

The traditional method of calculating this CWSI using skilled manpower needs several days to prepare CWSI maps whereas the drone consumes just 1 hour to record the values. Accuracy is also comparatively higher. Drones can fly closer to the soil, collect samples, capture pictures, and collect spectral reflectance data. These data are available immediately for further analysis. Bellvert *et al.* calculated CWSI as:

$$CWSI = (T_c - T_a) - (T_c - T_a)_{LL} / (T_c - T_a)_{UL} - (T_c - T_a)$$

where T_c is tree canopy temperature; T_a is the air temperature; LL is the lower limit of $(T_c - T_a)$; UL is the Upper Limit of $(T_c - T_a)$.

14.5.2   Drones to Monitor Water Resources

Drones are excellent flying vehicles over water bodies to monitor water storage levels, floating weeds, dam conditions, and so on (see Figure 14.8).

Figure 14.8 Drones in water sampling (Source: TechCrunch).

Blockage of water path, loss of water due to leakage, and spillage can be easily identified instantaneously. Water stored in the dams can be regulated using drone imagery. In 2014, McCabe reported that drones are effective for surveillance of water bodies and water sampling. Water sampling at different pick-up spots is used to analyze nutrients and contaminants.

Most irrigation uses water sprinklers to supply water to plants. It is important for the farmers to check the sprinklers frequently for effective distribution of water. A drone with a multispectral camera can capture 3D images of the entire form in one go. Drones can fly closure to the sprinklers and take close-up pictures to monitor the clogging or leakage in sprinklers [11].

14.5.3   Drones to Design an Irrigation System

In Australia, agricultural drones are used for the design of irrigation systems. Its 3D imaging technique of showing topography helps to design and construct irrigation pipes and channels to water the crops. Geo-referenced 3D pictures from drones help to detect drought and flood areas in the irrigation field. Due to this advancement, many private agencies are now operating agricultural drones and their population is increasing.

The traditional satellite imagery can provide information on a large area such as districts and states and lacks resolution. But drones can overcome all challenges with affordable cost of service. Thus, drones are useful in helping agriculture farms by aerial imaging, mapping, monitoring, and controlling the crop fields.

14.6   Crop Disease Identification

One of the key objectives for enhancing food output in agriculture is the identification of crop diseases. Using a drone to identify the disease is a difficult task. For this, a variety of unmanned aerial vehicles and sensors have been employed. The platforms and peripherals of UAVs have their own limitations when it comes to diagnosing plant diseases precisely. There are various kinds of vignetting and orthorectification image processing software available. Important aspects of data analysis include training and validating datasets. Various machine learning methods and architectures are currently employed to categorize and identify plant diseases. In order to comprehend results, these models aid in feature extraction and image segmentation. In order to produce results, researchers also use the values of vegetative indices, such as Normalized Difference Vegetative Index, Crop Water Stress Index, etc., gathered from various multispectral and hyperspectral sensors [12].

Because imaging sensors' own spectrum bandwidth, resolution, the background noise of the picture, etc., are limitations, there are still a variety of drifts in the autonomous identification of plant diseases. Drone crop health monitoring should incorporate a gimble with numerous sensors, big datasets for training and validation, the creation of site-specific irradiance systems, and other techniques in the future.

14.6.1   Monitoring and Identification Using Different Drone Platforms and Peripherals

Nowadays, as indicated previously, several agricultural activities have been carried out by Drones. Unmanned aerial vehicles (UAV) are one of them and are utilized more and more for disease monitoring and detection. They collaborate with many parts, including cameras, sensors, motors, rotors, controllers, and others. One of the basic purposes of UAVs is to capture photographs. The information contained in images is retrieved and turned into valuable information using image processing and deep learning algorithms [13].

Electromagnetic spectra also provide useful information, which is used to make judgments on plant physiological stress. In order to evaluate the state of the plants in the field in real-time, it is helpful to compare the spectroscopic in the area. In a UAV-captured image, physiological abnormalities brought on by foliar reflectance in the near-infrared region of the spectrum can be used to detect crop disease. Additionally, reflectance in the red wavelength range is utilized to investigate the disruptions in crop photosynthetic activity brought on by certain illnesses. There are different UAV types for diverse agricultural operations. These UAV-related constructions are referred to as platforms.

There are primarily two types of agricultural UAV platforms: fixed wings and rotating wings.

A fixed-wing UAV is utilized for wide-area coverage and is relatively larger in size. Rotary-wing UAVs are further classified into two types:

1. Helicopter
2. Multirotor

Helicopters are employed for aerial photography and spraying and have a huge propeller at the top of the aircraft. Similar to this, there are various types of multirotor depending on how many motors the aircraft has. The quadcopter, hex copter, and octocopter are three different varieties of multirotor. Both fixed-wing and rotary UAVs have been employed for the monitoring and identification of plant diseases.

14.6.2   Disease Symptoms

14.6.2.1   Sheath Blight

The fungus that causes the soil-borne disease sheath blight can remain active in the soil for up to two years as sclerotia. After the sclerotia float out of the soil with irrigation water, the illness begins when the sclerotia come into touch with leaf sheaths at or just above the water line. Initial symptoms appear as greenish-grey, round, oval, or ellipsoid, water-soaked patches on the leaf sheaths (Figure 14.9a). The lesions grow larger and join together to produce larger lesions with uneven borders and greyish-white centers (Figures 14.9a and b). Over time, entire leaves wither as lesions on the sheaths merge. The edges of lesions on leaf blades might be dark green, brown, or yellow-orange.

The lesions can spread widely and consolidate on some or all of the leaf blades, giving the appearance of rattlesnake skin (Figure 14.9c). Sheath blight propagates both steeply and straight over the field. In severe cases, the illness may spread up the plants and infect panicles and flag leaves (Figure 14.9d). By developing its runner hyphae from tiller to tiller, from leaf to leaf, and from plant to plant, the fungus travels throughout the field, causing damage in a circular pattern. Early sheath infections can cause the fungus to penetrate into the culms, weakening them so that tillers lodge

Figure 14.9 Sheath blight lesions on the sheaths. (a) Water soaked patches on the leaf sheaths, (b) uneven borders and greyish-white centers, (c) appearance of rattlesnake skin, (d) infect panicles and flag leaves, (e) blight lesions on the sheaths, and (f) fungus causing damage in a circular pattern.

and fall off. Plants that are unwell produce less grain, especially in the lower part of the panicles. Increased accommodation generally results in more severe yield losses [14].

14.6.2.2 Narrow Brown Leaf Spot

The principal inoculum for the NBLS fungus is infected rice plant waste and infected seed, both of which persist year after year. Conidia, the infection-causing structures, are produced by the fungus. The conidia germinate, enter the host plant tissue through the stomata, and begin to proliferate intercellularly in the tissue. This is the beginning of the infection. Figure 14.10a shows how the fungus target leaves. Figure 14.10b shows how it affects sheaths, internodes, panicle branches, and glumes (Figure 14.10d). It results in small, narrow, brown lesions parallel to the leaf veins on leaf blades (Figure 14.10a).

A noticeable brown spot, or "net blotch," that occurs from infection of the leaf sheaths is produced by the browning of the leaf veins (Figure 14.10c). The "neck blight" caused by the fungus results in a light brown to tan coloration of the internodal area above and below the node at the base of the panicle (Figure 14.10e). The lower portion of the panicle's kernels fail to fill, and the damaged tissue area dies.

Figure 14.10 Shows lesions of narrow brown leaf spot (NBLS) on leaves and sheaths, symptoms of NBLS net blotch on the sheath, lesions on panicle branches and glumes, and "neck blast" lesions at the base of the panicle (E). (a) How the fungus target leaves, (b) narrow brown leaf spot (NBLS) on sheaths, (c) symptoms of NBLS on sheath, (d) lesions on panicle branches and glumes, and (e) neck blast lesions at the base of the panicle.

14.7 Pest Control Management

Pest insects are known to devastate entire regions and drastically reduce the output of food grains. Over 37% of the losses that the FAO has predicted are attributable to pests and illnesses. Invasive pests like the autumn army-worm have recently attacked crops grown in India, and they significantly damaged crops in 2018 and 2019. To guarantee effective pest management, plant protection measures must be implemented on a community level. Since more than 80% of farmlands in India fall under the category of small and marginal, controlling invading pests is exceedingly challenging. Pests just shift their food to the nearby fields if one farm is sprayed. Drones are necessary in order to address this.

14.7.1 Drones Offer a Sustainable Pest Control Solution

Koppert Biological Systems is the source of the insects and biocontrol chemicals that UAV-IQ distributes by drone. For both food crops and aesthetic plants, Koppert creates environmentally friendly farming techniques. They contend that the employment of bumblebees for natural pollination, natural enemies to control insect infestations, and biostimulants to boost and strengthen crops both above and below ground can make agriculture healthier and safer [15].

Figure 14.11 New pest management strategy (Source by: Entomological Society).

Figure 14.11 shows the spraying of pesticides using drone. Drone-based aerial biocontrol has the following features and advantages:

- Kill pests even when they hide where chemicals can't reach
- More effective distribution than conventional application approaches
- Significantly lower consumption of chemical pesticides
- Lower labor expenses

14.8   Agricultural Drones to Improve Crop Yield Management Efficiency

1. Monitoring Field Conditions: Drones are used to map fields and offer data on their irregularities and elevation. If there are any dry places or drainage patterns, knowing about them will help you irrigate crops more effectively. In order to determine the proper amounts of fertilizers needed to preserve the health of the soil, specialized drones also allow for the monitoring of the nitrogen content in the soil.

2. Plant health monitoring: Drones employ onboard cameras to keep tabs on crop health. They operate unaffected by clouds and poor lighting since they soar close to the earth. Drones employ the Normalized Difference Vegetation Index, which evaluates plant health via color-coded data. As a result, the farmers get rapid information about diseases or pests and can take fast action to manage their machinery and equipment.

3. Spraying on Crops: Drone sprayers prevent prolonged exposure to dangerous chemicals by removing the requirement for physical intervention while spraying chemicals on crops. They may also access places where it is difficult for people to work, such as high elevations. By spraying on specific regions, these devices improve efficiency while decreasing waste and chemical expenses.

4. Planting Seeds: Drone seeders can also plant seeds in inaccessible places that are risky or difficult for people to access. As a result, these seeders assist in lowering the costs associated with labor and equipment while ensuring the producers' workplace safety.

5. Protecting Livestock and Crops with Drones: Drones are used to secure livestock and crops, especially high-value crops like cannabis, rather than employing a large number of security workers. Additionally, farmers can use these helpful tools to find agricultural animals that may have strayed a great distance.

6. Irrigation: Drought conditions have become more prevalent in many locations as a result of climate change. Therefore, it's crucial to increase irrigation efficiency in these places. Drones use microwave sensing technology to gather precise data on the condition of the soil and its present moisture content. The right amount of water may then be distributed, helping to preserve natural resources.

14.9   Issues and Challenges

Agricultural drones are widely used across the world. Their role in agriculture is numerous and research is going on to operate in different aspects of agriculture. Despite several advantages, there are a few issues and challenges which are needed to be looked into it.

1. Power source
2. High Capital cost
3. Capacity of the tank to carry fertilizer and water for spraying
4. Lack of Technical skills to operate, repair, and service
5. Job loss of existing farm workers

14.9.1   Power Source and Flight Time

Agricultural drones need the power to work, and it needs frequent charging and recharging. Generally, Lithium polymer batteries are used for powering drones. These batteries have high energy density and less weight. The capacity of these batteries is limited in such a way the drone can operate

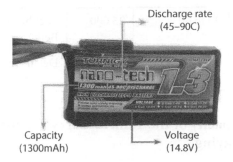

Figure 14.12 LiPo battery (Source: Tytorobotics).

from a few minutes to one hour. Due to this limited flight time, the work may be intermittent [16]. In the forest, it will be difficult to get a power source for charging/recharging. Hence backup power such as a battery bank or alternate LiPo batteries is needed. Figure 14.12 shows the structure of Lithium Polymer battery.

14.9.2   High Capital Cost

The capital cost of the agricultural drone is very high. Due to its high cost, it is challenging for small-scale farmers to adopt these drones and their technology. Some of the few drones and its cost are shown in the table below.

Drone model	Battery usage in minutes	Cost in US$	Manufacturer
AG550	0 to 30	3,000	Aerial Technology, Portland
Ag Drone UAS	30 to 60	9,995	Honeycomb Inc. Wilsonville, USA
Ag Eagle	30 to 60	13,500	AgEagle Inc. Neodesha, KS, USA
eBee Ag	30 to 60	25,000	Sense Fly Inc. Lausanne, Switzerland
Lancaster Hawk Eye-III	30 to 60	25,000	Precision Hawk Inc. Raleigh, NC State, USA

Quad Indago	30 to 60	25,000	Lockheed-Martin Inc. Bethesda, MD, USA
Crop Mapper DT-18	60 to 90	37,700	Delair-Tech, Toulouse, France

14.9.3   The Capacity of the Tank to Carry Fertilizer and Water for Spraying

The fertilizer or water spraying drone has a maximum capacity of 25 L, whereas another traditional way of spraying starts from 25 L to 250 L. Hence, for a large area, the tank of the drone must be refilled several times and becomes quite complex.

14.9.4   Lack of Technical Skills to Operate, Repair, and Service

Most farmers and farm workers have insufficient knowledge of operating agricultural drones. Skilled farm workers have challenges to repair and service the drones. Hence private agencies come in as an intermediate. Improper maintenance of drones leads to malfunction and accidents. Hence proper technical up-gradation of farmers is essential to learn the drone technology.

14.9.5   Job Loss of Existing Farm Workers

The agricultural drone can perform most of the tasks performed by the farm workers such as soil sampling, seeding, watering, crop health monitoring, weeding, surveillance, and so on. Due to this advancement, existing farmworkers may lose their jobs. It is reported that several drone technologists may replace existing farm workers.

14.10   Conclusion

The application of drones in agriculture is inevitable in the future. Therefore, to improve our economy and modernize agriculture, drones should be used. It is also helpful where human interventions are not possible for spraying chemicals on crops and where there is a scarcity of labor. With the advancement of technology in the future, the production of drones is expected to be economical. Drones also assist the farmers in knowing their land quality, what type of crops can be planted, proper usage of fertilizers, and pest control. Drones in precision agriculture are at an early stage. In the near future, with the use of drones, everyone expects that agriculture will get flourished and the quality of farming will be increased.

References

1. Kim, J., Kim, S., Ju, C., Son, H.I., Unmanned aerial vehicles in agriculture: A review of perspective of platform, control, and applications. *IEEE Access*, 7, 105100–105115, 2019.
2. Spoorthi, S.B., Shadaksharappa, S.S., Manasa, V.K., Freyr drone: Pesticide/fertilizers spraying drone—An agricultural approach, in: *Proceedings of the 2nd International Conference on Computing and Communications Technologies (ICCCT)*, pp. 252–255, Chennai, India, 23–24, February 2017.
3. Hartanto, R., Arkeman, Y., Hermadi, I., Sjaf, S., Kleinke, M., Intelligent unmanned aerial vehicle for agriculture and agroindustry. *IOP Conf. Ser. Earth Environ. Sci.*, 335, 012001, 2019.
4. Giordan, D., Adams, M.S., Aicardi, I., Alicandro, M., Allasia, P., Baldo, M., De Berardinis, P., Dominici, D., Godone, D., Hobbs, P. *et al.*, The use of unmanned aerial vehicles (UAVs) for engineering geology applications. *Bull. Int. Assoc. Eng. Geol.*, 79, 3437–3481, 2020.
5. Singh, K.K. and Frazier, A., A meta-analysis and review of unmanned aircraft system (UAS) imagery for terrestrial applications. *Int. J. Remote Sens.*, 39, 5078–5098, 2018.
6. Koh, L.P. and Wich, S.A., Dawn of drone ecology: Low-cost autonomous aerial vehicles for conservation. *Trop. Conserv. Sci.*, 5, 121–132, 2012.
7. Chen, P.-C., Chiang, Y., Weng, P.-Y., Imaging using unmanned aerial vehicles for agriculture land use classification. *Agriculture*, 10, 416, 2020.
8. Shafi, U., Mumtaz, R., García-Nieto, J., Hassan, S.A., Zaidi, S.A.R., Iqbal, N., Precision agriculture techniques and practices: From considerations to applications. *Sensors*, 19, 3796, 2019.
9. Fernández-Guisuraga, J.M., Sanz-Ablanedo, E., Suárez-Seoane, S., Calvo, L., Using unmanned aerial vehicles in postfire vegetation survey campaigns through large and heterogeneous areas: Opportunities and challenges. *Sensors*, 18, 586, 2018.
10. Yamamoto, K., Togami, T., Yamaguchi, N., Super-resolution of plant disease images for the acceleration of image-based phenotyping and vigor diagnosis in agriculture. *Sensors*, 2017, 17, 2557.
11. Stojcsics, D., Domozi, Z., Molnár, A., Automated evaluation of agricultural damage using UAV survey. *Acta Univ. Sapientiae Agric. Environ.*, 10, 20–30, 2018.

12. Hashimoto, N., Saito, Y., Maki, M., Homma, K., Simulation of reflectance and vegetation indices for unmanned aerial vehicle (UAV) monitoring of paddy fields. *Remote Sens.*, 2019, 11, 2119.
13. Qin, W., Xue, X., Zhang, S., Gu, W., Wang, B., Droplet deposition and efficiency of fungicides sprayed with small UAV against wheat powdery mildew. *Int. J. Agric. Biol. Eng.*, 11, 27–32, 2018.
14. Ni, J., Yao, L., Zhang, J., Cao, W., Zhu, Y., Tai, X., Development of an unmanned aerial vehicle-borne crop-growth monitoring system. *Sensors*, 17, 502, 2017.
15. Ayhan, B., Kwan, C., Budavari, B., Kwan, L., Lu, Y., Perez, D., Li, J., Skarlatos, D., Vlachos, M., Vegetation detection using deep learning and conventional methods. *Remote Sens.*, 12, 2502, 2020.
16. Nhamo, L., Magidi, J., Nyamugama, A., Clulow, A.D., Sibanda, M., Chimonyo, V.G.P., Mabhaudhi, T., Prospects of improving agricultural and water productivity through unmanned aerial vehicles. *Agriculture*, 10, 256, 2020.

15

Network Intrusion Detection of Drones Using Recurrent Neural Networks

Yadala Sucharitha[1*], Pundru Chandra Shaker Reddy[2] and G. Suryanarayana[3]

[1]*Department of Computer Science & Engineering, VNR Vignana Jyothi Institute of Engineering & Technology, Hyderabad, India*
[2]*Department of Computer Science and Engineering Geethanjali College of Engineering and Technology, Hyderabad, Telangana, India*
[3]*Department of Computer Science & Engineering, Vardhaman College of Engineering, Hyderabad, India*

Abstract

Flying Ad Hoc Network (FANET) has obtained a great deal of interest over recent times because of their significant applications. Thus, various examinations have been led on working with FANET applications in different fields. FANET's distinctive properties have made it intricate to reinforce its safeguard next to steadily varying security dangers. Nowadays, progressively more FANET appliances are carried out into common airspace, yet the enlargement of FANET protection has remained unacceptable. However, FANET's unusual roles ended it intricate to help arising dangers, particularly interruption recognition. This research explores FANET intrusion-detection threats by presenting a real-time data-analytics structure utilizing on deep-learning. The system comprises of Recurrent-Neural-Networks (RNN) as a foundation. It likewise includes gathering information from the network and breaking down it utilizing enormous information examination for inconsistency discovery. The information assortment is carried out through a specialist operating inside every FANET. The mediator is tacit to log the FANET real-time information. Furthermore, it includes a stream handling module that gathers the drone's correspondence data, with intrusion recognition associated data. This data is feed into 2-RNN components for data examination, trained for this function. First RNN inhabits in the FANET itself, and the second dwells in at the base-station. The investigations are directed for huge scale in light of different

Corresponding author: suchi.yadala@gmail.com

Sachi Nandan Mohanty, J.V.R. Ravindra, G. Surya Narayana, Chinmaya Ranjan Pattnaik and Y. Mohamed Sirajudeen (eds.) Drone Technology: Future Trends and Practical Applications, (375–392) © 2023 Scrivener Publishing LLC

datasets to assess the effectiveness of the presented model. The outcomes affirmed that the proposed model is better than other existing works.

Keywords: Flying ad hoc network (FANET), intrusion detection, deep learning, drone-ad-hoc network, RNN

15.1   Introduction

During previous years, Unmanned Aerial Vehicles (UAVs) are drawing in increasingly further consideration. The utilization of UAVs has lots of benefits against traditional manned airplanes particularly as far as functional cost, administrator's wellbeing, operability in troublesome/unsafe conditions and openness for common applications [1]. Present technical developments have made it more straightforward than any time in recent memory to arrangement an Unmanned-Aerial-System with intricate geography to accomplish refined assignments which were already inconceivable lacking genuine human contributions. The fast progressions and weighty inclusion of Information Technology tremendously affect the way which drone networks take to foster potential UAV frameworks. The present decentralized innovation advances appropriation of mission and relating assets [2]. This approach permits overt repetitiveness as far as basic parts and work on the general strength of the framework. In any case, a large portion of the present progressions in the area of organization appended UAV armada are zeroing in on the way to accomplish a drone set-up as portrayed in [3]. Slight has been believed for the digital protection of the drone set-up frameworks leaving even the vital condition of-workmanship drone set-up frameworks powerless alongside different safety dangers. A few investigates [4] have been directed depicting exhaustively the conceivable security dangers a UAV armada can be looking during its not unexpected activity. Now we attend to one sort of FANET safety concerns which are network-intrusion in a remote specially appointed network. As portrayed in [5] there are numerous danger models connected with network intrusion like over-burden, streak swarms, worms, port sweeps and sticking assaults. Among these strange examples, streak swarms badly affect the armada of UAVs in light of the fact that they make blockage and diminish fundamentally the Quality-of-Service of the whole organization. This is a significant difficulty for UAV certificate and incorporation into common airspace. Therefore, vindictive inconsistency recognition is a significant issue these days [6].

In [7] an outline is given assessing numerous research regions and application spaces. Network peculiarities and safety-related issues (like Distributed Denial of Service (DDoS) assaults) are significant concerns for the discovery of dynamic security dangers. An assortment of instruments for inconsistency identification is chiefly founded on information parcel signature. This conduct is known to be extremely viable for managing notable DDoS assaults. Notwithstanding, this instrument is wasteful when another kind of assault is performed. Consequently, we frame in this paper another sort of IDS ready to recognize various kinds of DDoS. The proposed intrusion discovery model is two-level components which initially describe the traffic by utilizing a factual sign and afterward select an exact assessor model to remake the assault traffic [8].

At present Scenario, analysts have proposed diverse ML & DL models for effective NIDS to recognize malevolent assaults. In any case, the ever-evolving expansion in network traffic and its security dangers has shown a few provokes for the NIDS strategies to recognize goes after productively. The IDSs target recognizing interlopers. In IoT, these gatecrashers go about as hordes endeavoring to get to different hubs without a permit. A NIDS comprises of three fundamental elements: a specialist, an investigation motor, and a reaction module [9, 10]. The fundamental capacity of the specialist is to gather data from the organization by observing occasions. Interestingly, the examination motor and a reaction module are liable for following the indications of intrusion, producing alarms, and dealing with the outcomes got from the investigation motor, individually. NIDSs have become more helpful and effective without fail, yet trespassers have likewise evolved different assault strategies to conquer these identification advancements. What's more, conventional NIDS is not appropriate for the complicated organization layers in IoT [11, 12]. The intricacy increments when UAV is concerned where drone tasks are more complicated and oblige more safety than different organizations because of their basic applications. As a matter of fact, effective drone assaults might prompt presenting individuals' life to risk. Besides, the idea of the drone's structures requires different security courses of action, particularly intrusion identification [13].

The main aim of the study is to give an intelligent circulated system to ramble intrusion recognition in light of DL strategies, in particular, LSTM-RNN structure. The primary thought is to have appropriated units of RNN where all drones will have single unit that attempts to recognize any assault on the actual drone. One more centralized LSTM-RNN unit dwells at the base station that affirms the assault and settles on the choice, advising different drones of specific assaults. The drones' LSTM-RNN manages just the traffic coming from its specialized gadget. Nonetheless, the LSTM-RNN base station inspects the entirety of the drones' traffic.

The objectives of the proposed research are:

- An intelligent data analytics system for exploring FANET intrusion discovery strings has been designed.
- Using DL techniques for drone networks intrusion identification.
- Experiments are conducted on huge scale to investigate the performance of the presented system.

The research work is coordinated as pursues: Section 15.2 is a survey of the related work while Section 15.3 is a depiction of the proposed system; the directed tests are point by point in Section 15.4 alongside the pre-owned assessment measurements; results and discussions. At long last, the paper finishes up and future work is given in Section 15.5.

15.2   Related Works

The intrusion discovery approach designed and implemented in this study is the consequence of a coordinated effort amid two innovative fields. The initial is connected with traffic portrayal. It involves phantom examination to create a particular traffic signature. Next is connected with programmed control techniques concerned for traffic reproduction. It utilizes powerful regulator techniques to examine the traffic and revamp its attributes and behavior [14].

The common utilizations of UAVs have encountered an extremely fast progression somewhat recently. Close by their new accomplishments and capable future, its problems are simply surfacing [15]. Different explores have been led to deal with previously mentioned difficulties, for example, GPS ridiculing, snooping and signal sticking, [16] yet just until late years, the organization intrusion location for FANET has at long last grabbed the eye from the drone local area. [17] has recommended an IDS configuration in light of conviction way to deal with recognize the excellent guess of gatecrasher inside a FANET. This process has a better bogus positive rate beside mischievous activities of an individual from the organization. Notwithstanding, this strategy is zeroing in on broad mischievous activities of a part and expecting earlier information on all potential ways of behavior of the framework. In opposite, we are proposing another IDS technique in view of both otherworldly mark examination and hearty control/assessment hypothesis pointing explicitly at distinguishing network intrusion inside the structure. The two strategies are thought of as free to the plan of FANET IDS frameworks [18].

The authors [19, 20] have exhibited how Long-Range-Dependence (LRD) can be a proficient boundary to measure the degree of fluctuation of Internet traffic. They have fostered a particular technique (counting a Matlab tool compartment) to handle information traffic. This cycle utilizes wavelet examination (see [21–23] for subtleties) which is a proficient instrument to get the fluctuation level of any information series for various time scales and various snapshots of investigation. Authors will utilize an upgraded adaptation of the technique presented in [24, 25] created by Hehri Wendt all the more as of late called Wavelet-Leader Multi-fractal (WLM) investigation toolbox.

Between the non-direct techniques [26, 27] depicted in the writing, the Super-Twisting Algorithm (STA) is the most broadly utilized for chatting aversion while distinguishing abnormalities. Its standards depend initially on the non-direct liquid technique applied on TCP elements and furthermore on sliding types [28, 29] which are regularly utilized to plan strong non-straight spectators regulations. Tragically, expanding ahead this particular spectator accommodates limited input-limited state (BIBS), limited time strength just [30, 31]. Therefore, this assertion confines the use of this onlooker to the class of the frameworks for which the higher bound of the underlying circumstance may be assessed ahead of time. Such a methodology can be extremely nonprecise for complex powerful frameworks, for example, the TCP model for an armada of UAVs. One more applicable technique proposed in the writing depends on time-defer straight state assessment. Such a methodology [32–34] illustrates on both Lyapunov-Krasovskii practical and energetic way of behaving of TCP (Transmission Control convention) to utilize a Luenberger eyewitness to adapt to inconsistency discovery. An AQM comprises of changing information stream rates send by the UAV during the organization. The standard comprises of dropping a few parcels before the cradle soaks. Thus, the assessor should be related with a strong AQM to play out its conclusion. The investigation of blockage control in a period defer framework structure isn't new and has been effectively exhibited in [35]. An applicable valuable calculation [36] has been proposed.

15.3   Drone Intrusion Detection Methodology

The idea of the UAV network is unique in relation to other conventional structures while they need to interact with the earth base station(s). Besides, drones shift starting with one spot then onto the next, contingent upon the application [37]. To that end the proposed structure comprises of two parts, the drone, and base station. Furthermore, here is no requirement for various RNN units where a concentrated unit is all that could possibly be needed in conventional structures. Then again, the drone structures could require greater than one RNN, 1 on the actual drone and another at the base station [38].

In this study, we expected that the base station is just a single station, and it can discuss straightforwardly with every one of the drones. DIDS could exertion in 2-distinct paths, cluster or stream, contingent upon the pre-owned innovation. For example, assuming that MapReduce is utilized as a concentrated part for the dynamic interaction, the handling must be run in groups, which takes a period to shape the bunch [39]. Then again, if different structures, like Apache Kafka, are used, the location cycle should be possible at the execution. Apache-Kafka is liked in this research because of its conveyance attributes, particularly when the huge information stream is imminent on. This imitates the continuous analysis by taking care of the information into a type of a stream to the RNN units [40]. The structure, displayed progressively examination by taking care of the information into a type of a stream to the RNNT and it is displayed in Figure 15.1.

As publicized in Figure 15.1, drone is supposed to have an enhanced version of LSTM-RNN. Experiencing the same thing, the enhanced variation

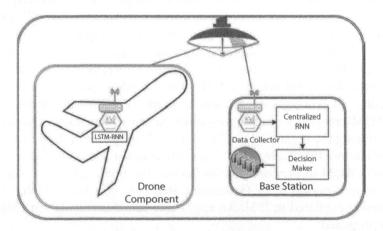

Figure 15.1 Structure of drone intrusion detection system.

implies a structure with less traffic in regards to getting ready and runtime traffic. Thusly, it is acknowledged that every drone will have light correspondence and less traffic compared to base station. Therefore, the drone LSTM-RNN unit doesn't necessitate comparable proportion of data assessment as the base station LSTM-RNN, where the base station LSTM-RNN unit ought to obtain all of the drone traffic for examination [41].

15.3.1   Drone RNN

The RNN with DL involves hubs connected to one another. These hubs might store data consecutively, however the information components are handled each in turn, and they can deal with information and result independently. RNNs are utilized really for various applications, for example, discourse union, time series expectation, video handling, and regular language handling [42]. RNN utilizes a multi-facet perceptron plan. It additionally has a circling structure that fills in as the fundamental way to empower information move starting with one phase then onto the next. Figure 15.2 proves the collapsed RNN layers where the RNN circles are extricated [43].

Because of many causes, RNNLSTM is utilized in this study for DID. In RNNLSTM, the inclination might move toward zero while duplicating values somewhere in the range of nothing and one or surpassing one. This infers that the network probably won't interface past contributions to their yields. Taking a gander at the RNNLSTM structure, 4-layers are deemed as the principle parts of the structure. Those layers speak with each other and that is displayed in Figure 15.3. RNNLSTM will be working on the drone. Nonetheless, the structure needs to become familiar with the typical way of behaving to recognize the variation. It gets the boundaries separated from the drone traffic. These boundaries are conceded to the forecast work that attempts to track down the traffic oddity. When there is a strange way of behaving, the framework will alert the regulator for direction [44, 45].

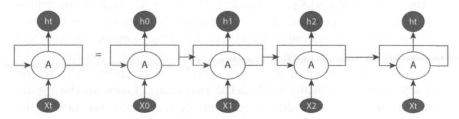

Figure 15.2 Folded RNN.

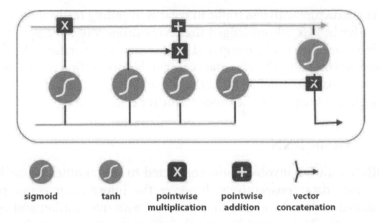

sigmoid   tanh   pointwise   pointwise   vector
                           multiplication   addition   concatenation

Figure 15.3 Internal structure of LSTM.

15.3.2   Data Collector

The data collector's essential capacities are to deal with the information prior to taking care of it to the RNNLSTM. It likewise has the obligation to hoop the information bundles and concentrate the important elements, for example, transmission rate, gathering rate, transmission to gathering proportion, sourceIP, objectiveIP, transmission mode, and movement span. Since our system is intended to work in the two information modes, cluster and stream, as referenced prior, the information gatherer is relegated this undertaking. In this manner, as found in Figure 15.1, two authority units are proposed in this structure; an information gatherer modules lives in each drones part, and another lives at the base station part. Accordingly, in cluster information handling, the gatherer is designed to keep the information in a cradle prior to fleeting it to the RNNLSTM. The cradle range relies upon the group volume to be handled, remembering the drone restrictions. On account of the information stream-mode, which is the situation of the current research, the information gatherer will be answerable for taking care of the information as stream to the RNNLSTM module [46]. As such, since we are reproducing the robot's exercises, the information authority imitates the constant information handling and sets it up likewise. Nonetheless, in actual robots, which aren't true in this field, the information authority will be answerable for capturing the correspondence module's information and setting it up to fit the RNNLSTM necessities. Likewise, the robot's information authority module is additionally answerable for moving the gathered information to the base station gatherer module alongside the robot's RNNLSTM choice [47].

The base station information authority module gets the entirety of the drone's information and their choices. It routes generally got information and leaves them to the brought together RNNLSTM on the base station for choice check. Then, it will advance an ultimate choice to the chief module for additional handling [48].

15.3.3 Centralized-RNN

RNNLSTM is sent on the base station. Yet again this unit could chip away at clumps. It gets drone travel from the information gatherer module either in clusters or a stream in light of the arranged mode. The unified RNNLSTM will settle on a worldwide choice in view of the by and large gathered information to check which robot is concessioned [49]. The incorporated RNNLSTM sends its choice to the leader unit. The incorporated RNN will be further prepared than the robots' RNN because of the travel impending from the various drones.

15.3.4 Decision-Maker

It is a unit, which essential job is to dissect the drones gotten and the incorporated RNN unit choices. The investigation is done during a ballot cycle where three admonitions:

Red: It caution implies that the intrusion is sure and is recognized by the 2-RNN units. For this situation, every one of the systems is reported, and the uncovered drone is separated.

Yellow: It caution implies that one of two modules distinguished the intrusion while the other one didn't perceive the intrusion. They just permitted sending the got posts to the concentrated RNN unit. In this time, the general network is informed to quit managing the uncovered drone for a specific timeframe. The present circumstance goes on until additional warning comes from the dynamic module.

Green: It intends that there is no requirement for any caution where the robots should be protected. For this situation, both the RNN units affirm no intrusion on any of the robots.

15.4   Results and Discussion

The assessment platform is described, and various arrangements of investigations are acquainted with analyze the productivity of the proposed approach is discussed. We imitated the drone's motions by presenting the "Data Collector" unit to stream the information to the RNNLSTM. It is contrasted with Linear-Regression & KNN calculations. One more arrangement of tests is led in this segment to think about between our proposed AI approach and non-machine learning draws near, signature-based approach.

The issue of interruption recognition in drones is still new, and as far as anyone is concerned, there is no particular dataset connected with drone assaults. Accordingly, various principles datasets are used to analyze the structure's presentation. Those datasets are painstakingly decided to cover countless assaults and inspect the proposed calculation utilizing different assault WSN-DS strategies [50]. All datasets has unique legitimacy and an alternate number of records; subsequently, choosing the specific rate for all of the datasets won't be proper much of the time. Along these lines, we needed to change the rates as indicated by the quantity of records in the dataset as well as the idea of the assaults and their rate in the datasets.

15.4.1   Model Assessment

To judge the proposed model efficiency, few notable measurements are utilized in comparable circumstances. The tuples in the dataset are grouped into ordinary and assaults. Subsequently, the accompanying boundaries are believed for the suitable quantifier to the exhibition of the designed arrangements: True Positive (TP), True Negative (TF), False Positive (FP), and False Negative (FN). The efficiency of the proposed model is assessed by following measurements: Accuracy, Precision, True Positive Rate and F1-Score.

15.4.2   Performance Analysis

However the initial bunch of examinations inspected the exactness of the LSTMRNN with the diverse example estimates, this was finished utilizing these datasets because of an enormous amount of tuples. The dataset was taken on to fit the presented system for drones. Figure 15.4 explains the recognition exactness expanded with the increment of the example size. The normal precision was 96%, and now and again, it came to 98%.

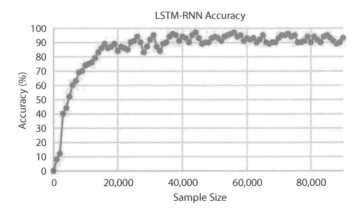

Figure 15.4 Accuracy of LSTMRNN with sample size.

15.4.3   LSTM_RNN Performance over UNSW-NB15 Dataset

Figure 15.5 presents the comparison among the LSTMRNN, LR, and KNN in paired characterization. In this investigation, 10% of dataset was utilized for preparing. As displayed in the Figure 15.4, LSTMRNN efficiency displayed a greatly improved exhibition than LR&KNN calculations. The exactness of the LSTMRNN came to 97%, while the Precision, Recall, and F-score were 95%, 98%, and 97%, individually. Next, KNN presents fine with, 92% Accuracy, Recall, and F-score pursued by LR with practically 89%.

Figure 15.6 displays another characterization execution in view of multiclass assaults. As should be visible, the LSTMRNN and KNN had relative

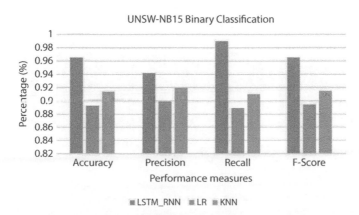

Figure 15.5 Performance comparison binary classification for of UNSW-NB15 dataset.

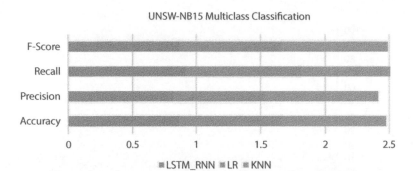

Figure 15.6 UNSW-NB15 multiclass classification performance contrast.

execution with regards to Accuracy and Precision. In any case, the Recall of LSTMRNN &LR was obviously superior to that of KNN, where the genuine positive worth was far superior to KNN. On the opposite side, LR actually has poorer accuracy, exactness, Recall, & F-score than LSTMRNN.

Here is one more arrangement of examinations focusing on the three calculations' presentation in light of the UNSW-NB15 itemized assaults. As displayed in Figure 15.7, LSTM_RNN showed an incredible by and large execution of the assaults introduced in the UNSW-NB15 dataset. On overall, LSTMRNN better than KNN by 2% accuracy while it superior than LR, on average, by 5% exactness.

One more conceivable perspective on the past figures' information is the normal outcomes assumed control over the entirety of the parallel and multiclass orders. The Critical Distance Diagrams (CDDs) from the posthoc Nemenyi test were applied to the normal outcomes to more readily

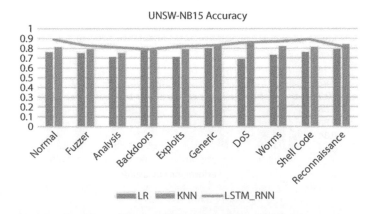

Figure 15.7 UNSW-NB15 multiclass classification exactness contrast.

pass judgment on the calculation's general execution on the utilized data-sets. To do as such, the paper adhered to similar methodology and rules introduced in to carry out the test in R-Tool. As should be visible in Figure 15.8, Figure 15.9, the crest line in the graph addresses the hub along which the normal position of every Spectro-worldly component is contrived, beginning from the majority reduced positions (generally significant) on the left to the most noteworthy positions (least significant) on the right. The Figure 15.8 works by gathering the calculations with comparative mea-surable qualities; those calculations are associated. As displayed in Figure 15.8, in double grouping, LSTMRNN performed greatly improved in all cases, which is the reason it is furthest left with a huge distance among it and different calculations. What's more, the Figure 15.8 illustrates that

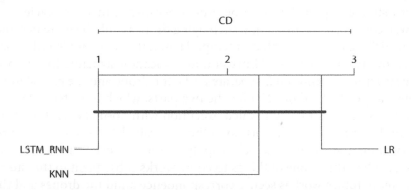

Figure 15.8 CDD graph for average binary classification.

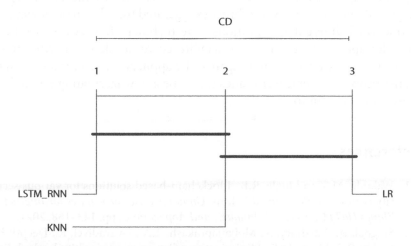

Figure 15.9 CDD graph for average multiclass classification.

KNN is still in front of the LR, with a little exhibition distinction amid them. Additionally, Figure 15.9 affirms the presentation of the LSTMRNN calculation over different calculations. Simultaneously, it proves that KNN is in the center among LR and LSTMRNN with practically a similar distance in the middle of them. In any case, LR appears to drop behind both KNN and LSTMRNN in the multiclass grouping.

15.5   Conclusion

The field of FANET has acquired massive consideration in recent scenario. It has been utilized in numerous regular citizen and military applications. Be that as it may, there is as yet a gap in their protections. Current cryptography strategies probably won't be proper because of their intricacies and power-cost. Also, intrusion into such flying drones could be more perilous than sticking or message block attempt. Hence, this study is viewed as one level towards an intellectual intrusion identification structure for Drones. The structure depends on 2-separate parts: the drone and the base station. It uses a specific kind of RNN for the two parts, which is RNNLSTM. The outcomes demonstrated unrivaled execution with normal execution, very nearly 15 on mean, over a portion of the new calculations. One of the primary limitations of this research is applying the proposed structure above real FANET. This is one of things to come works subsequent to this study. One more future work is secure correspondence amid the drones and the base station as well as among the actual drones. In our future work, we expect to expand the proposed structure in this study by executing a lightweight calculation in view of a direct exaggerated tumultuous arrangement of fractional differential conditions. The turbulent framework is fundamentally applied to drones and satellite correspondence. Furthermore, investigating the cyber assaults against the application-layer interaction of drones utilizing diverse techniques might be a new fascinating research to be executed in a future.

References

1. Wakode, M.S. and Ingle, R.B., Blockchain-based solutions for various security issues in UAV-enabled IoT, in: *Unmanned Aerial Vehicles for Internet of Things (IoT) Concepts, Techniques, and Applications*, pp. 143–158, 2021.
2. Singhal, A., Varshney, S., Mohanaprakash, T.A., Jayavadivel, R., Deepti, K., Reddy, P.C., Mulat, M.B., Minimization of latency using multitask scheduling

in industrial autonomous systems. *Wireless Commun. Mobile Comput.*, 2022, 10pp, 2022.

3. Sujihelen, L., Boddu, R., Murugaveni, S., Arnika, M., Haldorai, A., Reddy, P.C., Feng, S., Qin, J., Node replication attack detection in distributed wireless sensor networks. *Wireless Commun. Mobile Comput.*, 2022, 11pp, 2022.

4. Reddy, P.C.S., Yadala, S., Goddumarri, S.N., Development of rainfall forecasting model using machine learning with singular spectrum analysis. *IIUM Eng. J.*, 23, 1, 172–186, 2022, https://doi.org/10.31436/iiumej.v23i1.1822.

5. 4.Yahuza, M., Idris, M.Y.I., Ahmedy, I.B., Wahab, A.W.A., Nandy, T., Noor, N.M., Bala, A., Internet of Drones security and privacy issues: Taxonomy and open challenges. *IEEE Access*, 9, 57243–57270, 2021.

6. Sucharitha, Y., Vijayalata, Y., Prasad, V.K., Predicting election results from twitter using machine learning algorithms. *Recent Adv. Comput. Sci. Commun. (Formerly: Recent Pat. Comput. Sci.)*, 14, 1, 246–256, 2021.

7. Altawy, R. and Youssef, A.M., Security, privacy, and safety aspects of civilian drones: A survey. *ACM Trans. Cyber-Phys. Syst.*, 1, 2, 1–25, 2016.

8. Balamurugan, D., Aravinth, S.S., Reddy, P., Rupani, A., Manikandan, A., Multiview objects recognition using deep learning-based wrap-CNN with voting scheme. *Neural Process. Lett.*, 21, 1–27, 2022 Apr.

9. Shvetsova, S.V. and Shvetsov, A.V., Ensuring safety and security in employing drones at airports. *J. Transp. Secur.*, 14, 1, 41–53, 2021.

10. Abualigah, L., Diabat, A., Sumari, P., Gandomi, A.H., Applications, deployments, and integration of internet of drones (IoD): A review. *IEEE Sens J*, 21, 25532–25546, 2021.

11. Ahmad, N., Chaturvedi, S., Masum, A., Unregulated drones and an emerging threat to right to privacy: A critical overview. *J. Data Prot. Priv.*, 4, 2, 124–145, 2021.

12. Sucharitha, Y., Prasad, V.K., Vijayalatha, Y., Emergent events identification in micro-blogging networks using location sensitivity. *J. Adv. Res. Dyn. Control Syst.*, 11, 596–607, 2019.

13. Dey, V., Pudi, V., Chattopadhyay, A., Elovici, Y., Security vulnerabilities of unmanned aerial vehicles and countermeasures: An experimental study, in: *2018 31st International Conference on VLSI Design and 2018 17th International Conference on Embedded Systems (VLSID)*, IEEE, pp. 398–403, 2018, January.

14. Shaker Reddy, P.C. and Sureshbabu, A., An enhanced multiple linear regression model for seasonal rainfall prediction. *Int. J. Sens. Wirel. Commun. Control*, 10, 4, 473–483, 2020.

15. Yaacoub, J.P., Noura, H., Salman, O., Chehab, A., Security analysis of drones systems: Attacks, limitations, and recommendations. *Internet Things*, 11, 100218, 2020.

16. Jeler, G.E. and Alexandrescu, G., Analysis of the vulnerabilities of unmanned aerial vehicles to cyber attacks. *Rev. Air Force Acad.*, 2, 17–26, 2020.

17. Reddy, P.C.S., Nachiyappan, S., Ramakrishna, V., Senthil, R., Sajid Anwer, M.D., Hybrid model using scrum methodology for softwar development system. *J. Nucl. Energy Sci. Power Gener. Technol.*, 10, 9, 2, 2021.
18. Alhawi, O.M., Mustafa, M.A., Cordiro, L.C., Finding security vulnerabilities in unmanned aerial vehicles using software verification, in: *2019 International Workshop on Secure Internet of Things (SIOT)*, IEEE, pp. 1–9, 2019, September.
19. Bekmezci, İ., Şentürk, E., Türker, T., Security issues in flying ad-hoc networks (FANETS). *J. Aeronaut. Space Technol.*, 9, 2, 13–21, 2016.
20. Lakew, D.S., Sa'ad, U., Dao, N.N., Na, W., Cho, S., Routing in flying ad hoc networks: A comprehensive survey. *IEEE Commun. Surv. Tutorials*, 22, 2, 1071–1120, 2020.
21. Oubbati, O.S., Atiquzzaman, M., Lorenz, P., Tareque, M.H., Hossain, M.S., Routing in flying ad hoc networks: Survey, constraints, and future challenge perspectives. *IEEE Access*, 7, 81057–81105, 2019.
22. Tan, Y., Liu, J., Kato, N., Blockchain-based key management for heterogeneous flying ad hoc network. *IEEE Trans. Ind. Inf.*, 17, 11, 7629–7638, 2020.
23. Maakar, S.K., Khurana, M., Chakraborty, C., Sinwar, D., Srivastava, D., Performance evaluation of AODV and DSR routing protocols for flying ad hoc network using highway mobility model. *J. Circuits Syst. Comput.*, 31, 01, 2250008, 2022.
24. Du, W., Tao, J., Li, Y., Liu, C., Wavelet leaders multifractal features based fault diagnosis of rotating mechanism. *Mech. Syst. Signal Process.*, 43, 1–2, 57–75, 2014.
25. Lakshminarayana, D.H., Philips, J., Tabrizi, N., A survey of intrusion detection techniques, in: *2019 18th IEEE International Conference on Machine Learning and Applications (ICMLA)*, IEEE, pp. 1122–1129, 2019, December.
26. Khraisat, A., Gondal, I., Vamplew, P., Kamruzzaman, J., Survey of intrusion detection systems: Techniques, datasets and challenges. *Cybersecurity*, 2, 1, 1–22, 2019.
27. Karatas, G., Demir, O., Sahingoz, O.K., Deep learning in intrusion detection systems, in: *2018 International Congress on Big Data, Deep Learning and Fighting Cyber Terrorism (IBIGDELFT)*, IEEE, pp. 113–116, 2018, December.
28. Musa, U.S., Chhabra, M., Ali, A., Kaur, M., Intrusion detection system using machine learning techniques: A review, in: *2020 International Conference on Smart Electronics and Communication (ICOSEC)*, IEEE, pp. 149–155, 2020, September.
29. Alzahrani, A.O. and Alenazi, M.J., Designing a network intrusion detection system based on machine learning for software defined networks. *Future Internet*, 13, 5, 111, 2021.
30. Fang, W., Tan, X., Wilbur, D., Application of intrusion detection technology in network safety based on machine learning. *Saf. Sci.*, 124, 104604, 2020.

31. Prasad, R. and Rohokale, V., Artificial intelligence and machine learning in cyber security, in: *Cyber Security: The Lifeline of Information and Communication Technology*, pp. 231–247, Springer, Cham, 2020.

32. Lew, J., Shah, D.A., Pati, S., Cattell, S., Zhang, M., Sandhupatla, A., Ng, C., Goli, N., Sinclair, M.D., Rogers, T.G., Aamodt, T.M., Analyzing machine learning workloads using a detailed GPU simulator, in: *2019 IEEE International Symposium on Performance Analysis of Systems and Software (ISPASS)*, IEEE, pp. 151–152, 2019, March.

33. Gleeson, J., Gabel, M., Pekhimenko, G., de Lara, E., Krishnan, S., Janapa Reddi, V., RL-Scope: Cross-stack profiling for deep reinforcement learning workloads. *Proceedings of Machine Learning and Systems*, vol. 3, pp. 783–799, 2021.

34. Liang, J., Chen, J., Zhu, Y., Yu, R., A novel intrusion detection system for vehicular ad hoc networks (VANETs) based on differences of traffic flow and position. *Appl. Soft Comput.*, 75, 712–727, 2019.

35. Huang, S. and Lei, K., IGAN-IDS: An imbalanced generative adversarial network towards intrusion detection system in ad-hoc networks. *Ad Hoc Networks*, 105, 102177, 2020.

36. Kowsigan, M., Rajeshkumar, J., Baranidharan, B., Prasath, N., Nalini, S., Venkatachalam, K., A novel intrusion detection system to alleviate the black hole attacks to improve the security and performance of the MANET. *Wirel. Pers. Commun.*, 1–21, 2021.

37. Parameshwarappa, P., Chen, Z., Gangopadhyay, A., Analyzing attack strategies against rule-based intrusion detection systems, in: *Proceedings of the Workshop Program of the 19th International Conference on Distributed Computing and Networking*, 2018, January, pp. 1–4.

38. Asad, H. and Gashi, I., Dynamical analysis of diversity in rule-based open source network intrusion detection systems. *Empir. Software Eng.*, 27, 1, 1–30, 2022.

39. Vaidya, A. and Kshirsagar, D., Analysis of feature selection techniques to detect DoS attacks using rule-based classifiers, in: *Applied Information Processing Systems*, pp. 311–319, Springer, Singapore, 2022.

40. Sucharitha, Y., Vinothkumar, S., Rao Vadi, V., Abidin, S., Kumar, N., Wireless communication without the need for pre-shared secrets is consummate via the use of spread spectrum technology. *J. Nucl. Energy Sci. Power Gener. Technol.*, 10, 9, 2, 2021.

41. Al-Jarrah, O.Y., Siddiqui, A., Elsalamouny, M., Yoo, P.D., Muhaidat, S., Kim, K., Machine-learning-based feature selection techniques for large-scale network intrusion detection, in: *2014 IEEE 34th International Conference on Distributed Computing Systems Workshops (ICDCSW)*, IEEE, pp. 177–181, 2014, June.

42. Nazir, A. and Khan, R.A., A novel combinatorial optimization based feature selection method for network intrusion detection. *Comput. Secur.*, 102, 102164, 2021.

43. Selvakumar, B. and Muneeswaran, K., Firefly algorithm based feature selection for network intrusion detection. *Comput. Secur.*, 81, 148–155, 2019.

44. Krishnaveni, S., Sivamohan, S., Sridhar, S.S., Prabakaran, S., Efficient feature selection and classification through ensemble method for network intrusion detection on cloud computing. *Clust. Comput.*, 24, 3, 1761–1779, 2021.

45. Hakim, L. and Fatma, R., Influence analysis of feature selection to network intrusion detection system performance using NSL-KDD dataset, in: *2019 International Conference on Computer Science, Information Technology, and Electrical Engineering (ICOMITEE)*, IEEE, pp. 217–220, 2019, October.

46. Reddy, P.C.S., Pradeepa, M., Venkatakiran, S., Walia, R., Saravanan, M., Image and signal processing in the underwater environment. *J. Nucl. Energy Sci. Power Gener. Technol.*, 10, 9, 2, 2021.

47. Almomani, O., A feature selection model for network intrusion detection system based on PSO, GWO, FFA and GA algorithms. *Symmetry*, 12, 6, 1046, 2020.

48. Al-Yaseen, W.L., Othman, Z.A., Nazri, M.Z.A., Multi-level hybrid support vector machine and extreme learning machine based on modified K-means for intrusion detection system. *Expert Syst. Appl.*, 67, 296–303, 2017.

49. Besharati, E., Naderan, M., Namjoo, E., LR-HIDS: Logistic regression host-based intrusion detection system for cloud environments. *J. Ambient Intell. Humaniz. Comput.*, 10, 9, 3669–3692, 2019.

50. Moustafa, N., A new distributed architecture for evaluating AI-based security systems at the edge: Network TON_IoT datasets. *Sustainable Cities Soc.*, 72, 102994, 2021.

16

Drone-Enabled Smart Healthcare System for Smart Cities

Subasish Mohapatra[1]*, Amlan Sahoo[1], Subhadarshini Mohanty[1] and Sachi Nandan Mohanty[2]

[1]Department of Computer Science & Engineering, Odisha University of Technology and Research, Bhubaneswar, India (Formerly College of Engineering and Technology) [2]School of Computer Science & Engineering, VIT-AP University, Amaravathi, Andhra Pradesh, India

Abstract

Nowadays, smart healthcare system is very much essential for human life as it is more simplified and trustworthy in comparison to the traditional healthcare system. Drone technology gives a boost to smart health as various cutting-edge technologies are used such as infrared kinds come with technologies, including intrared cameras, GPS, and lasers (consumer, commercial and military UAVs). Remote ground control systems operate these (GSC). It functions in conjunction with the vehicle's sensors and a GPRS (GPS). Out of the many Benefits of Drones, telemedicine is a major aspect through which it can easily save lives. With the help of IoT analytics, the objective can be achieved suitably. In this chapter, a three-layered architecture is proposed in which the first layer consists of the wearable device which is responsible for the acquisition of body parameters. The sensor nodes are in charge of tracking bodily functions like temperature, heart rate, blood pressure, etc., and sending this information, together with the patients' geolocations, to the cloud via a sink node. The sink node's built-in Wi-Fi module connects it to the cloud. In the second layer, the acquired body parameters will be stored in cloud storage and analyzed via various supervised and unsupervised learning techniques. A Cloud database is mainly responsible

Corresponding author: smohapatra@outr.ac.in

Sachi Nandan Mohanty, J.V.R. Ravindra, G. Surya Narayana, Chinmaya Ranjan Pattnaik and Y. Mohamed Sirajudeen (eds.) Drone Technology: Future Trends and Practical Applications, (393–424) © 2023 Scrivener Publishing LLC

for storage, analysis, and decision-making in real-time. Further in the third and final layer, the IoT cloud will be coordinating with the drone and the medical store which is nearest to the patient and deliver the medicines at the desired location. There is a subscription-based model where the patient can log in to the UI system. After logging in he/she can able to view his/her body parameters, can make payment, and get the medicine remotely. The IoT along with the help of drone technology facilitates the integration of heterogeneous sensing components which are capable of providing real-time analyzed data through the internet anytime anywhere. Based on these data a patient can get assisted in a better and more feasible way in case of any urgency without bothering about time, speed, cost, and the rigidity to traverse physical barriers which make drones handy to adopt.

Keywords: Healthcare, GPS (global positioning system), GSC (ground control system), UAV (unmanned aerial vehicle), sink node

16.1   Introduction

In this modern era, digital and mobile technologies are being used to develop smart healthcare solutions for smart cities all around the world. Health care will be significantly impacted by the Internet of Things (IoT) revolution that has swept the infrastructure, security, and transportation industries. A new era in tackling today's health concerns will be ushered in by a unified healthcare system, data collecting and sharing, analysis, and research methods. Several revisions to this communication technology in resilient cities have already been completed. Smart cities would be safer and more resilient if drones or aerial vehicles (AVs) were integrated. The drone technology embedded along with the internet of things is a very new concept. Drones may connect to and gather data from IoT devices and sensors after they are deployed and operational, resulting in a more efficient and optimal data collecting, aggregation, and offloading process. Drones may be operated remotely, and the acquired data can be observed in real-time also. Data can be sent to the cloud and monitored at a later date [1]. Programs for various situations may be designed and implemented, and when they are fulfilled, they can inform the person monitoring the area for any irregularities or concerns. All of these smart cities are said to be built on technology and the Internet of Things [2]. From traffic distribution and rerouting to ensure easy passage to the management of street lights and traffic lights, to the implementation of various types of sensors and actuators to sense the environment and turn lights on and off to save energy, to the implementation and maintenance of low emission zones to reduce inner-city traffic, population, and air quality. Everything that happens inside a smart city requires a communication system, data analysis, and at times artificial intelligence to conserve resources and at the same time manage and improve the lives of the dwellers of the city. In this chapter, a detailed conceptual amenity of the basic drone system is being discussed and one of the best-suited deploying frameworks is being proposed at the end concerning the evasion of the challenges faced by today's smart cities. We would endeavor to implement the smart city idea with the integration of the aerial vehicle (drone) in such a manner that the cities' growth and development would be sustainable since sustainability is the need of the hour. These smart cities lean towards investing in research and innovation on different applications of recent Information and Communication Technology (ICT) along with the evolution of various strategies to plan out towards improving quality of life and sustainability.

Drones are routinely alluded to as unmanned aerial vehicles or UAVs. This remotely steered airplane (drone) framework is a more worldwide and ordinary term while the remotely directed vehicle is most frequently connected with military conditions. Fundamentally, a drone is a type of robot intended to be flown by controlling it in a good way by utilizing a PC or gadgets with programming-controlled flight plans customized in them. It is embedded with worked-in sensors and a global positioning system (GPS). The automated airborne vehicles are typically associated with the military, wherein past, they have been utilized all the more antagonistically as a stage for weapons as well as in military insight for social occasion data about a specific substance and for against airplane target. Earlier, drones have been utilized in a broad scope of exercises outside military jobs with the line up from observation, traffic, and weather conditions checking, search and salvage, firefighting, photography or videography, farming, conveyance administrations also for individual and other business purposes. This automated airplane (drone) is fundamentally comprised of an impetus framework, an air edge, and a route framework which comprise a broad scope of supporting hardware and airplane format settings that works with a few application programs. It recently started to arrive at the importance of regular organizations notwithstanding not being another innovation by conveying a savvy, swifter, and ideal decision contrasted with the monitored airplane. As of now, an illness brought about by the infection known as COVID-19 is right now assaulting the world at a disturbing speed because of the staggering ability of the causative creature (SARS-CoV-2). By utilizing drones in the healthcare battle or to support medical services laborers are constantly being presented with this pandemic situation without thinking twice [3]. Drones have set off a quick headway in an assortment of business, sporting, and modern administrations with the ability to assemble constant information cost as well as transport air freight. Nonetheless, its leap forward in medical care has been slower contrasted with different areas where drones have been utilized widely. Furthermore, drones can develop clinical consideration and move headway in the well-being business. It is against this background that our plan to make mindful of the significance of drones in medical care conveyance especially in the period of COVID-19 while checking its current and future applications [4].

With regards to brilliant urban communities, the insightfulness estimation is connected with personal satisfaction, medical care, public security, fiasco the executives, natural perspectives, (for example, energy proficiency, air quality, traffic checking, and so forth), and administrations. Subsequently, highly appreciated present-day Information and Communication Technologies (ICT), Artificial Intelligence (AI), and advanced mechanics assume fundamental parts in making urban areas more astute. These innovations assist with working on the foundation of brilliant urban areas and make savvy administrations conceivable to work on the general personal satisfaction of the inhabitants.

Drones are independent robots that fly overhead and are associated with various applications in regular citizen society [5]. Drones along with the Internet of Things technology support increasing the smartness of smart cities in various ways, for example, correspondence, transportation, farming, healthcare, security and surveillance activities, etc. Drones are ready to turn into an essential piece of brilliant urban communities and further develop generally educational involvement with the feeling of checking contamination, mishap examination, putting out fires, bundle conveyance, supporting person on-call exercises, conveying medication, observing traffic, and regulating building destinations. Drone innovation can additionally prompt tremendous auxiliary advantages like diminishing power utilization, saving assets, decreasing contamination, getting too risky and war zones, and expanding readiness for crises. Propels in innovations in the space of sensors, information handling, and battery-powered resources have made drones more reasonable. Drones can likewise be emulated to convey correspondence administrations (both uplink and downlink) for the supporters via the ground base station on the ground. Dexterity and line-of-sight (LoS) are the highlights that have made drones assume a fundamental part in the Internet of Things (IoT) structure [6]. The appearance of IoT has empowered critical advances in shrewd urban communities' autonomous applications like brilliant homes, savvy roads, savvy stopping, shrewd power matrices, and so forth. The primary thought of drone technology is that everything can associate and convey over the positioning system and take care of the healthcare application along with various other claims which are applicable in the worldwide foundation for a data-driven society [7]. The improvement of such applications has become basic to our ways of life, the economy, and the climate. Aside from monetary development, the general advancement of a geological substance is additionally determined by green innovations which safeguard the climate from unsafe emanations and dangerous squander, monitor regular assets, moderate the outcomes of environmental change, and lessen contamination and power utilization [8].

This paper is partitioned into five segments. In Section 16.1, the authors give a concise prologue about technological evolution. In Section 16.2, the subsequent related works are discussed concerning drone technology. Section 16.3 discusses the various applications, communication paradigms, and basic components of a drone. In Section 16.4, an integrated framework is depicted concerning the healthcare applications of drone technology. Some of the issues and challenges are highlighted in Section 16.5. Finally, this chapter incorporates the conclusion and future extent of this contemporary technology in Section 16.6.

16.2   Related Works

Because of the various happening events of catastrophic events, crisis administrations and medical services suppliers, specifically, should give close consideration to somewhere safe and secure [2–5]. For this reason, the remote correspondence linkage assumes an imperative part in surveying a harmed region, gathering information on provisions, assisting police with observing the areas of episodes in shrewd urban communities, planning salvage groups, and helping group exercises, saving individuals' lives, and representing missing individuals. To lay out a productive and powerful remote correspondence network for conveying information in a war zone range width is strongly suggested [9]. Notwithstanding, earthly remote correspondence advances could be missing, inaccessible through the blockage, or harmed [10]. The basic elements for crisis correspondence network arrangements and debacle recuperation are fast sending, prompt accessibility, and unwavering quality. In this way, space advances address the best answer for calamity recuperation, public wellbeing, SAR, and crisis administrations [11]. Space innovations are utilized for gathering data expected to safeguard humans and diminish financial misfortunes. As the result of a debacle, a satellite is a solid correspondence arrangement [12], however, the shortcoming is a period deferral and sending-off cost. Hence, utilizing an aeronautical stage can be an improved arrangement, since it has the benefits of both space and earthbound remote correspondence frameworks. It contended the utilization of flying stages for catastrophe and crisis circumstances and showed the meaning of salvage groups during the calamity. The capacity of an ethereal stage to convey correspondence administrations, for example, this type of work with SAR tasks is talked about in [14]. The benefits of flying stages are the capacity and stable inclusion region, survivability [15], the moderated impedance that happens in the remote correspondence [16], and the capacity to oversee traffic [17]. They offer an important choice to help crisis correspondences after a catastrophe [18]. Debacle expectation inclusion for relief of calamity sway a from the low-height stage (LAP) is examined in [19]. Drones have a place with the LAP family [20, 21] and are viewed as space robots. The huge benefits of robots are organization cost, view (LoS), low proliferation delay, fast arrangement, fixed station, and use in a fiasco. The spread models, versatility, and situating of robots for a correspondence network are examined in [22]. Drones are alluring for crisis correspondence as a result of the chance of quick organization and clients working them from their current portable handsets in misfortune zones. Subsequently, drones address the best answer for calamity recuperation and crisis administrations since they can be utilized to help alleviate and salvage groups in playing out their errands proficiently. Drones assume a fundamental part in associated gadgets in brilliant urban communities. Henceforth, the creators of [23] evaluated the different parts of robots connected with protection, digital protection, and public security in shrewd urban areas. Robots, IoT devices, and humans all convey favorably. Particularly, they can maintain the accessibility of now operating very remote firms like mobile and broadband organizations. The advantage of using drones as flying base stations over earthbound base stations is their ability to provide on-the-fly exchanges and to lay out correspondence that links to ground clients. Notwithstanding, the inclusion region for the sending of robot base station was considered for limiting energy cost, further developing the inclusion range and ideal elevation of robots [24]. For sure, one more significant utilization of robots is in IoT situations [25, 26], because gadgets frequently have little transmission power and will most likely be unable to impart over a long reach. Vignesh *et al.* [27] summed up the different participation approaches for the arrangement of robots. Likewise, Ganesan *et al.* [28] fostered a situation for traffic the executives and participation of robots and hubs on the ground to give consistent information moves and organization adjustment by utilizing ad hoc innovation. The objective applications and innovative ramifications of IoT-supported mechanical technology were talked about [29]. Besides, Mustapha *et al.* [30] tended to the organization security upgrade of IoT-supported mechanical technology in an intricate climate. Notwithstanding crafted by the creators in [30], the creators of [31, 32] evaluated the combination concerning network conventions, designs, and inserted programming for IoT advanced mechanics for shrewd urban areas. The cooperation of mechanical and IoT gadgets was researched [33, 34]. Along these lines, AI, robots, and IoT will give the up-and-coming age of IoT applications [35]. Also, sustainable power collecting was examined for the energy Internet of robot correspondence and organizations [36]. Consequently, drone innovation and data and correspondence advancements play a crucial job in brilliant urban areas' decreased asset utilization and expenses. Greening data and correspondence advances empowered the green IoT [37] by lessening energy utilization, contamination, and dangerous discharges in brilliant urban areas. The majority of the difficulties in regards to energy proficiency, obstruction, and correspondence networks are talked about top to bottom [38–40], alongside wise strategies for handling information. Moreover, various research articles, it is being investigated the expected helpfulness of the IoT to improve public well-being and talked about the difficulties and chances of utilizing the IoT to help public security organizations and SAR. The primary thought of the primary responder was to show up at the war zone before the specialists on call. The work was separated into two sections, the assortment of information, and the utilization of neighborhoods looking to observe the ideal situation in which the robot could convey correspondence administrations to the people in question. Notwithstanding, keeping a network connection between the prime responder and different responders was not talked about. The cooperation of drones and IoT assumes a critical part in open security. The proposed network design, that is to say, the reconciliation of drones, IoT, and savvy wearable gadgets, offers various administrations like supporting fiasco help group to save human lives, significant distance correspondence, greening correspondence, and so forth.

16.3 Applications of Drones

Construction and Infrastructure
One of the most common applications of drone technology is in the planning and management of construction projects. The software has been created to analyze building progress, material quality, elapsed time, and other factors. When dispensaries open or the deadline is missed, an alert is issued

to the authorities with the details. Drones also do ground and surrounding surveying, which is an important element of the process. Buildings, high rises, bridges, and monuments are all monitored by flying cameras for any failures. They can warn authorities ahead of time about probable problems, perhaps preventing tragedies and saving lives and money.

Emergency/Disaster Relief

Drones can be quite useful in the event of a disaster. Floods, hurricanes, landslides, earthquakes, and other natural disasters cut communities off from the rest of the world. Drones can be used to discover who is causing the damage, locate victims, and, most importantly, send critical supplies to them. Food, clothing, and medications, among other things, can be transported and supplied to victims or those who have been forced to live in unfavorable situations. Due to the current state of affairs, the People's movement is extremely difficult due to the COVID 19 breakdown. Drones can be quite useful in the delivery of goods. Food and medications are examples of necessities. They can be used to keep an eye on certain regions and see if they are safe. There is a proper lockdown and quarantine in place. They can also be used to squirt disinfectants throughout the metropolis.

Smart City Monitoring

The widespread acceptance of IoT enabled the transformation of legacy cities into smart cities. However, now is the moment for drone technology. Drones are the smart city's future. With criminal detection, drones will make smart cities more secure and safe. As needed, search and rescue, tracking of vehicles, people, or other things. Because stationary cameras may not be able to record the entire scene and all of the details, flying drones will be more durable and secure. Every corner of the city will be covered by drones flying above, making monitoring a breeze. It will also aid in the monitoring of traffic on the roads and the reporting of incidents. Drones can assist in reading electric, water, and gas meters without the need for human intervention.

Smart Agriculture

Due to little or no connectivity in rural areas, using drone technology in agriculture remains a difficulty. Drones will assist farmers in gathering large amounts of data from their farms, automating some procedures, and increasing efficiency through sound decision-making. Drones can be deployed to monitor the farm, allowing for early detection of animal infiltration. Cattle tracking may be done quickly and effectively. Cattle health and behavior can be effectively handled. Sensors placed in the soil and surrounding the field can communicate data to drones, allowing for faster and more accurate monitoring and decision-making. Drones can be used to monitor the health of crops and trees regularly. Drones can be used to disperse seeds and spray insecticides and pesticides from the air.

Wildlife Conservation

The two main factors affecting the health of wildlife around the world are climate change and poachers. The World Wildlife Fund estimates that thousands of species are predicted to become extinct every year. Conservationists are seeking new and inventive ways to better study and protect different ecosystems, wildlife, and their habitats to assist solve this problem. Drones can effortlessly monitor and follow animals in their natural habitats and settings without causing them any distress or anxiety. Drones can be used to track endangered species and relay any pertinent information to their caretakers and researchers. Drones will also aid in the fight against poachers, allowing them to be identified and punished if necessary.

Healthcare

The prevention of diseases, the extension of human life expectancy, and the improvement of living conditions have all been significantly impacted by modern medicine. However, due to a lack of infrastructure, such as hospitals and highways connecting them to cities, many rural areas around the world do not have access to high-quality healthcare. Drones can transport medical supplies to such areas quickly and efficiently, with no delays or issues. Drones can also be used to quickly carry blood, small to medium-sized equipment, and pharmaceuticals. They can also be used to transport patient samples to hospitals for testing purposes.

Weather Forecasting

Data collection for weather forecasting is now carried out by stationary structures or geospatial imaging satellites. Scientists are always attempting to improve hardware and software solutions for data collection and worldwide weather forecasting. Drones can be extremely useful in this situation. They provide a variety of climatic data collection tools. They can travel to remote areas or deep-water locations to gather data for weather forecasting. They can also be set to follow various weather patterns as they develop and move. Drones can also be used to collect data on the ocean and atmosphere from above the water's surface.

Energy

Although alternative energy sources are becoming more and more popular, fossil fuels continue to be the world's main energy source. Since it is necessary to ensure compliance with laws and standards, routine inspection and observation of the infrastructure used to extract, process, and transport oil and gas are key aspects of the sector. Drones can effortlessly check and monitor the property from a distance. They can detect various leaks and cracks faster when equipped with various sorts of sensors. Some faults can be diagnosed remotely and quickly with a high-resolution camera.

Mining

Mining is a field that necessitates constant measurements and evaluations of the physical materials on the job site. Substances like ore, rock, and minerals are kept in stockpiles that must be tested for radiation, quantity, and amount. Drones are capable of obtaining volumetric data on stockpiles using high-quality and unique cameras. They can also be used to assist with mining activities from the air. This drastically decreases the risk of having persons on the ground survey the stocks. They can also be utilized to improve the area's security.

Insurance

An insurance claim involves a lot of inspection. Drones can be utilized in a variety of ways for this. Drones can be used by insurance companies to scale and observe structures and cars that they insure. Drones with high-quality cameras can provide precise assessments that can be used to accept or deny claims without involving people on the road. Drones can also be used to inspect infrastructure for flaws and damages regularly. In the case of vehicles, drones can be used to examine vehicles regularly to determine patterns of damages and dangers, which would aid insurance firms in determining whether or not to provide insurance to users.

Aerial Vehicle Applications in Healthcare

We live in a digital age where technology advances; a drone is one component of this technology that has yet to be adopted in the medical field. Drone technology can help the healthcare sector improve care delivery and safety by quickly delivering medications, vaccines, and other medical supplies to the locations where they are required, preventing the spread of harmful communicable illnesses. Every day, medical professionals from paramedics to other health workers at primary health care centers to large hospitals encounter a variety of problems, including the ability to reach patients in dire need of rapid medical attention. It might mean the difference between life and death for these patients and the medical staff. Drones can assist in providing effective health care to patients promptly. A drone may be the sole way to deliver blood, vaccines, drugs, snakebite serum, and other medical supplies to towns and rural areas without access to health facilities. Drones have the ability to fly over difficult obstacles like buildings and bridges, as well as deliver cargo to places with no roads, where manned planes cannot. Drones can help transfer medications to patients' bedsides and courier blood for transfusion between hospitals promptly within hospital walls. Unmanned aerial vehicles (UAVs) provide the healthcare business with a wide range of cost-effective and life-saving possibilities. Medications, communication equipment, temporary shelter, mobile technology, and other necessities are among the many products that could be delivered quickly by drone to locations that have been rendered inaccessible by traditional land and air delivery due to disasters, pandemics, or other tragedies. The capacity to transport laboratory samples, blood products, drugs, and other requirements over extensive distances in a short amount of time is a significant advantage.

These benefits of unmanned aircraft systems are the reason why academics, businesspeople, and other organizations are experimenting with drone technology to create applications that improve productivity and patient outcomes. The inventors of the patent have offered a thorough explanation of how to use unmanned aerial vehicles (UAVs) to observe livestock. People throughout the world have struggled to access laboratories and transport specimens for laboratory examination as a result of poor infrastructure, traffic, and inaccessibility. This applies to both rural and urban places, particularly in Nigeria, where the so-called big towns, shockingly, have awful roads and are constantly congested. Drones are required to provide access to health care services in these areas, particularly in remote and rural areas. In terms of test outcomes, the medical laboratory is an important part of health care. Some medical centers are largely collecting sites, with a substantial number of specimens being gathered in one area and then transported to another for analysis and diagnosis. Some laboratory examinations are only conducted in specific locations, necessitating the transfer of such specimens over considerable distances. It is crucial to act quickly because if specimens are left neglected for a long time, they will degrade and become less usable for analysis. Sending test results to the doctor as soon as possible for disease diagnosis and treatment is another requirement of in-patient care. A control room receives real-time data from UAVs flying over feedlots while monitoring and assessing each animal's health based on core body temperature.

Communication Protocols and Technology

Numerous physical items are being connected via various communication technologies to create smart services in this Internet of Things era. Aerial vehicle-assisted networks frequently use noisy and lossy communication channels, yet the vehicle nodes should operate with very little power.

Aerial vehicle-aided networks in IoT contexts may use Wi-Fi, Long Term Evolution-Advanced (LTE-A), WiMAX, ZigBee, Long Range Wide Area Network (LoRa-WAN), and 5G as examples of communication protocols [6].

One of the most popular communication technologies is Wi-Fi which can access smart devices over a range of up to 100 meters. This communication technology (IEEE 802.11a/b/g/n/ac) has two distinct operating modes known as infrastructure mode and ad hoc mode, which can be used for communication between an aerial vehicle and a ground base station.

In comparison to IEEE 802.11n, according to the study in [7], IEEE 802.11ac-based aerial vehicle-assisted IoT networks perform better in terms of throughput. However, further research is needed to properly understand this type of behavior. Additionally, the majority of commercial aerial vehicles are intended to be verified by the indications of the Wi-Fi, enabling users to control drones or other aerial aircraft using their mobile devices.

Because AI's long-term viability is dependent on access to vast amounts of data, it can accelerate its development by collecting and analyzing data. As one of the most popular Internet access methods is Wi-Fi, which enables mobile users to rely on wireless services based on context information like location. Together, AI and Wi-Fi technology may create trustworthy, predictable, and quantifiable wireless operations by allowing Wi-Fi data to be sent to the cloud and then analyzed by AI.

The Universal Mobile Telecommunications System (UMTS) and the Global System for Mobile Communications (GSM) are the foundations of the universal technology known as LTE (UMTS). Also, LTE-A is an advanced technology that offers lower latency, better speed, and improved coverage. In and around locations with a lot of macro cells and small cells, tracking the plan out for streaming real-time video surveillance from vehicles to a control center can be built using cellular infrastructures like 5G,

4G LTE, and LTE-A. This kind of structure might provide early disaster relief and crime prevention. Cellular networks, on the other hand, can leverage AI as a tool to comply with 5G standards requirements. For example, performance estimation in LTE small cells is evaluated using statistical regression and machine learning methods [12].

The results demonstrate that learning powered by AI can accomplish very significant gains. WiMAX is another potential drone-based communication technology. WiMAX technology, which is based on IEEE 802.16e, can be used to expand the service area and the number of open communication channels between service providers using communication equipment on aerial vehicles. Users having a nifty device prepared to receive advice can communicate with WiMAX's ability to deliver data transfer over Voice over IP (VoIP) services. Vehicles can be outfitted with a variety of gadgets that enable them to act as nodes in a network and properly route communications. Furthermore, it is critical to forecast WiMAX network traffic to monitor performance and provide improved network management. In this situation, WiMAX network performance can be predicted using AI-based approaches like ANN (Artificial Neural Networks) based on the maximum number of online users.

Based on IEEE 802.15.4, the wireless communication standard along with Low-power and the low-data-rate connection is offered by ZigBee. It can allow communication for a 50-meter distance via smart IoT devices. IEEE 802.15.4 allows for communication between moving vehicles and stationary ground stations in a very short range of communication. Additionally, a range of applications and scenarios can make use of AI techniques in conjunction with ZigBee technology.

In the coming decade, it is anticipated that LoRaWAN, a low-power, low-bit-rate, long-range wireless technology for the Internet of Things, would provide connectivity for millions of intelligent IoT devices. As it supports low power consumption and can spread across a great range, thus, it can be utilized as the communication protocol between aerial vehicles and terrestrial base stations. When utilized with specific AI techniques that don't require a lot of parallel processing across the nodes, LoRaWAN can be cost-effective. LoRaWAN can also be the basis for machine learning powered by Artificial Intelligence (AI). AI has several consequences for operators, including changes to network architecture and the need for big data analysis [13].

In addition, LoRaWAN technology can be connected with AI-based approaches like ANNs to predict its propagation. ANNs are less computationally intensive and more accurate than deterministic and nondeterministic algorithms. The future of wireless networks, or 5G, will adhere to the Internet of Things (IoT) smart devices with extremely low latency, high data rates, and widespread coverage. Unmanned aerial vehicles (UAVs) are flying objects that will be used in the upcoming generation of wireless networks. The flying platform will be in focus as the 5G networks support in terms of cost and energy efficiency. For instance, in the 5G future, cars can be connected to function as drone base stations (BSs) to increase capacity and network coverage.

Therefore, in the age of the Internet of Things, 5G is among the most crucial communication technologies for networks with aerial vehicle assistance. To solve security and privacy concerns in 5G networks, AI-based drone communications solutions may also be deployed. Additionally, when integrated with UAV systems, 5G networks and AI can interact in real-time with the location of people, bicycles, and vehicles, reducing the likelihood of mishaps.

Communication Architecture

This section briefs about the communication architecture of aerial vehicles, which describes how data is exchanged between a drone and a base station or between drones. The architecture can be divided into two categories: Centralized Communications and Decentralized Communications.

In the centralized architecture depicted in Figure 16.1, all drones are connected to a base station, which is situated at ground level. In this architecture, vehicles are not directly connected. Instead, command and control

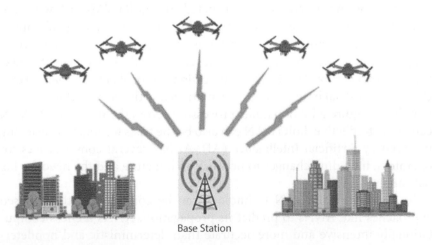

Base Station

Figure 16.1 Centralized aerial vehicle architecture.

information is transmitted directly from the base station to each vehicle. Furthermore, the central node is in charge of communication between the two media. As the data has been routed through the ground base station, this causes a rather considerable delay for information to be delivered between two vehicles. Furthermore, because communications between drones and the base station are frequently long-distance, vehicles in the centralized design require high transmission power along with compatible media. Because of their size and payload limitations, tiny and medium vehicles are not ideal for this. Additionally, if the terrestrial base station fails, the entire aerial vehicle network will be offline.

The centralized architecture is not robust or trustworthy as a result. Vehicles can connect directly or indirectly without the use of a central node or a ground station, in contrast to centralized architecture. The three different types of decentralized communication architecture are aerial vehicle ad hoc networks, multigroup aerial vehicle networks, and multilayer aerial vehicle ad hoc networks. Each vehicle participates in data promotion in an aerial vehicle ad hoc network, as depicted in Figure 16.2. In this architecture, the base station is connected to just one vehicle, which serves as a gateway for communication between the base station and additional cars.

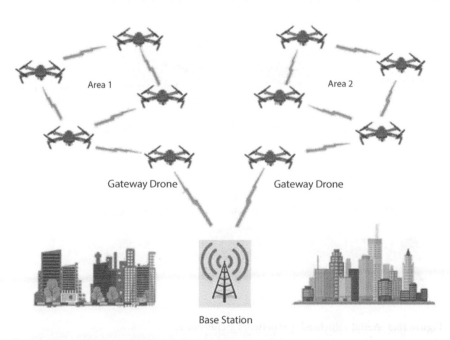

Figure 16.2 Multigroup aerial vehicle network architecture.

The network's coverage area may be significantly increased as a result. The gateway vehicle must use one radio transmission medium to communicate with the ground base station and another to communicate with all other aerial vehicles connected to the network.

This architecture works well for medium- and small-sized vehicle networks since the flying vehicles are so close to one another. As a result, radio transmission equipment can be inexpensive and lightweight. However, to assure network connectivity, all aerial vehicles must have identical mobility patterns, such as speed, heading direction, etc. As a result, this design is appropriate for a variety of related vehicle activities such as ongoing surveillance.

A multigroup aerial vehicle network combines unified architecture and an ad hoc network. A set of vehicles and their corresponding gateways constitute an aerial vehicle that is linked to the base station available at the ground level, as shown in Figure 16.3.

For communication inside the same group, the aerial vehicle ad hoc network described earlier in this section is employed, whereas the gateway vehicle and the terrestrial base station are used for communication between two groups. It should be emphasized that this communication architecture works best when a large number of aerial vehicles are engaged, each having a different communication or flying capability. However, due

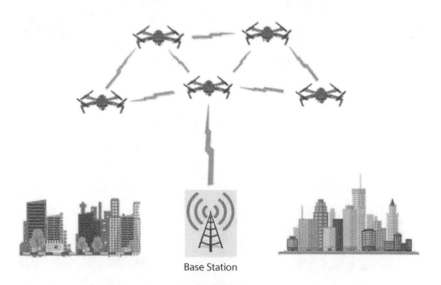

Base Station

Figure 16.3 Aerial vehicle ad hoc network architecture.

to the semi-centralized nature of communication, this communication framework is not yet a vigorous design. As illustrated in Figure 16.4, a multilayer aerial vehicle ad hoc network comprises numerous groups of assorted aerial vehicles. In this type of communication architecture, the gateway vehicles of various ad hoc networks are linked together.

Furthermore, only one of the gateway vehicles is directly linked to the base station on the ground, which grips data communication for the station. The gateway aerial vehicles handle communication between two vehicle groups, so it doesn't have to go through the base station directly. This reduces the station's communication and processing load dramatically. The multilayer aerial vehicle ad hoc network has no single point of failure, making it a reliable architecture.

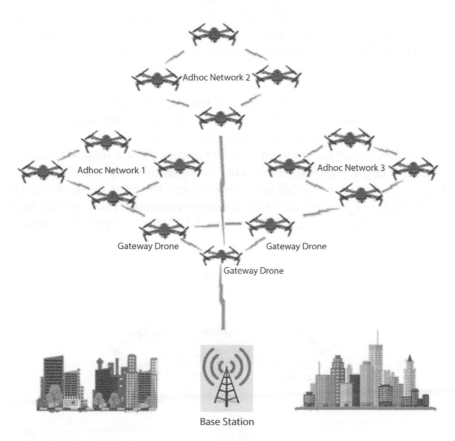

Figure 16.4 Multilayer aerial vehicle ad hoc network architecture.

Components

Various components of a basic drone are addressed in this section, which aids in understanding its design and mechanism to create an enabling environment that warrants time and funds saved in daily commercial engagements. A detailed diagram is depicted in Figure 16.5. These components are divided into two categories i.e., Primary Components and Secondary Components. Primary components are those that are most vital in the design of the drone system, while secondary components are less critical but still important in the design of an efficient aerial vehicle system.

Primary Components
GPS Module: The GPS module is a critical component that provides the aerial vehicle with its related spatial information like latitude, longitude, and elevation points. It enables the vehicles to fly further distances and control certain areas of the surroundings. It also aids in the safe return of the aerial vehicle to the controller, even if the link to the main control unit is lost.

Battery: The battery is a component of the aerial vehicle that allows it to effectively perform all of its functions. The vehicles would be unable to fly without the battery. The battery needs a different vehicle's contrast. Because of their low power requirements, smaller drones require smaller batteries. Similarly, larger vehicles, on the other hand, require larger batteries with higher capacity to power all of the operations. The pilot can keep an eye on the battery's performance using a remote-controlled monitor mounted at the controller which displays battery data. Also, an application-based

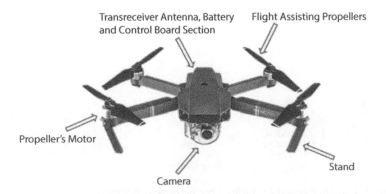

Figure 16.5 Basic components of a drone.

user interface is provided by the manufacturer to monitor the status of the aerial vehicle.

Camera: The camera is one of the most important components of an aerial vehicle. They assist in the capture of photographs from above, which is a major use for drones. Some aerial vehicles come with cameras built-in, while others can be equipped with a variety of camera types available on the souk.

Controller: This is essentially the flight's brain or motherboard. The flight controller is in charge of all directions given to the vehicles. It decodes the GPS modules, receivers, and sensor inputs. It is also in charge of the vehicle's control and the electronic speed controller's regulating of motor speeds. In addition, the flight controller is in charge of the vehicle's autopilot mode and other autonomous tasks.

Secondary Components
Propellers: The propellers are usually mounted on the top of the aerial vehicle in each direction, vary in size, and differ in material utilized in their construction. Some are constructed of plastic, which is good for smaller vehicles, but the more expensive propellers are made of carbon fiber. They are in charge of the vehicle's movement and direction.

Transceiver: The transceiver is the component that transmits and receives radio signals back and forth from the vehicle. During the flight, the transceiver communicates with the drone via radio transmission. Each radio broadcast has a unique code that allows it to be distinguished from other signals in the air.

Solar Panel: Due to certain limitations in the cell of the aerial vehicles, there is a chance to add a portable, renewable energy source, like a solar panel, which can significantly increase the efficiency and the flying time of the vehicle. It can also do away with the requirement to use an external power source to recharge the vehicle's battery. Due to this, the airborne vehicle might be ideal for long-distance travel without landing.

16.4   Suggested Framework

A few applications, as well as certain model-making frameworks, will be discussed in this section. Out of the various applications discussed earlier, the suggested model will focus on the smart healthcare application deployed in smart cities.

In the case of the various intelligent applications of aerial vehicles, it is very difficult to present every possible situation and use case when discussing the widespread deployment of flying vehicles. As a result, a few prospective applications and use cases will be recommended in this section, and projected outcomes will be discussed in detail. The proposed model leverages a three-stage system that consists of wearable health monitoring devices in the first stage, a web application, and a cloud storage medium in the second stage, and a network of drones will be required in the third or final stage. The detailed proposed model complies with various stages as described in Figure 16.6. Wearable sensors, including a temperature sensor, pulse sensor, and sink node, make up the wearable gadget. The sensors can regularly gather the body's parameters, which are then sent to the cloud via the sink node. The sink node is equipped with a built-in wireless communication chip that allows it to communicate with the server remotely. To begin measuring a patient's body vitals, the wearable Internet of Things gadget must first be engaged. The apparatus measures the patient's temperature and pulses as soon as it is activated. The sensor-gathered data is delivered to the server at regular intervals. The cloud database containerizes the aberrant data for further examination and distribution. The cloud is being used to implement a variety of prediction algorithms, which aid in defining the doctors' recommendations.

The concerned medical actors will receive an alert message in the very next step of decision-making. In this stage, we have attached a temperature sensor and a well-designed plug-and-play heart-rate sensor called a pulse sensor to the human body to collect physiological parameters regularly. We have utilized the DS18B20 Sensor and the SN0203 Pulse Sensor together as temperature sensors.

Furthermore, a sink node will transmit the sensor data to the cloud server. The Nodemcu (microcontroller) has been employed in this case as the sink node or gateway. It contains an integrated Wi-Fi module (ESP8266) that may wirelessly communicate data to the server. These sensors can able to collect vital body signs or parameters at a regular interval of time. These signs can able to transfer into the cloud database via an intermediate microcontroller and a sink node. The microcontroller can control the flow of the sensor data via an internal program. This program has a certain interval time, which can be able to control the proceeding of the sensor data. The acquired data is sent to the cloud via a sink node when it has been gathered at the microcontroller.

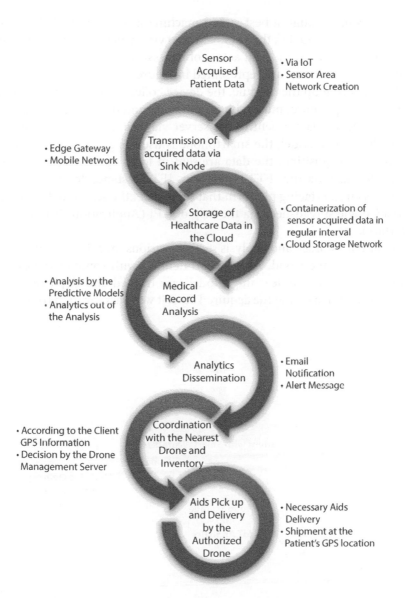

Figure 16.6 A detailed flowchart of the suggested framework.

It has direct access to the cloud via the mobile network. For further research and distribution, the collected data are containerized in a cloud database. The cloud is being used to implement a variety of prediction algorithms, which aid in defining the doctors' recommendations.

Afterxpassing through a best-suited machine learning model the output will be communicated in the very next step via an alert message which will notify the concerned medical actors of necessary precautions. The details of the proposed model are represented in Figure 16.7.

After booting up the module, the device collects different body parameters like temperature, pulses, blood pressure, and oxygen saturation. The sensor-acquired data is sent to the server via a sink node at regular time intervals. In this model, the smartphone is used as a sink node. After the sensor data acquisition, the data are sent in form of numerical values to a Google sheet via the IFTTT platform. IFTTT stands for IF This Then That. It is an interfacing platform that can collect the sensor data and send it to a Google spreadsheet via the help of API (Application Programming Interface).

In the data storage and analysis section, various machine learning models are used in the cloud. After experimenting with some of the existing data sets the machine learning models are trained and deployed in the cloud. After normalizing the acquired data, it will pass through the existing

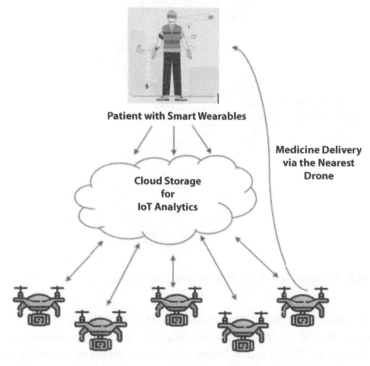

Figure 16.7 Proposed model.

best-suited predictive model for better accuracy. The outcome will be noti-
fied via text messages or email to the concerned medical actors.

An alarm message is sent to the patient, caregivers, and doctor if the
server detects any irregularities so they can provide the patient with the
necessary care. Doctors can use the credential to log into the cloud-hosted
portal after getting the mail. Doctors review patient data stored in the cloud
and recommend any necessary assistance. Once the suggestion is posted in
the portal the drone management network will automatically be activated
and the patient's location will be tracked and recorded. The drone management server will automatically activate and deploy the nearest localized
drone according to the patient's location, to a medicine store.

Further, once it is picked up by the authorized drone it will get delivered
to the desired patient's location. In case of a laboratory test is required at the
patient side, the nearest drone can able to get the sample from the patient
by the requisition of the caretaker at the web portal. The patient can enter
the desired portal and receive the thorough information supplied by the
related medical actors. In this way, the remote and smart healthcare system
is achievable in a smart city by the use of drone technology. Concerning
saving precious time, cost and most importantly life, the implementation
of this technology is highly appreciable.

16.5 Challenges

Drone-enabled smart healthcare is currently classified as a developing
invention. The most severe test is probably the continuous monitoring of
a network in real-time without any interruption. Various security threats
have been examined previously. Thus, a few of the issues are further noted
as follows.

I. Adjusting the burden dynamically and at a lower level: When compared
to how much information is created and transferred, memory, storage,
energy, and deferrals are still falling short, resulting in asset corruption and
low effectiveness. As a result, supporting assignments with low overheads
is still a challenge to work on. If done well, dynamic burden adjustment
can be deemed a compelling arrangement. To ensure the throughput of the
sensitive acquired data from the IoT gadgets and improve UAV competence, advancement of common regions of its operating frequency closely
together, expected IoT gadgets, and the amount of data expected to be
stored.

II. Longevity and Survivability: Drones are battery-powered; thus, they should have returned to the station before the battery died. Unfavorable indicators and starving conditions, which lead to asset depletion, exacerbate survivability. Examining assets without increasing the weight and size of the robots necessitates the involvement of the examination networks. During vehicle surveillance, determining the exact ID of multiple aerial vehicles moving at the same time is also a challenge. During take-off and landing in severe atmospheric conditions, fluctuating heat indexes, and within the sight of an even and lopsided formed construction, a review and estimation mission drone to drone channel, drone to the ground hub, and drone control joins is exceptionally expected to empower business and crisis aerial vehicle activity.

III. Low operating costs and high throughput: Cost is likely the most important factor in the organization and acceptance of drone technology. Drones are often offered at a higher price than the acknowledgment rate. Following that, low-cost and effective costing procedures for upgrading and increasing the dispatching of these robots should be found.

IV. Execution and uncompromising quality: The efficiency with which a task is completed and the efficiency with which assets are used to determine performance. Researchers should like to find more efficient and dependable ways to transport and run aerial vehicles in more congested environments. Better execution will result from optimal asset use. One of the major limitations of aerial vehicles is their ability to handle information and communicate across long distances. Adapting to the aforementioned challenges will enable vehicles to go beyond the point of view (PoVs) improvements and subsidize the steady growth of the drone-based sector.

V. Low dissatisfaction rates: Drone Enabled smart healthcare system is designed to run specific errands. Nonetheless, the rate of disappointment continues to be exceedingly high. Disappointment with a web connection, batteries, cameras, and asset fatigue are just a few examples. Disappointment taints the smart system's overall presentation. Steps should be taken to avoid disappointments like this and make the framework more powerful.

VI. Needs better security options: Recent advances in this technology have focused on information rate and security. New approaches must be implemented to deal with the various types of attacks that could be launched against robots and the organization. The security of the acquired data is one of the most important and pressing concerns. The programmers can access aerial vehicle information and operate its items due to the unavailability of the essential components. Different picture streamlining procedures, such as information combining, sewing of flying photographs, and so on, should be used to defend the information obtained by drones. To confirm the data security of unmanned aerial vehicles, various applications of blockchain-like technology should be explored.

Furthermore, this automated elevated framework comes with a few drawbacks, which is true of every industry that has been influenced by this invention. According to various researchers, these issues could range from battery life to payload limits and recommendations. These rules limit the amount of testing that may be done on drones' health-related delivery services across the globe, even though similar research is being done in various countries. Although the Federal Aviation Administration (FAA) has made improvements to the Part 107 guideline, with the release of a new Part 107 guideline that removes a few restrictions, there are still vulnerabilities. Various firms will be more interested in this innovation in the future as these obstructions are removed, with the ventures benefiting from the advantages associated with this application, which range from transportation to more advanced administrations. The government's avionics organization remains the biggest roadblock to this innovation, but there have been improvements. There are concerns about security breaches, as the drone innovation could enable remote spying by unknown entities.

There are also concerns that the airspace will be clogged with planes, obstructing growth and perhaps resulting in a crash among monitored and automated planes. Other concerns include drones colliding with structures and flying over or falling on people. The previous U.S. extraordinary activities' Command head for automated ethereal frameworks recommended proposals on the compelling activity of drone innovation. According to their belief system, ground-to-air interfaces should be implemented to prevent programmers from taking control of unmanned flights and to prevent cybercrime. Flying aerial vehicles past the administrator's perspective will also add to the confusion, as well as increase the cost of pilot-to-ramble. Thirdly, integrating robots into the public air control framework will necessitate some form of a remote organization. Finally, drones designed for therapeutic usage should be able to carry out campaigns more efficiently than what the US currently demands from typical beginning drones.

16.6   Conclusion

In the predicted future, the Internet of Things along with drone technology will change the way of handling the smart healthcare framework. Here in this chapter, an attempt has been proposed to advance the idea along with the change in a new wording of drones in smart healthcare. The model incorporates the cloud, predictive analysis of past health records, shrewd sensors, actuators, and so on. It can perform outrageous undertakings, while the accurate acquisition of the sensor data. In this approach, a total organization of various communicating drones associated with various applications, particularly healthcare surveillance is being emphasized. Still a great deal of work must be done in the field, as there are various weaknesses and issues in the framework. Cost, unwavering quality, and support are the first spot issues on the list.

Drone convenience is on the ascent as it offers an assortment of invigorating open doors. Conveyance of clinical and lab supplies and blood items, including biopsy for crisis medical procedures is currently at the very beginning stage. Drones used for clinical supplies are the guide and eventual fate of far-off regions that have been delivered distant by traditional land and air transport because of fiascos, devastation, awful framework, war, and different misfortunes and, an alleviation for individuals confronted with severe weather conditions and pandemics-needing administrations yet are left with severe lockdown arrangements. In various spots around the world, wellbeing experts, for example, clinical researchers, doctors, attendants, and other clinical workforce are beginning to encounter drone utility in delivering administrations. The exploitability of drone innovation is unascertained and broad. Drones can develop clinical consideration as well as impel headway in the well-being business. Drones can be a bonafide method of conveyance in the clinical business for Laboratory tests, drugs, antibodies, crisis clinical gear, and rescue vehicle administrations. Government organizations overall ought to put drone use on a worldwide plan particularly thinking about the current reality. With limitations in development and the absence of admittance to medical services offices during pandemics as in COVID-19, robots will be an exceptionally viable method for commending as well as giving locally established medical services. There ought to be a requirement of speeding up the research drives in the space of wellbeing and airspace for the executives to guarantee serviceable traffic within the board framework. For security concerns, severe standards on enlistment and authorizing of robots at a reasonable rate ought to be urged to empower close observing, following, and recognizable proof if there should arise an occurrence of thought violations and interruption. There ought to likewise be expanded public mindfulness and makers ought to investigate producing drones that can oblige greater payload and rambles with better battery duration.

The joining of healthcare along with the drone technology is thought of as quite possibly the most encouraging arrangement in regards to the improvement of shrewd conditions as they can be used to upgrade innovations and to manage intricacy and large information, as well as to give high precision and fast handling. In this regard, through explorative research on the utilization of drone technology, this paper is being emphasized various applications and correspondence conventions of the aerial vehicle-assisted smart healthcare in smart cities and gave different plan variables and prerequisites which guarantee an effective and dependable framework with greater Quality-of-Services (QoS). Besides, a detailed outlined based grouping and picture-based procedures that are utilized for aerial vehicle correspondences are discussed. At last, a few uncluttered issues and research challenges concerning drone technology in healthcare are being outlined which need to have some sort of solution-centric approaches are being highlighted concerning the drone-based smart healthcare system.

Future Scopes

There is a definite expectation for drone technology in achieving higher precision in healthcare applications. Out of enormous applications, one of the major outcomes can be remote patient monitoring which can be provided securely without risk and effortlessly. After the consultation by the medical actors' samples of the patients can be taken from a remote location for laboratory analysis. After diagnosis, the prescribed medicines also can be delivered to the concerned person's end.

The revolution of AI (Artificial Intelligence), ML (Machine Learning), and all related predictive support-based learning are assuming a huge part in the plan of cutting edge-based approaches that are at present being explored to tackle issues like aerial vehicles' direction and obligation pattern of sent ground hubs, battery limit, detecting handling and numerous unmanned aerial vehicle-based applications, e.g., accuracy agribusiness, vehicles following, climate correspondence, crash evasion issue, etc. Very small-scale wave correspondence-based innovation can be utilized in existing aerial vehicle organizations to propose quality information.

There are various hopeful assumptions for the innovation and implementation of the drone technology fit for conveying the necessary aids from the drug store to the patient's bedside. The future will encounter more short-term care and maybe locally situated care recently in the emergency clinic. Giving locally established care safely without risk and easily for various conditions can be accomplished utilizing drones. After consulting an appointed doctor on a patient's real-time data, based on analysis taken from past health records the results can be conveyed without a moment's delay using a robot to the centralized server for dissemination. After receiving the test results, prescriptions, and other handling requested by the doctor, it can likewise be moved back to the home of the end-user i.e., the patient through a drone delivery system. Additionally, by the virtue of this technology individuals residing in offices will presently return to the isolation center for a delayed time frame empowering them to rely less upon others for help. Drones can be utilized to convey food to individuals who are sick and can't set up their suppers as well as it can able to monitor isolated patients with ailments that disrupt their everyday work. With the development of this innovation, a ton of imminent clients have gone past the review and individual verification stage with various enterprises grappling with the truth of drone proficiency and advantages as they go to this innovation for secure and financial plan amicable missions. This anyway will assist with moving quick unrest and support progression in their respective businesses. Additionally, with limitations in development and tracking the home isolated patients during pandemics as in COVID-19, drones will be exceptionally powerful for conveyance as well as giving locally situated medical services in a contact freeway which can be a great asset in the pandemic situation.

References

1. Al-Turjman, F., A novel approach for drone positioning in mission-critical applications. *Trans. Emerging Telecommun. Technol.*, 33, 3, e3603, 2022.
2. Al-Turjman, F., Zahmatkesh, H., Shahroze, R., An overview of security and privacy in smart cities' IoT communications. *Trans. Emerging Telecommun. Technol.*, 33, 3, e3677, 2022.
3. Ullah, S., Kim, K.I., Kim, K.H., Imran, M., Khan, P., Tovar, E., Ali, F., UAV-enabled healthcare architecture: Issues and challenges. *Future Gener. Comput. Syst.*, 97, 425–432, 2019.
4. Krey, M. and Seiler, R., Usage and acceptance of drone technology in healthcare: Exploring patients and physicians perspective, in: *52nd Hawaii International Conference on System Sciences*, Grand Wailea HI, USA, 8–11 January 2019, HICSS, pp. 4135–4144, 2019.

5. Bravo, R.Z.B., Leiras, A., Cyrino Oliveira, F.L., The use of UAV s in humanitarian relief: An application of POMDP-based methodology for finding victims. *Prod. Oper. Manage.*, 28, 2, 421–440, 2019.

6. Wazid, M., Bera, B., Mitra, A., Das, A.K., Ali, R., Private blockchain-envisioned security framework for AI-enabled IoT-based drone-aided healthcare services, in: *Proceedings of the 2nd ACM MobiCom Workshop on Drone Assisted Wireless Communications for 5G and Beyond*, 2020, September, pp. 37–42.

7. Zhang, S., Zhang, H., Di, B., Song, L., Cellular UAV-to-X communications: Design and optimization for multi-UAV networks. *IEEE Trans. Wireless Commun.*, 18, 2, 1346–1359, 2019.

8. Al-Turjman, F., Ever, E., Fahrioglu, M., UAVs in intelligent IoT-cloud spaces, in: *Drones in IoT-Enabled Spaces*, pp. 1–6, CRC Press, Boca Raton, 2019.

9. Sharma, V., You, I., Kumar, R., Chauhan, V., OFFRP: Optimised fruit fly based routing protocol with congestion control for UAVs guided ad hoc networks. *Int. J. Ad Hoc Ubiquitous Comput.*, 27, 4, 233–255, 2018.

10. Hussain, S., Mahmood, K., Khan, M.K., Chen, C.M., Alzahrani, B.A., Chaudhry, S.A., Designing secure and lightweight user access to drone for smart city surveillance. *Comput. Stand. Interfaces*, 80, 103566, 2022.

11. Jaiswal, R., Agarwal, A., Negi, R., Smart solution for reducing the COVID-19 risk using smart city technology. *IET Smart Cities*, 2, 2, 82–88, 2020.

12. Gupta, R., Kumari, A., Tanwar, S., Fusion of blockchain and artificial intelligence for secure drone networking underlying 5G communications. *Trans. Emerging Telecommun. Technol.*, 32, 1, e4176, 2021.

13. Mbunge, E., Chitungo, I., Dzinamarira, T., Unbundling the significance of cognitive robots and drones deployed to tackle COVID-19 pandemic: A rapid review to unpack emerging opportunities to improve healthcare in sub-Saharan Africa. *Cognit. Rob.*, 1, 205–213, 2021.

14. Hoque, M.A., Hossain, M., Noor, S., Islam, S.R., Hasan, R., IoTaaS: Drone based Intrenet of Things as a service framework for smart cities. *IEEE Internet Things J.*, 9, 14, 12425–12439, 2021.

15. Sharma, K., Singh, H., Sharma, D.K., Kumar, A., Nayyar, A., Krishnamurthi, R., Dynamic models and control techniques for drone delivery of medications and other healthcare items in COVID-19 hotspots, in: *Emerging Technologies for Battling COVID-19*, pp. 1–34, Springer, Cham, 2021.

16. Chowdhry, B.S., Shaikh, F.K., Mahoto, N.A. (Eds.), *IoT architectures, models, and platforms for smart city applications*, IGI Global, United States, 2019.

17. Alsamhi, S.H., Ma, O., Ansari, M.S., Almalki, F.A., Survey on collaborative smart drones and internet of things for improving smartness of smart cities. *IEEE Access*, 7, 128125–128152, 2019.

18. Sharma, M., Drone technology for assisting COVID-19 victims in remote areas: Opportunity and challenges. *J. Med. Syst.*, 45, 9, 1–2, 2021.

19. Jin, Y., Qian, Z., Gong, S., Yang, W., Learning transferable driven and drone assisted sustainable and robust regional disease surveillance for smart healthcare. *IEEE/ACM Trans. Comput. Biol. Bioinform.*, 18, 1, 114–125, 2020.

20. Magaia, N., Ribeiro, I.D.L., de Aguiar, A.W., Fonseca, R., Muhammad, K., de Albuquerque, V.H.C., An artificial intelligence application for drone-assisted 5G remote e-health. *IEEE Internet Things Mag.*, 4, 4, 30–35, 2021.

21. Kumar, A., Sharma, K., Singh, H., Naugriya, S.G., Gill, S.S., Buyya, R., A drone-based networked system and methods for combating coronavirus disease (COVID-19) pandemic. *Future Gener. Comput. Syst.*, 115, 1–19, 2021.

22. Abid, A., Cheikhrouhou, S., Kallel, S., Jmaiel, M., Temporal constraints in smart contract-based process execution: A case study of organ transfer by healthcare delivery drone, 2021.

23. Das, S., Mohanta, B.K., Jena, D., IoT commercial drone and it's privacy and security issues, in: *2020 International Conference on Computer Science, Engineering and Applications (ICCSEA)*, IEEE, pp. 1–4, 2020, March.

24. Kumar, A., Sharma, K., Singh, H., Srikanth, P., Krishnamurthi, R., Nayyar, A., Drone-based social distancing, sanitization, inspection, monitoring, and control room for COVID-19, in: *Artificial Intelligence and Machine Learning for COVID-19*, pp. 153–173, Springer, Cham, 2021.

25. Munawar, H.S., Inam, H., Ullah, F., Qayyum, S., Kouzani, A.Z., Mahmud, M.A., Towards smart healthcare: UAV-based optimized path planning for delivering COVID-19 self-testing kits using cutting edge technologies. *Sustainability*, 13, 18, 10426, 2021.

26. Gera, U.K., Saini, D.K., Singh, P., Siddharth, D., IoT-based UAV platform revolutionized in smart healthcare, in: *Unmanned Aerial Vehicles for Internet of Things (IoT) Concepts, Techniques, and Applications*, pp. 277–293, 2021.

27. Sangeetha, D., Rathnam, M.V., Vignesh, R., Chaitanya, J.S., Vaidehi, V., MEDIDRONE—A predictive analytics-based smart healthcare system, in: *Proceedings of 6th International Conference on Big Data and Cloud Computing Challenges*, pp. 19–33, Springer, Singapore, 2020.

28. Maheswari, R., Ganesan, R., Venusamy, K., MeDrone-A smart drone to distribute drugs to avoid human intervention and social distancing to defeat COVID-19 pandemic for Indian hospital. *J. Phys.: Conf. Ser.*, 1964, 6, 062112, 2021, July, IOP Publishing.

29. Ren, X., Vashisht, S., Aujla, G.S., Zhang, P., Drone-edge coalesce for energy-aware and sustainable service delivery for smart city applications. *Sustainable Cities Soc.*, 77, 103505, 2022.

30. Mohammed, M., Hazairin, N.A., Al-Zubaidi, S., AK, S., Mustapha, S., Yusuf, E., Toward a novel design for coronavirus detection and diagnosis system using IoT based drone technology. *Int. J. Psychos. Rehabil.*, 24, 7, 2287–2295, 2020.

31. Zagrai, A. and Hassanalian, M., Drones as a driving force for smart towns: Technology and accessibility, in: *AIAA Propulsion and Energy 2020 Forum*, p. 3967, 2020.

32. Kumar, K., Kumar, S., Kaiwartya, O., Kashyap, P.K., Lloret, J., Song, H., Drone assisted flying ad-hoc networks: Mobility and service-oriented modeling using neuro-fuzzy. *Ad Hoc Netw.*, 106, 102242, 2020.

33. Li, J., Goh, W.W., Jhanjhi, N.Z., A design of IoT-based medicine case for the multi-user medication management using a drone in elderly centre. *J. Eng. Sci. Technol.*, 16, 2, 1145–1166, 2021.

34. Anand, P., Arjun, P., Kumar, N.B., Gowtham, K., Drone ambulance support system. *Int. J. Eng. Tech.*, 4, 2, 369–373, 2018.

35. Çalhan, A. and Cicioğlu, M., Drone-assisted smart data gathering for pandemic situations. *Comput. Electr. Eng.*, 98, 107769, 2022.

36. Khan, N.A., Ahmad, M., Alam, S., Siddiqi, A.M.U., Ahamad, D., Khalid, M.N., Development of medidrone: A drone based emergency service system for Saudi Arabian healthcare, in: *2021 International Conference on Computational Intelligence and Knowledge Economy (ICCIKE)*, IEEE, pp. 407–412, 2021, March.

37. Ahmad, K.A.B., Khujamatov, H., Akhmedov, N., Bajuri, M.Y., Ahmad, M.N., Ahmadian, A., Emerging trends and evolutions for smart city healthcare systems. *Sustainable Cities Soc.*, 80, 103695, 2022.

38. M. Johnston and C. Hoffmann, Hybrid MAC protocol, based on contention and reservation, for mobile ad hoc networks with multiple transceivers. U.S. Patent 10,251,196, 2 Apr 2019.

39. W. Sham, Facilitating communication with a vehicle via a UAV. U.S. Patent 10,454,564, 22 Oct 2019.

40. T.D. Erickson, C.A. Pickover, M. Vukovic, K. Weldemariam, Unmanned aerial vehicle-based system for creating foam splint. U.S. Patent Application 15/222,574, 1 Feb 2018.

17

Drone Delivery

V. Sakthivel[1,2]*, Sourav Patel[2], Jae Woo Lee[1] and P. Prakash[2]

*[1]Konkuk Aerospace Design-Airworthiness Institute, Konkuk University,
Seoul, South Korea
[2]School of Computer Science and Engineering, Vellore Institute of Technology,
Chennai, Tamil Nadu, India*

Abstract

A delivery drone is an unmanned aerial vehicle (UAV) that delivers goods and packages. Drones are autonomous and controlled remotely using software. Using drone delivery, we can transport a product or item to a specific spot. Consumers who cannot handle big objects to transport may use drones to provide mechanical assistance. Delivery drones can be used by retailers and transportation organizations around the world. Drones can also transport medicinal items such as blood, vaccinations, medications, and medical samples. In remote areas where access is not easy, we can use drones for medical and other deliveries.

With the technology being new to the market, these drones are much more expensive. While delivering, they perpetually exhaust their batteries; thus, in case a delivery is interrupted in between due to a power shortage, this will affect an organization's delivery system. Delivery drones are also equipped with cameras. The cameras perpetually record customers' location and property, and corrupt drone users may exploit customers' recorded information. While in air, a drone can also be easily stolen if someone can disconnect its power supply and take it away when it drops. Furthermore, as they are operated using different software, anyone who can access or breach the software can misuse the drone.

Moreover, drones can be used in defense systems to improve the country's security. Organizations must be careful of every consumer's security when making drone deliveries. Additionally, drones were made in such a way that they could tolerate various weather conditions and be more aerodynamically precise. For further enhancement, drone manufacturers should build drones or install safety features to reduce the risk of harm. Companies are also working very

*Corresponding author: mvsakthi@gmail.com

Sachi Nandan Mohanty, J.V.R. Ravindra, G. Surya Narayana, Chinmaya Ranjan Pattnaik
and Y. Mohamed Sirajudeen (eds.) Drone Technology: Future Trends and Practical Applications,
(425–440) © 2023 Scrivener Publishing LLC

hard to develop a better visual representation for drones that can monitor even smaller objects.

Keywords: Drone module, corrupt drone users, outdoor test

17.1   Introduction

The drone comes from the word drān, which signifies a male bee and continuous persistent noise (see Figure 17.1). A drone is an aircraft or a ship that can travel on its own. Also stated as [3]

- UAVs (Unmanned Aerial Vehicles)
- UAS (Unmanned Aircraft Systems)
- RPAS (Remotely Piloted Aircraft Systems)

UAVs, especially drones, have become one of the most popular technological toys in recent years, but what exactly are these flying robots, and what can they do?

Unmanned Aerial vehicles (UAVs) are vehicles that do not contain humans on board. They can be operated by a person or by a computer. UAVs were first utilized in the military to complete missions that were either too complex or dangerous for humans to achieve. Drone surveillance and drone assaults are two prominent military applications.

Civilian unmanned aerial vehicles (UAVs) are becoming increasingly prevalent. Several companies were currently pushing drone delivery services. Delivery drones can be used by Retailers and transportation organizations around the world. It can transport Medicinal items such

Figure 17.1 Drone.

as blood, vaccinations, medications, medical samples, etc. In the remote area where we cannot access it quickly, we can use medical deliveries using the drone. Photography and racing are two recreational drone hobbies.

Drones were originally designed for war usage and were primarily utilized as attack weapons. The use of drones has dramatically extended due to their ability to move quickly throughout the sky. Drones are increasingly being used in logistics services, drone photos, surveillance, news reports, and many making use of their sensing capabilities via cameras as well as their accessibility to challenging terrains [4]. For example, Amazon, December 2013, launched "Amazon Prime Air," which uses drones known as "Octocopters" to transport different products, particularly those that are small in size and weigh around 55 pounds, within 30 minutes.

DHL, a multinational logistics firm, created third-generation drones and verified their suitability for the delivery purpose that includes better aerodynamic technology, precise shipment, and autonomous performance. Delivery organizations, where speed is crucial, have also tried the use of drones for distribution. We will discuss various pros and cons of drones in this report. Let us take an insight into the history of drones.

17.2 History of Drones

Drones have been known for quite some time. A question arises who invented these human-less aerial vehicles well answer is Nikola Tesla and William Crozier invented these in the early 1900s. Since then this technology has improved drastically the surveillance and personal drones becoming more famous [1].

In the autumn of 1900, the Wright Brothers made their historic Kitty Hawk flight, and after a few years, Britain built the first pilotless aircraft which was named 'The Ruston Proctor Aerial Target' and it was based on Nikola Tesla's ideas. It was operated by radar, like today's drones. Nikola Tesla invented this vehicle that can be operated remotely using a radar beam over short distances. Tesla's remarkable invention is considered the first human-less automated aircraft [1].

"The first quadcopter was built by the Brequet brothers. The Gyroplane No. 1 was its name. Although it doesn't fly far in the sky, it flies only a few feet, but it was the first rotary-wing craft to lift by itself (see Figure 17.2). The gyroplane was said to have been able to fly several times." [5].

Figure 17.2 History of drones.

Initially, drones were designed for military use, but as time passed, they were widely used for commercial operations, and the demand for consumer drones increased significantly (see Figure 17.3). Aside from military applications, they were widely used for day-to-day applications. Commercial drones can be traced back to Otto Lilienthal, who is credited with the first operational human-crewed glider (1891).

Drones stretch back to 1849 when Austria waged war on Venice and they used unmanned balloons filled with explosives. The history of toy drones is essentially trial and error, with each decade seeing very tiny steps forward. Both military and toy drones made significant progress in the early 1900s. Percy Sperry piloted a radio-controlled kite plane in 1907, and in 1917, Archibald Low attempted to build an autonomous aerial vehicle. The first drone assault took place in 1918 when the US Postal Service launched drones against German targets.

Figure 17.3 History of drones. (Source: Bill Larkins/Wikimedia Commons.)

The royal army of the Netherlands developed the first modern drone for military use as an alternative to weaponry. Most drones were formerly operated remotely by a pilot on the ground. These revolutionary 'Fairy Rotors' were able to fly autonomously and employed gyroscopes to aid flight control. The De Havilland DH.82B Queen Bee is an aircraft designed by De Havilland. Many consider it the first modern drone, while there are other competitors (see Figure 17.4). The Navy began using the QF-2 drone in 1940, and the United States Army Air Forces fell into line throughout WWII. FIREBEES were the name given to these drones.

Drone technology began for civilian purposes in the 1980s, and it became increasingly common in many sectors. Drones had already been fitted with cameras at this stage. The drone was then used for surveillance purposes and then continuously spreading in all fields like spying and photography [1].

Due to technological advancements in aeronautics and astronautics, there has been a resurgence of interest in unmanned aerial vehicles (UAVs) throughout the last decade. UAVs' number of applications continues to

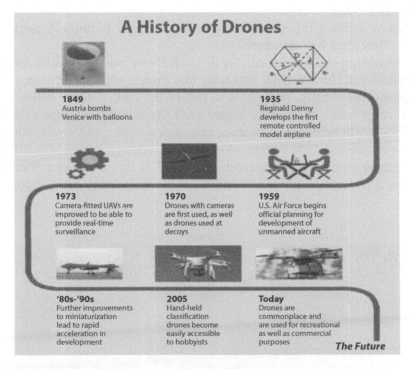

Figure 17.4 History of drones. (Source-http://www.bestrccopters.com/images/historyofdroneinfographic.jpg)

grow as sensors, motors, and other control instruments get more precise, smaller, and lighter. It's no wonder that UAV research has been embraced by both industry and academics, with applications ranging from military reconnaissance drones to emulating biologically inspired flight tactics and formations. UAVs provide a low-risk, adorable, and repeatable flight environment. As a result, UAVs can serve as an initial platform for novel technologies that can eventually be used for commercial applications.

Let us now focus on our topic [19, 20]. Several startups were established that used drones for delivery purposes and get enormous popularity in the transportation industry.

Before we explore this interesting topic deeper, we must first address the question, "What is drone delivery?" and "How does drone delivery work?" [7] Drone delivery, as the name implies, comprises the delivery of items to customers using unmanned drones. Drones can be controlled in one of two ways. The first is using artificial intelligence (AI). Drones can also be operated manually by pilots stationed at a nearby distribution center.

While drones are not currently mature enough to transport products over large distances at a cost-effective rate, they can help speed up the delivery process.

Drone delivery [8] is now being used by several prominent corporations. Here are a few examples:

Calculating the precise cost of drone delivery can be difficult. Many corporations, like Amazon, anticipate that drones will be cost-effectively implemented shortly. The goal is to reduce the cost of drone delivery of each cargo to $1 or less (see Figure 17.5). As a result, package delivery by drones will be significantly more economical than traditional modes of transportation.

Because commercial drone delivery is still in its early stages, there's a lot we don't know about how things work at all. Delivery using drones is

Amazon   Walmart   DHL

Figure 17.5 Delivery methods using drones.

far better than delivering traditionally because of its speed accuracy and time-efficient. Suppose a person ordered something from any organization, after ordering company loads that package on a drone that knows the location given by the operator or it may be operated by using software, flown in the sky, and brought to the door of that customer (see Figure 17.6).

The package delivery [9] method has been improved thanks to technological advancements. For example, FedEx and Amazon have both started experimenting with package delivery robots. Drones were introduced into delivery systems [16]. They may be willing to facilitate delivery firms with last-mile deliveries.

Consider an Amazon distribution center in a city. As usual, Amazon's delivery network may transfer products, Customers order what they want on Amazon, Walmart, or another retailer's website, and a drone picks it up and delivers it to their house or business. In Cambridge, UK, Amazon has begun testing its Amazon Prime Air service. From the time the order is being placed to the time it is delivered to the consumer, Amazon is striving to achieve a 30-minute turnaround time. This is unsurprising given Amazon's role as a pioneer in the drone delivery industry since the technology's inception.

E-commerce businesses are experimenting with new ways to be inventive and efficient in their operations. One example is the use of unmanned aircraft systems (often known as "Drones" or "UAS") to deliver products quickly. Drones are beneficial and productive when utilized commercially. Drone delivery offers several promising applications in a variety of industries, including healthcare, food, postal delivery, and freight.

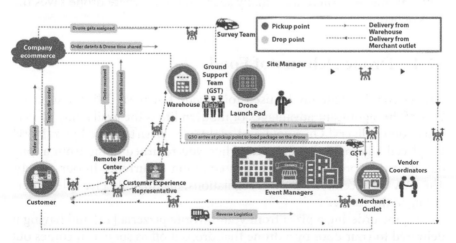

Figure 17.6 Amazon delivery method.

17.3   Drone Delivery in Healthcare

Drones can be used to deliver life-saving medical supplies. This includes things like:

- Vaccines
- Pharmaceuticals
- Samples
- Plasma and blood
- Aerial disinfection

Medical supplies can be delivered to remote locations or sections of the world that are unreachable to humans or airplanes using drones. Using drones to carry supplies is also a lot more cost-effective solution than using planes or helicopters [2]. As a beautiful tribute to these discoveries, it was published that an autonomous drone assisted as a lifesaver of a man who suffered from OHCA. According to the press release, "A 71-year-old man was shoveling snow in his driveway when he suffered an (OHCA)."

Drone for telehealth: University of Cincinnati researchers developed a telehealth drone with cameras and a display screen to enable two-way communication — with drone assistance. The drone may also transport samples or medication, making it an ideal solution for isolated locations worldwide [25]. According to an article by a telehealthcare director at the University of Cincinnati's College of Nursing, "We identified a need for telehealthcare delivery drones to give healthcare in the home and in regions where access to treatment was not readily available when the COVID-19 outbreak started," There were many such examples where drone saves the lives of people.

17.4   Drone Delivery of Food

Drones could potentially be used to provide fast meal delivery services. One of the significant advantages of modern life is the availability of personal food delivery! On a cold, wet night, who hasn't ordered pizza? [21] Or placed a grateful dinner order when you returned home from a long day at work? Food delivery [10] is a common occurrence in our society. The FAA has recently relaxed its regulations, making this a feasible undertaking in the future.

Imagine ordering a pizza from your favorite pizzeria [13] and having it delivered to your door by a drone that drops it off as soon as it comes out of the oven. The restaurant prepares the meals and put on the drones once

Figure 17.7 Delivery of food.

after placing the order [14] by the consumer. Then these drones fly in the sky and drop the delivery at the designated location (see Figure 17.7).

17.5   Drone Delivery in Postal Service

Drone delivery demonstrations have already been carried out by postal operators in several nations. These experiments are being used to see if large-scale postal delivery through drones is possible. For quite some time, the USPS has been experimenting with drone deliveries. Jeff Bezos, the CEO of Amazon, has long had a great vision: he wants his e-commerce company to use drones to deliver orders in under 30 minutes, as long as the packages are under 5 lb.

When it comes to mail delivery, Amazon isn't alone in experimenting GeoPost, Alibaba, the Asian e-commerce equivalent of Amazon, has also successfully tested mail deliveries using drones.

17.6   Delivery of Goods

Delivery time has become a critical factor in online users' purchase intention. The need for speed has outpaced the ability of railroads, trucks, ships, and planes to provide it. As a result, a new type of technology, drones, has been added to the mix.

Businesses are now looking to the skies to enhance delivery times, expand delivery radiuses, and gain a competitive sales advantage. In 2019, the global Drone delivery market [22] size in the United States from 2018 to 2030 is given below and is foretold with a CAGR of 44.2 percent to USD

9.51 billion by 2027 (see Figure 17.8). Drones are pilotless aircraft that can deliver packages to particular locations. Drones that deliver packages are either self-contained or operated from a base station. By simultaneously monitoring, many drone controllers can keep track of all packages delivered.

Drone deliveries [12] have the potential to be extremely beneficial to the shipping business. Drones are not just a smart alternative for corporations like Amazon, but they can also aid global logistics firms. Drones can refill cargo ships and conduct other activities that were previously carried out by small drones.

Now what makes the drone overcome other means of transport (deliveries)?

The advent of delivery drones has sparked debate over what the technology can achieve to reduce vehicle trip time, safety, and sustainability (see Figure 17.9).

Below are some of the potential [15, 17, 18] advantages:

- Reduced congestion on the road: Amazon estimates that 87 percent of its parcels [21] weigh less than five pounds, making them ideal for the type of drone delivery that the FAA has approved. Consider how much space is taken up on every Amazon delivery truck by-products light enough to be carried by a drone from a warehouse—a that's a lot of volumes to be freed up, especially as drone technology allows

Figure 17.8 Drone package delivery system market size.

Figure 17.9 Drones in the city.

the devices to travel longer distances. As a result, larger parcels can be crammed onto a smaller fleet of human-driven vehicles for hand delivery. Distributors can reduce the size of their road-bound delivery fleets by taking advantage of the open skies above [6].

- Accuracy: Drones were more effective and consistent when it comes to delivering the product to the customers. They are more accurate than humans at delivering products to the correct customer. As they were operated by automated software so there will be fewer chances of failure of delivery.

- Enhances time management: As drones use automated software to deliver products, the algorithms of such devices accurately locate the delivery place, which allows for faster delivery. This can be done so precisely with fewer marginal errors. Delivery drones [23] assist workers in saving energy by reducing the amount of effort they have to put in while delivering goods. This makes workers less exhausted as compared to normal deliveries. As a result, faster deliveries were made in sustainable surroundings.

- Safety: Consider a traditional delivery that requires big delivery trucks and vehicles to deliver products. Such vehicles can cause traffic, accidents, greenhouse gas emissions, and many other problems. However, if deliveries were made using drones that use air as a mode of transportation, these vast problems would be solved simply and effectively. They also provide more route flexibility. Less use of big vehicles means less expenditure on roads.

Every coin has two faces; let's flip the coin and look for insight into the cons of drone delivery.

1. Expensive device: Because the technology is new to the market, these drones are much more expensive. According to the site Drone-blog, "The cost of a drone varies from $50 for small drones to $1000 for large drones, while a camera drone costs $300-$500 as a starting price." Due to the high cost of using drones for delivery, only a few retailers use them.

2. Drone battery defects: While delivering things to a specific place, delivery drones quickly consume their batteries and they might be completely depleted in just a few minutes. Thus, if delivery fails, it may result in market complaints. So one must be careful if we are using drones and must be capable of enough power to operate. "We're committed to making our goal of delivering packages by drones in 30 minutes or less a reality," Amazon says on its Prime Air website. Our existing battery technologies are one of the reasons for the need for such a short turnaround. The average delivery drone's battery requirements could cause it to lose power in as little as 20 minutes. That may not be enough time for the product to arrive at its intended destination - or for the drone to return to the company's headquarters. Client complaints, product expenses, and insurance premiums would all rise if the delivery drone failed to meet its obligations to a customer. Hence battery of drones must be sustained and capable.

3. Requires technical understanding [24]: As drone technology is new to the market so few people understand its works. Flying a delivery drone for a normal person will be difficult (see Figure 17.10). The operator must be well-known with its handbook (how to operate) and its procedure. People must be trained before executing them in the field. Before this service takes off, a tremendous amount of time and effort would be invested.

4. Defective drone: Delivery drones with defective equipment months after purchase have been reported; this is the result of a manufacturing flaw that the maintenance staff at work

Figure 17.10 Drone directions.

went unnoticed. Ominous drones can cause delivery services to be disrupted. If one of these flaws occurs during delivery, the company's logistics will face many issues, additional fees will be incurred, and other possible complications will arise.

5. Data insecurity: Delivery drones are occupied with cameras when delivering items to the customers. The camera perpetually records wherever it goes then in such it also records the customer's privacy things like property, and his/her location. Corrupt drone users may exploit customers' recorded information.

6. Drones can be easily stolen by any person as they use air as a medium for delivering products; anyone able to disconnect the drone's power supply can take it away. Also, as they are operated by using different software anyone who can able to get access to the software or breach the software can misuse the drone.

7. Unemployment: In traditional deliveries, the company required a person that delivers the product to the customers but if deliveries were done using automated drones [11], then the company didn't need those delivery people and thus drone delivery would worsen the situation for such employees.

Conclusion: In 1980, political scientist Langdon Winner stated, "If there is a particular path that modern technological progress has taken, it is that technology goes where it has never been." Drone development has, without a doubt, followed this pattern. In an essay on the effects of new technology,

Winner concluded that the most critical questions are: "How are we to live together?" "How can we live in grace and justice?" These may appear to be very generic questions at the end of a book about drones. However, the goal conveyed in this book is that the evidence obtained by drones will, in some minor way, assist in answering these questions.

With improvement in the penny, things may increase the vitality of drone as a delivery purpose, before operating the drone; delivery employees should configure the batteries. Perpetually check its functioning and maintenances. After every shift, keep a record of everything you did with the drone (see Figure 17.11). Before distributing materials, double-check that the client, product, and location are correct. If you see any problems with the drones, notify your supervisors.

When delivering goods, keep an eye on the drones. Disciplined control of the delivery drone is required. To make drone deliveries, the organization must be careful regarding the security of every consumer. Drones were made in such a way that they could tolerate weather conditions and be aerodynamically more precise. Drone manufacturers should build drones or install safety [26] features to reduce the risk of harm. Companies are working 24 hours to develop a better visual representation for a drone that can monitor even smaller objects. Drone manufacturers should build drones or install safety features to reduce the risk of harm. Companies must work for 24 hours to develop a better visual representation for a drone that can monitor even smaller objects.

With the modern and futuristic technologies like Big Data, Internet of Things (IoT), web 3.0, and Artificial Intelligence (AI) pushing efficiency to new heights, Autonomy has become a possibility in cargo drone designs. This is reflected in the technology's dominance in this area, and the vast potential for developments in ADVs has the potential to counter the whole drone delivery sector.

Figure 17.11 Drone package delivery.

Acknowledgements

This research was supported by the Basic Science Research Program through the National Research Foundation of Korea (NRF) funded by the Ministry of Education (No. 2020R1A6A 1A03046811).

References

1. Gross, R.J., Complete evolution & history of drones: From 1800s to 2022, December 31, 2021, Retrieved April 10, 2022, from https://www.propelrc. com/history-of-drones/.
2. Balasubramaniam, S., Drones may become 'The next big thing' in healthcare delivery. January 9, 2022. Retrieved April 16, 2022, from https://www.forbes. com/sites/saibala/2022/01/09/drones-may-become-the-next-big-thing-in-healthcare-delivery/?cv=1&sh=7a5d3a681e9b.
3. Andreas, Introduction to drones - drone guide for dummies!, AirBuzz.One Drone Blog, September 2, 2018, Retrieved April 10, 2022, from https://air-buzz.one/introduction-to-drones/.
4. Park, J., Kim, S., Suh, K., A comparative analysis of the environmental benefits of drone-based delivery services in urban and rural areas. *Sustainability*, 10, 3, 888, 2018, https://doi.org/10.3390/SU10030888.
5. Mario, The history of drones (timeline from 1907 to 2021), Drone Tech Planet, 2018, Retrieved April 10, 2022, from https://www.dronetechplanet. com/the-history-of-drones-timeline-from-1907-to-2019/.
6. What are the benefits of drone delivery?, DRONEDEK | The Mailbox Of The Future, 28 February 2021. Retrieved April 10, 2022, from https://www. dronedek.com/news/benefits-of-drone-delivery/.
7. Dunham, J., What is drone delivery and how does it work?, Reveel, May 22, 2019, Retrieved April 10, 2022, from https://reveelgroup.com/ drones-an-in-depth-guide-to-success/.
8. Future of drone delivery: Can drones take over package delivery ecosystem?, 29, November 2019, iotdesignpro.com.
9. Pinguet, B., The role of drone technology in sustainable agriculture, May 25, 2021, (Introduction to drones - Drone guide for dummies! - AirBuzz.One drone blog, n.d.), PrecisionAg.
10. Mario, What are the disadvantages of drones?, Drone Tech Planet.
11. https://www.Dronedek.com/news/benefits-of-Drone-delivery/.
12. https://grindDrone.com/info/pros-cons-delivery-Drones.
13. https://www.whatnextglobal.com/post/food-delivery-Drones.
14. https://www.smartsoftusa.com/news/the-future-of-mail-is-with-Drones. html.
15. https://dispatchninja.com/the-future-of-Drone-delivery/.

16. https://www.Dronedek.com/news/benefits-of-Drone-delivery/.
17. https://www.wipro.com/business-process/the-future-of-delivery-with-Drones-contactless-accurate-and-high-speed/.
18. https://ratings.freightwaves.com/what-is-Drone-delivery/.
19. https://www.reveelgroup.com/Drones-an-in-depth-guide-to-success/.
20. https://www.Dronezon.com/Drones-for-good/Drone-parcel-pizza-delivery-service/.
21. https://timesofindia.indiatimes.com/blogs/voices/last-mile-delivery-by-Drones-an-assessment-of-market-viability-citizen-accessibility/.
22. https://grindDrone.com/info/pros-cons-delivery-Drones.
23. https://en.wikipedia.org/wiki/Delivery_Drone.
24. https://www.coursehero.com/file/128567799/Since-its-first-acknowledgement-at-the-World-Economic-Forumdocx/.
25. https://www.safetyfabrications.co.uk/news/droning.

Index

Printed and bound by CPI Group (UK) Ltd, Croydon, CR0 4YY

27/10/2024

14580128-0004